Water, Agriculture and the Environment in Spain: can we square the circle?

T0225381

Water, Agriculture and the Environment in Spain: can we square the circle?

Editors

Lucia De Stefano &
M. Ramón Llamas

Water Observatory of the Botín Foundation;
Complutense University of Madrid, Spain

CRC Press
Taylor & Francis Group
Boca Raton London New York Leiden

CRC Press is an imprint of the
Taylor & Francis Group, an **informa** business

A BALKEMA BOOK

Cover illustration: Morning twilight on the Gallocanta wetland (Aragón, Spain), Photographer: Miguel Ángel Sentre Domingo

First issued in paperback 2017

CRC Press/Balkema is an imprint of the Taylor & Francis Group, an informa business

© 2013 Taylor & Francis Group, London, UK

Typeset by V Publishing Solutions (P) Ltd, Chennai, India

Published by: CRC Press/Balkema
              P.O. Box 447, 2300 AK Leiden, The Netherlands
              e-mail: Pub.NL@taylorandfrancis.com
              www.crcpress.com – www.taylorandfrancis.com

Library of Congress Cataloging-in-Publication Data

Water, agriculture and the environment in Spain: can we square the circle?/
    Editors: Lucia De Stefano & M. Ramón Llamas. -- 1st ed.
      p. cm.
    Includes bibliographical references and index.
    ISBN 978-0-415-63152-5 (hardback : alk. paper) -- 1. Water resources
    development--Spain. 2. Water--Government policy--Spain. 3. Water-supply--
    Spain. I. De Stefano, Lucia. II. Llamas, Manuel Ramón.

    HD1697.S7W37 2012
    333.9100946--dc23

                                     2012027481

ISBN 13: 978-1-138-07607-5 (pbk)
ISBN 13: 978-0-415-63152-5 (hbk)

Table of contents

PART 5
Case studies

Foreword

The book now presented here is preceded by about twenty books and many other scientific and working papers on water resources that the Botín Foundation (BF) has produced since 1998. A detailed description of all of them can be found and downloaded from the website of the BF [http://www.fundacionbotin.org/agua.htm]. A brief history of the interest and activities of the Botín Foundation on water follows.

In 1998 the BF, in response to the proposal of its board member Emilio Botín O'Shea, initiated a multidisciplinary analysis of the problems of groundwater in Spain. The study, led by Academician Prof. M.R. Llamas, culminated four years of intensive work with a presentation at the Third World Water Forum (Kyoto, March 2003). The monograph PAS no. 13[1] is a summary of that project, which had, among other relevant outputs, a clear impact on the regulations related to groundwater management in the Spanish National Hydrological Plan, approved by the Government in 2001. In the history of Spanish research on groundwater problems, the project carried out by the BF is perhaps the most influential landmark.

During the years 2004 to 2008, the research programs of the BF related to water resources focused mainly on the organization of the Second and Third International Water Workshops held in Santander (Spain) in 2005 and 2007. The papers presented in these Workshops were published in two books: *Water Crisis: Myth or Reality?*[2] and *Water Ethics*[3], which can be freely downloaded from the website of the BF. The book *Water Ethics* was presented at the Fifth World Water Forum (Istanbul, March 2009).

At the end of 2005, Prof. Llamas delivered a speech in the opening session of the academic year of the Royal Academy of Sciences about the new concepts concerning the colours of water and virtual water[4]. The BF resumed investigations into water resources, in order to disseminate these new ideas in Spain. This decision crystallized in the formal constitution of the Water Observatory of the Botín Foundation

1 Llamas, M.R. (2003). *El Proyecto Aguas Subterráneas: resumen, resultados y conclusiones*. Papeles del Proyecto Aguas Subterráneas (PAS, no. 13). Mundi-Prensa Ed. and Marcelino Botín Foundation. Madrid, Spain.
2 Rogers, P.; Llamas, M.R. & Martínez-Cortina, L. (eds.) (2006). *Water Crisis: Myth or Reality?* Taylor & Francis Group. London, UK.
3 Llamas, M.R.; Martínez-Cortina, L. & Mukherji, A. (eds.) (2009). *Water Ethics*. CRC Press.
4 Llamas, M.R. (2005). Los Colores del Agua, el Agua Virtual y los Conflictos Hídricos. [Water Colours, Virtual Water and Water Conflicts] Opening lecture of Academic Year 2005–2006. *Revista de la Real Academia de Ciencias Exactas, Físicas y Naturales*, 99(2): 369–389.

(BF-WO), which became a program attached to the Observatory of Trends of the BF. Unlike the previous project on groundwater, which was performed in an office of the BF and with BF staff, in this case it was decided to establish two separate agreements with the University Complutense of Madrid (UCM) and the Technical University of Madrid (UPM), both located in the International Moncloa Campus of Excellence, a strategic alliance of both Universities launched in 2010. All research staff of the BF-WO is recruited through these universities, which are also the venue for the Secretariat and offices of the researchers. The Director of the BF-WO is Professor Llamas, Emeritus Professor of UCM, and the Deputy Director is Professor Garrido, Professor of Agricultural Economics and Natural Resources of UPM, and Director of the CEIGRAM, UPM Research and Development Centre. Initially, the research of the recently established Water Observatory was primarily to apply and disseminate the methodology of the water footprint, involving also estimates of the virtual water trade in Spain. The activities of the BF-WO broadened soon, currently including three main lines: 1) a think-tank for innovation in water resources management; 2) an exchange place for relevant experts in this field; and 3) the transfer of new knowledge to society at large, mainly to water decision-makers.

The activities of the BF-WO had an outstanding policy impact, which resulted in the decision of the Ministry of Environment to include the use of the water footprint method in its regulation for preparing the River Basin Management Plans (BOE, no. 229, September 22nd 2008), in line with the European Union (EU) Water Framework Directive (WFD) (2000/60/EC). Since 2008 the BF-WO has published eight monographs (PAV papers), four monographs called SHAN papers, and a book with the title *Water Footprint and Virtual Water Trade in Spain*[5]. In a certain way, the present book is a continuation of that one. By the time it was completed it was not possible to take into account some important data, especially on the role of groundwater, due to delays in the elaboration of River Basin Management Plans required by the EU Water Framework Directive.

At the same time, the Observatory continued its international relations and activities. In 2008, the Fourth Botín Foundation Water Workshop on *Re-thinking Water and Food Security* took place in Santander, at the BF headquarters. The book *Re-thinking Water and Food Security*[6] was presented at the 2010 Stockholm Water Week. Afterwards, in 2010, the Fifth Botín Foundation Water Workshop entitled *Water and Food Security in a Globalized World: Cooperation vs. Confrontation*, was held in Santander. In this case, the contributions to that Workshop led to the elaboration of a Special Issue in the scientific journal *Water Policy* (Volume 14, Supplement 1, 2012), presented at the Sixth World Water Forum in Marseilles (March 2012).

This book is the result of the joint effort of the BF-WO and several other water experts. The Water Observatory is a small think-tank made up of 14 people, including six postdoctoral researchers, three university professors who are advisors, and five doctoral students. The aim of this group is to analyse problems related to water

5 Garrido, A.; Llamas, M.R.; Varela-Ortega, C.; Novo, P.; Rodríguez-Casado, R. & Aldaya, M.M. (2010). Water Footprint and Virtual Water Trade in Spain: Policy Implications. Springer, New York, USA.
6 Martínez-Cortina, L.; Garrido, A. & López-Gunn, E. (eds.) (2010). *Re-thinking Water and Food Security*. Taylor & Francis Group, London, UK.

resources with independence of thought and following an interdisciplinary approach. The main and initial focus has been on the Spanish case. However, in this globalized world water issues in Spain are not detached from world problems. Therefore, from the early work in 1998, the BF-WO has always considered the Spanish case with an international perspective and since 2011 has expanded its work in Latin America in partnership with national research institutions. The six countries participating in the project are Argentina, Brazil, Chile, Costa Rica, Mexico and Peru. This research project was presented the first time in Stockholm at the 2010 World Water Week. A side-event on the preliminary results of the project took place at the Sixth World Water Forum in Marseilles (March 2012). It is expected that each country will complete its analysis in 2013. The BF-WO will produce a report summarizing the main findings for the whole project.

Since the early stages of the BF water programs, with freedom of thought and without submission to the *politically correct*, the BF-WO has demanded high professionalism, scientific rigor and a concerted effort to find solutions for Spanish society in the first place, but also for the *global village*, which today is this blue planet. This book shows how these objectives are being achieved so far.

In order to achieve scientific rigor this book and the previous one (*Water Footprint and Virtual Water Trade in Spain*) were submitted for peer review to a group of well-known experts: Prof. A. Allan, Prof. A. Hoekstra and Prof. M. Shechter for the first book; and Dr. J. Bogardi, Mr. J. Corominas, Dr. T. Estrela, Prof. E. Fereres and Prof. J. Lundqvist for the second book. The authors of this book are very grateful to the comments and suggestions received from the last five reviewers. The editors also thank Dr. L. Martínez Cortina for his help and advice throughout the editorial process.

Madrid, May 2012
The Botín Foundation Water Observatory

Authors' biographies

MAITE ALDAYA

Maite Aldaya is a postdoctoral researcher at the Water Observatory of the Botín Foundation, and consultant for the Sustainable Consumption and Production Branch of the Division of Technology, Industry and Economics of the United Nations Environment Programme, Paris, France. Maite has a PhD in Ecology and MSc in Environmental Policy and Regulation from the London School of Economics and Political Science (UK). She has worked for several international organizations such as the Agriculture and Soil Unit of the European Commission, and the Land and Water Development Division of the Food and Agriculture Organization of the United Nations. She has developed her research on water accounting, footprint and efficiency at different organizations, such as the University of Twente, the Netherlands; Complutense University of Madrid, Spain; and Technical University of Madrid, Spain.
E-mail: Maite.Aldaya@unep.org

ENRIQUE CABRERA

Prof. Enrique Cabrera has been a Mechanical Engineer since 1971 and obtained a Physical Science degree in 1974. In 1976 presented his PhD on Hydraulic Transients at the Technical University of Valencia, Spain, where he has been full professor of Fluid Mechanics since 1982. His main fields of interest are urban hydraulics and those related to the management of water, with special emphasis on its efficient use. Professor Cabrera has published widely in the urban hydraulic field, including 40 papers in refereed journals, and over 200 papers at national and international conferences. He has edited 26 books and participated in around 60. Eleven of the 26 edited books have been published in English, edited by well-known international editors. Dr. Cabrera has supervised 14 PhD dissertations and published over 80 newspaper articles on water and educational policies in the main Spanish daily journals.
E-mail: ecabrera@ita.upv.es

MARÍA DEL CARMEN CABRERA

Prof. María del Carmen Cabrera holds a PhD in Geology from Salamanca University (Spain), and is Associate Professor at the Department of Physics, University of Las Palmas de Gran Canaria, Spain. She has been working on Canary Islands hydrogeology since 1987, in the Geological and Mining Institute of Spain, the Water Authority of Gran Canaria, and currently at the University of Las Palmas de Gran Canaria. She lectures in the MSc Programme in Hydrology of the Universities of Alcalá de Henares and Málaga (Spain), and is responsible for the Teaching Support Centre of the International Course on Groundwater Hydrology (Barcelona, Spain) for on-line students. She is the main researcher in several European and Spanish hydrology projects and is author of numerous international and national articles and papers.
E-mail: mcabrera@dfis.ulpgc.es

JAVIER CALATRAVA

Javier Calatrava is Associate Professor of Agricultural Economics and Policy at the Technical University of Cartagena, Spain. He conducts his research in the field of agricultural and resource economics, mainly focused on the economics and policy of water resources and agricultural soil conservation, participated in 12 public research projects and supervised two doctoral dissertations. He has authored 40 academic references, including articles in indexed journals and chapters in international books, and has conducted consultancy work on water economics and soil conservation policy for several Spanish agricultural and environmental administrations, the European Commission and the OECD.
E-mail: j.calatrava@upct.es

DANIEL CHICO

Daniel Chico is an Agricultural Engineer, currently working as junior researcher at the Water Observatory of the Botín Foundation. He is also a member of the Research Centre for the Management of Agricultural and Environmental Risks (CEIGRAM) of the Technical University of Madrid, Spain. His interest lies in the evaluation of the water footprint and virtual water trade of agricultural products and the drivers of agricultural water use. He holds a MSc in Agriculture and Natural Resources Economics from the Technical University of Madrid (Spain), where he is doing a PhD in Agricultural Economics. Prior to his work at the Water Observatory he worked for an environmental consulting company on the economics of the agricultural biomass supply chain. During his undergraduate studies, Daniel Chico spent a year at the Institute for Agricultural and Urban Ecological Projects of Humboldt University in Berlin (Germany), where he collaborated in research on heavy metal pollution of urban waters.
E-mail: daniel.chico@upm.es

CARMEN COLETO

Carmen Coleto holds a PhD in Biology. She has been working in the water planning and water quality sectors for more than 15 years and in the Groundwater Project of the Marcelino Botín Foundation for four years as a scientific advisor. Afterwards she joined the Limnology Department at the consultancy firm United Research Services (URS) in Spain, as Project Director, where she conducted several projects on water quality control for seven years. Currently she is Head of the Area of Relationships with the European Union at the Spanish Ministry of Agriculture, Food and Environment, and her responsibilities include the coordination of the reporting activities for all European Directives related to water and the direction of some of the European Union expert groups on water, including water scarcity and droughts, water and climate change, and floods.
E-mail: CColeto@magrama.es

EMILIO CUSTODIO

Prof. Emilio Custodio, Dr. Industrial Engineer, is currently Professor Emeritus of Groundwater Hydrology at the Department of Geo-Engineering, Civil Engineering School, Technical University of Catalonia, Barcelona, Spain. His professional career started in 1965. He has been Professor of Nuclear Engineering and co-founder, former director and currently responsible for the Assessment Council of the Foundation International Centre for Groundwater Hydrology (Barcelona, Spain). He is advisor to the Water Observatory of the Botín Foundation. He has been Director General of the Geological and Mining Institute of Spain, Engineer at the former Water Authority of the Eastern Pyrenees basin (Spain), and collaborator of the former Public Works Geological Survey, Ministry of Public Works (Spain). He has been President of the International Association of Hydrogeologists and visiting professor at several Universities in Argentina. Doctor Honoris Causa by the University of Tucuman (Argentina). He has led numerous international and national research and professional projects on groundwater hydrology. He is author and co-author of 25 books and more than 500 papers.
E-mail: emilio.custodio@upc.edu

LUCIA DE STEFANO

Lucia De Stefano is a research fellow at the Water Observatory of the Botín Foundation, and Associate Professor at Complutense University of Madrid (Spain), where she teaches hydrology and hydrogeology. Her previous position was as a research assistant at Oregon State University, USA, working on global studies on water conflicts and resilience to climate-change-induced water variability, and regional water governance benchmarking in the Middle East and Northern Africa. Lucia has worked a policy officer for WWF, the international environmental NGO and as a water management specialist in the private sector. A hydrogeologist by training, she obtained her advanced degree in Geological Sciences from the University of Pavia (Italy), and holds a PhD on evaluation of water policies from Complutense University of Madrid, Spain.

She received formal training in water conflict management and negotiation at Oregon State University, USA; and UNESCO-IHE in Delft, the Netherlands. Her main fields of interest are policy evaluation, water planning, groundwater management and the assessment of good governance attributes.
E-mail: *luciads@geo.ucm.es*

AURÉLIEN DUMONT

Aurélien Dumont is a junior researcher at the Water Observatory of the Botín Foundation. He is a geological engineer of the École Nationale Supérieure de Géologie (INPL, Nancy, France) with a specialization on geological risks and hydrogeology. Aware of the political and social dimensions of environment-related issues, he enrolled in MSc studies in Geopolitics (Reims Champagne-Ardenne University) and chose to deal with Spanish water geopolitics in the frame of his MSc thesis. His PhD research (in progress) addresses the issue of groundwater use in Spain and, more particularly, how water footprint assessment tools contribute to an integrated water resources management.
E-mail: *adumont@geo.ucm.es*

ALBERTO GARRIDO

Prof. Alberto Garrido is Professor of Agricultural and Resource Economics at the Technical University of Madrid (Spain), Director of the Research Centre for the Management of Agricultural and Environmental Risks (CEIGRAM), and Deputy Director of the Water Observatory of the Botín Foundation. His work focuses on natural resources, water economics and policy, and risk management in agriculture. He has been consultant for OECD, IADB, European Parliament, European Commission, FAO, and various Spanish Ministries, Autonomous Communities, and private companies. He is the author of 123 academic references. His more recent books are: a) *Water Footprint and Virtual Water Trade in Spain: Policy Implications* (Garrido *et al.*, Springer, New York, USA, 2010); b) *Re-thinking Water and Food Security* (Martínez-Cortina, Garrido & López-Gunn [eds.], Taylor & Francis, Leiden, the Netherlands, 2010); c) *Water for Food in a Changing World* (Garrido & Ingram [eds.], Routledge, London, UK, 2011 [2nd ed.]).
E-mail: *alberto.garrido@upm.es*

LAURENT HARDY

Laurent Hardy is an expert in the water-energy relationships for public and private sectors. As a research Engineer in academic institutions, he has published several scientific articles, documents and book chapters, and made presentations in seminars and for MSc classes. He is interested in the study of the private energy sector and how water availability could be a constraint to energy systems development. Recently, he has started to work on his own analytical models to quantify the water-energy nexus.
E-mail: *laurent.hardy.mail@gmail.com*

NURIA HERNÁNDEZ-MORA

Nuria Hernández-Mora holds a BS in Economics and Business Administration from the Comillas Pontifical University in Madrid (Spain), and MSc degrees in natural resources policy from Cornell University and University of Wisconsin, Madison, USA. She has worked for non-profit organizations dedicated to land use and water resources policy both in the USA and Spain. She has participated in research projects focusing on water governance, risk management and vulnerability to extreme climatic events. She has worked as a consultant for non-profit and governmental organizations, including the Spanish Ministry of the Environment and the World Bank. She actively participates in Spanish water policy evaluation and analysis through her involvement with the Foundation for a New Water Culture (*Fundación Nueva Cultura del Agua*), which she currently presides. She has authored or co-authored multiple publications in the fields of water management, water governance, public participation and transparency. She is currently a collaborating researcher at the Technical University of Madrid, the Autonomous University of Barcelona and the University of Seville (Spain).
E-mail: nuriahernandezmora@gmail.com

ROSA HUERTAS

Rosa Huertas is a Law Graduate from the University of Valladolid (Spain), joining the civil service in 1993 as an official of the State Senior Civil Service Corps. Rosa worked as Deputy Secretary General in the Government Office of Castilla y León regional government (1993–2001). From 2001 she has worked at the Duero River Basin Authority (Spain), currently as Deputy Water Commissioner. Her work has focused on legal and water management issues, water administration, user communities, different instruments to protect the public water domain, and aspects related to river restoration. She has published a number of papers and chapters addressing issues related to the management and protection of water. She has also given courses, lectures, conferences, and is involved in training and courses run by the Ministry, as well as lecturing in MSc programmes. She is currently teaching the MSc in Environmental Engineering and Water Management at the EOI (Spain's School for Industrial Organisation), the International MSc on the Operational Safety and Management of Dams by Spancold and CICCP, as well as the MSc on Environmental management and stewardship run by the Biodiversity Foundation.
E-mail: rhg.ca@chduero.es

M. RAMÓN LLAMAS

Prof. M. Ramón Llamas is Emeritus Professor of Hydrogeology at the Complutense University of Madrid, Spain. Since 1986 he is a Fellow of Spain's Royal Academy of Sciences. He is also Fellow of the European Academy of Science and Arts (2005). He is author or co-author of nearly 100 books or monographs and almost 400 scientific papers. He has received the Cannes International Great Prize for Water in 2006 and is a Member of the French Academie de l'Eau (2006). He was President of the Inter-

national Association of Hydrogeologists (IAH) from 1984 to 1989. In 1992 he was appointed Honorary Fellow of the Geological Society of the United Kingdom. He has been coordinator of the UNESCO Working Group on the Ethics of the Use of Freshwater Resources (1998–1999) and has been member of the Scientific Advisory Committee of the International NGO Action Against Hunger (1999–2004). Since 1998 he is Director of the Water Observatory of the Botín Foundation.
E-mail: mramonllamas@gmail.com

ELENA LÓPEZ-GUNN

Elena López-Gunn is a Senior Research Fellow at the Water Observatory of the Botín Foundation at the Complutense University of Madrid (Spain), where she is currently leading a project studying groundwater governance, including collective action on groundwater. Dr. López-Gunn is also an Associate Professor at IE Business School (Madrid, Spain). She was an Alcoa Research Fellow at the London School of Economics Grantham Research Institute (UK), where she was engaged on applying a rights-based approach to water in Bolivia. Her interests focus on governance aspects of water management, particularly the institutional analysis of different aspects related to collective action dilemmas at different scales, and public policy analysis on water policy and politics. She has published on a range of issues, including issues related to self-regulation and the key role played by water user associations in water governance.
E-mail: elopezgunn@gmail.com

LUIS MARTÍNEZ-CORTINA

Luis Martínez-Cortina has a PhD in Civil Engineering at the University of Cantabria, Spain. He has worked as a researcher in several European Union research projects, and also for the four-year Groundwater Project, launched in 1999 by the Botín Foundation. He has carried out hydrogeological studies, groundwater flow numerical models, other empirical tools and its application to the practical management of water resources. He is co-author of five books and monographs, and is the author or co-author of about 40 scientific articles. He was co-editor of several books corresponding to the Botín Foundation Water Workshops, relating to groundwater, water ethics, water crisis, and water and food security. He was the Coordinator of the Spanish Association of Groundwater Users. He developed his work as a full researcher in the Geological and Mining Institute of Spain for six years. In 2012 he has joined the Department of Water Planning at the Spanish Ministry of Agriculture, Food and Environment.
E-mail: LMCortina@magrama.es

PEDRO MARTÍNEZ-SANTOS

Prof. Pedro Martínez-Santos obtained his Bachelor of Civil Engineering (Honours) and his MSc of Technology Management from the University of New South Wales,

Sydney, Australia. He completed his PhD on groundwater modeling and management at the Complutense University of Madrid (Spain), where he became a lecturer in 2008. Ever since, he has worked with different government institutions and private companies in the field of water resources. His research interests include the social and economic value of groundwater development, as well as new methods for environmental protection and modeling of hydrological processes. He is the author of over 30 papers and chapters in indexed journals and books.
E-mail: pemartin@geo.ucm.es

BEATRIZ MAYOR

Beatriz Mayor holds a MSc degree in Integrated Water Management from Wageningen University, the Netherlands. During her MSc studies she focused on water accounting issues, green water management and water productivity. In 2011–2012 she did an internship at the Water Observatory of the Botín Foundation and is now working at the Complutense University of Madrid (Spain) in a research project funded by the company Repsol. Beatriz holds a degree in Environmental Science from the Autonomous University of Madrid, Spain (2009), where she also had a 4-month internship at the Water Quality Institute.
E-mail: beamrchiki@hotmail.com

INSA NIEMEYER

Insa Niemeyer is a researcher at the Water Observatory of the Botín Foundation and in the Research Centre for the Management of Agricultural and Environmental Risks of the Technical University of Madrid, Spain (CEIGRAM). In the framework of the Water Observatory, she works in a project related to virtual water trade in Latin America. Her research work addresses the issue of sustainability aspects related to international agricultural trade. The focus lies on whether virtual water trade can contribute to achieving the goal of food security in a sustainable way. The results are applied for policy advice on a scientific basis. Insa Niemeyer is a Business Economist by training. During her MSc studies she majored in development economics at the University of Hamburg (Germany), after which she worked at the Potsdam Institute for Climate Impact Research (PIK) (Germany), in a joint project with the German Technical Development Cooperation (GTZ) for almost two years. Within the project she focused on adaptation and impact analysis of climate-change-induced water stress in various developing countries.
E-mail: insa.niemeyer@upm.es

DOLORES REY

Dolores Rey is a PhD student at the CEIGRAM (Research Centre for the Management of Agricultural and Environmental Risks, Technical University of Madrid, Spain). She holds an MSc in Agriculture and Natural Resources Economics. She has undertaken

several studies on impacts of climate change on Spanish agriculture, including crop yields, irrigation requirements and crop insurance. Currently she is working on a project related to water markets as a solution to water scarcity and drought in the Mediterranean region.
E-mail: dolores.rey@upm.es

MARTA RICA

Marta Rica is a PhD student at the Water Observatory of the Botín Foundation. In the framework of the Observatory, she is studying groundwater governance in Spain, focusing on the institutional analysis of collective action by resource users at different levels, and how it articulates within certain socio-ecological systems. She holds a degree in Environmental Sciences from the Autonomous University of Madrid, Spain (2008). She completed her MSc degree in June 2010 at Wageningen University (the Netherlands), specializing in Integrated Water Resources Management. As part of her studies, she did a practical period in Centro AGUA (Cochabamba, Bolivia) researching on drinking water management in rural areas, and institutional and policy reforms in the context of human rights to water.
E-mail: marta.rica85@gmail.com

JERÓNIMO RODRÍGUEZ

Jerónimo Rodríguez, born in Bogotá, Colombia, is an associate researcher at the Alexander von Humboldt Institute for Research on Biological Resources in Bogotá, working in the assessment of the environmental impact of national development policies specifically in the fields of agriculture and mining. He obtained his Economics Degree at the National University of Colombia-Bogotá in 2006, and his MSc Degree in Agricultural Economics at the Humboldt University in Berlin (Germany) in 2012. He carried out part of his research activity for the Colombian Palm Oil Research Center, in areas related to the social and environmental sustainability of the Colombian palm oil sector; and for the Colombian Coffee Growers Federation, in relation to the search for alternative income generation and social policy. He has also studied the social and environmental sustainability of the intensive horticultural sector in the Doñana area, in Andalusia, Spain.
E-mail: jerroe@hotmail.com

GLORIA SALMORAL

Gloria Salmoral is a Forest Engineer from the University of Córdoba (Spain), having studied two academic years of her degree at the University of Freiburg, Germany. In 2009 she enrolled in the MSc in Environmental Engineering and Project Management at the University of Leeds, UK. In 2011 she completed the MSc in Agro-Environmental Technology for Sustainable Agriculture at the Technical University of Madrid, Spain. Since March 2010 she has been a junior researcher at the Water Observatory of the

Botín Foundation, where she works on the water footprint assessment of agricultural products (olive oil and tomatoes) and of regions, particularly for the Guadalquivir basin. Currently, she is carrying out her PhD studies regarding the impacts of changing agricultural crop patterns on water quality in Spain considering technical, economic and political factors.
E-mail: gloria.salmoral@upm.es

NORA VAN CAUWENBERGH

Nora Van Cauwenbergh holds a PhD in participatory decision support for natural resources management. Nora has over six years of experience in river basin planning and governance in the semi-arid region of Almería, Spain. After concluding her PhD in 2008 at UCL, Belgium, she continues to investigate governance structures, stakeholder processes and the relevance, accuracy and use of intuitive, soft knowledge of stakeholders as compared to knowledge provided by modelling in decision support. She is the project leader of the 2009–2012 Altaguax project on participatory design of water management alternatives in the Andarax river basin, in Almería (Spain). The project set up a multi-stakeholder platform in the basin and delivered several online tools that aim to increase transparency on management options in the area. At UNESCO-IHE, The Netherlands, Nora lectures water resources planning and organizes different education activities. Her research and co-supervision of MSc theses has led to a number of publications in peer-reviewed journals and conference proceedings.
E-mail: N.vanCauwenbergh@unesco-ihe.org

FERMÍN VILLARROYA

Prof. Fermín Villarroya has a PhD in Geological Sciences and is a senior lecturer in Hydrogeology at Complutense University of Madrid, Spain. He teaches environmental geology, hydrogeology, and groundwater contamination to engineering students. He has worked in several European projects and has undertaken research activities in several areas of Spain, mainly the Upper Guadiana and Tagus basins. Currently he is involved in cooperation projects in Ethiopia, Haiti, Western Sahara, Mexico and Morocco. He has collaborations with Latin American Universities, including Sonora and Guadalajara in Mexico, La Plata in Argentina and Les Cayes in Haiti. Chairman of several Congresses and Symposia, he has been President of the Spanish Branch of the International Association of Hydrogeologists. Since 2011 he is an advisor at the Water Observatory of the Botín Foundation.
E-mail: ferminv@geo.ucm.es

BÁRBARA WILLAARTS

Bárbara Willaarts is a research fellow at the Water Observatory of the Botín Foundation, where she is working in a project on water and food security in Latin American and Spain. She holds a degree in Environmental Biology, a Master and a PhD in

Environmental Sciences. Her field of expertise is environmental planning, including the modeling and assessment of the hydrological consequences linked to land use changes, the spatial assessment and valuation of ecosystem services, and the analysis between food security, global trade and environmental sustainability. Besides her research activity she is also a lecturer in Global Environmental Change and Integrated Water Resources Management in several MSc programs and an Associate Professor at IE Business School (Madrid, Spain).

E-mail: barbara.willaarts@upm.es

PEDRO ZORRILLA-MIRAS

Pedro Zorrilla-Miras is a postdoctoral researcher at Terrativa Soc. Coop. Madrid (Spain), and at the Social-Ecological Systems Laboratory, Autonomous University of Madrid (Spain). Currently he is working on the Millennium Ecosystem Assessment of Spain, funded by the Biodiversity Foundation. He has a PhD in Ecology and Environment from the Autonomous University of Madrid, Spain. Pedro has worked in different research projects such as the groundwater management of an over-exploited aquifer in the Upper Guadiana Basin (Spain), evaluating the trade-offs between different management strategies through a participatory process using Bayesian networks; mapping and quantifying the loss of natural capital and land use changes in Doñana, southern Spain; and the biophysical accountability and mapping of ecosystem services in the transhumant Conquense Royal Drove, central Spain. His research interests are focused on the interaction between society and ecosystems, since the first need for a sustainable world is changing the functioning and relations of society with ecosystems.

E-mail: pedrozm@terrativa.net

Chapter 1

Introduction

M. Ramón Llamas[1], Lucia De Stefano[1],
Maite Aldaya[1], Emilio Custodio[2], Alberto Garrido[3],
Elena López-Gunn[1] & Bárbara Willaarts[3]

[1] *Water Observatory of the Botín Foundation; Department of*
 Geodynamics, Complutense University of Madrid, Madrid, Spain
[2] *Department of Geotechnical Engineering and Geo-Sciences,*
 Technical University of Catalonia, Barcelona, Spain
[3] *Water Observatory of the Botín Foundation;*
 CEIGRAM, Technical University of Madrid, Madrid, Spain

1 INTRODUCTION

The world's water problems are due to bad governance, not to physical water scarcity.
This book is inspired by this statement and explores whether it holds true in a specific
country, Spain, where climatic conditions - Spain is one of the most arid countries of
the European Union - could lead to the assumption that water problems are due to
physical water scarcity. In order to do so, this book builds on *Water Footprint and
Virtual Water in Spain* (Garrido *et al.*, 2010), where the Water Observatory of the
Botín Foundation focused on one of the pillars of good governance – accurate data
and information – and estimated how much of the resource is available, how much
is used, what for, and the extent of its economic productivity. The previous book
provided an analysis of the water footprint of Spain and explored the role of trade
in relation to the movement of virtual water – virtual water trade – in and out of the
country. Some of its main conclusions are particularly significant due to their policy
relevance and potential implications:

- 10% of the consumption of freshwater from rivers and aquifers produces around
 90% of the economic value of Spanish irrigated agricultural production.
- Spain imports water-intensive but low economic value products (mainly animal
 feeds and agricultural commodities), and transforms them into high-value export
 goods (mainly meat and elaborated products).
- Changes in virtual water trade cushion and respond to drops in agricultural supply caused by droughts.
- Green water (i.e. rainwater stored in the soil) has a very important share of Spain's
 water consumption. This is normally neglected by governments in traditional statistics and also in most national and river basin plans.
- The economy of water in Spain is very different now than it was 30 years ago,
 and current tensions around water often have little to do with old arguments and
 rationales centred on water scarcity.

The book of Garrido *et al.* (2010) was a breakthrough in understanding the relationship between the consumptive use of water and food production in Spain. However, some important aspects could not be treated with appropriate detail, both in terms of knowledge on water uses and also in terms of the complex factors that determine water policy and water-related decisions. In terms of knowledge of water uses, the following issues remained untouched or incomplete and are now addressed in the present book:

- The role of groundwater in the Spanish water sector and the economic value of its use in irrigated agriculture.
- The externalities of the economic uses of water, including agricultural diffuse pollution and the impact of water management decisions on the environment.
- The relevance and challenges associated with water uses in the urban, tourism and industrial sectors.
- The possible impacts of climate change on Spanish water resources and sectors.
- Other issues related to water in connection with energy, water institutions and collective management, and water footprint of specific basins, land uses and products.

Regarding the factors that determine water-related decisions, the book by Garrido *et al.* (2010) showed that a rational reallocation of water – based on economic considerations – could contribute to solving many of the problems of water scarcity (which is interpreted as an imbalance between demand and supply), but concluded that there are other crucial elements, beyond mere economic arguments, that need to be included in the *equation* of good water governance. As pointed out by the Declaration of the UNESCO World Commission on the Ethics of Scientific Knowledge and Technology (COMEST) on the conditions for the good governance of water resources, a key goal is reaching an equilibrium between the utilitarian and intangible values of water resources. Those *intangible* elements often determine the result of the equation, and if ignored or underestimated can sometimes bring water management to a deadlock.

The growing need for maintaining Spain's natural capital is added to the *equation* in the present book, as well as the human component: people that have needs, wishes, (vested) interests, aspirations, social and cultural considerations. This book thus takes a step forward in showing a more complex picture – probably closer to reality – of water governance in Spain.

2 STRUCTURE OF THE BOOK

In this book the reader will find chapters that contribute to shedding light on each of the elements of the *equation* of good water governance. Chapters are grouped into five sections (Table 1).

Section I (Political framework and institutions) characterizes the context of water governance in Spain. It starts by exploring what the concepts of water and food security mean for Spain (Chapter 2). It then outlines the main legal, administrative and economic elements that influence water management after the approval in year 2000

Table 1 Book structure.

Water, agriculture and the environment in Spain: can we square the circle?

Section I: Political framework and institutions
- Water and food security framework
- Legal, economic and administrative setting
- Politics and institutions

Section II: Metrification of water uses

- Spain's extended water footprint
- Groundwater resources
- Guadalquivir water footprint
- Guadiana water footprint
- Olive oil and tomato water footprint

Section IV: Possible mechanisms/enabling conditions
- Water markets
- Modernization of irrigation systems
- Collective action in groundwater
- Participation and transparency

Section III: Looking at the environment and sector uses
- Forests' water consumption
- Diffuse agricultural pollution
- Urban and industrial uses
- Water-energy nexus
- Potential impacts of climate change

Section V: Case studies
- Las Tablas de Daimiel National Park
- Irrigated agriculture in the North-west of Doñana
- The Canary Islands

of the European Union (EU) Water Framework Directive (WFD) (Chapter 3). Finally, Chapter 4 delves into the political and institutional issues that determine water decisions, how current institutions hinder the governability of water resources, and some potential venues for institutional reform.

Section II (Metrification of water uses) deals with metrification, which is undertaken calculating the extended water footprint (EWF) of the main water sectors at different scales. Chapter 5 sets the scene for this calculation by defining key concepts and showing the advantages and limitations of this approach in pursuing Integrated Water Resources Management (IWRM). Chapter 6 focusses on the metrification of water uses at the national level, by updating and upgrading the knowledge of water uses and their economic value, as provided in Garrido *et al.* (2010). Likewise, Chapter 7 presents the most updated data available on groundwater, the hidden half of *blue water* (surface and groundwater) and provides figures on the economic value of groundwater use in agriculture. The EWF is also applied to two river basins, the Guadiana and the Guadalquivir (Chapters 8 and 9), and to specific agricultural products (olive oil and tomatoes, Chapter 10), showing the added value of using this tool with a high level of data resolution.

Section III (Looking at the environmental and other sectorial uses) explores some of the challenges set by economic water uses, especially in terms of impacts on natural resources. Chapter 11 analyses how much water forests consume to maintain their ecological functioning, and how changes in land use and land cover impact water availability downstream. Chapter 12 presents the problem of diffuse water pollution due to agricultural activity, which is probably one of the most difficult and significant challenges for water sector over the next decades. Chapters 13 and 14 look at challenges related to urban water use and the water-energy nexus, showing the strong linkages between urban, energy and agricultural water uses. Finally, Chapter 15 provides an up-to-date picture of research findings on the potential impacts of climate change on water resources in Spain, how this could affect many water-dependent sectors and the different adaptation and mitigation options available.

Section IV (Possible mechanisms and enabling conditions) offers a closer look at specific mechanisms that are often portrayed as effective tools to solve the jigsaw of good water governance: water markets (Chapter 16), transparency in water management (Chapter 17), collective action in groundwater management (Chapter 18), and improved water use efficiency (Chapter 19).

Section V (Case studies) presents three specific cases where the socio-economic, political and environmental dimensions of water governance interact and lead to different challenges and opportunities for water management: Las Tablas de Daimiel National Park in the Guadiana basin (Chapter 20), an intensively irrigated agricultural area in the north-west of the Doñana National Park, in the Guadalquivir basin (Chapter 21), and the Canary Islands (Chapter 22).

3 BOOK'S MAIN FINDINGS

3.1 Political framework and institutions

Spain has a long, well-known, and globally acknowledged history in water management and planning, which has emerged in many ways as a response to its geographic location and climate. Water scarcity has been a regular and defining element for all Mediterranean countries and has triggered innovation throughout history, both in institutional terms and in water infrastructures, as adaptation measures to reduce risk. Looking to the future – at the start of the 21st century – Spain, like other countries across the world, will have to identify what water and food security mean for its domestic policy and also as part of a wider European and global international system, since this has important implications on water resource use, agricultural policy and the environment.

Water and food security are tags widely used nowadays, although these terms can hide very different meanings, depending on the countries' socio-economic context. Chapter 2 explains that in Spain, since production and market access is largely guaranteed, food security at the domestic level is closely linked to guaranteeing food safety and nutrition. In this context, the book shows how globalization has contributed to a shift in the dietary habits of Spanish consumers, with changes to the standard Mediterranean diet to include higher meat and processed food consumption, and a drop by half in the intake of cereals, legumes, fruits and vegetables. This dietary shift has been largely facilitated by globalization and virtual water imports from third countries. The book highlights some of the unexplored trade-offs, for example in terms of potential reduced food security due to poorer diets and also impacts from virtual water imports into third countries. In this context, Spain is also part and parcel to ongoing negotiations and debates over global food security, as a partner of one of the largest trading blocs in the world – the European Union – which is currently part of key agricultural food trade regulations in the Doha round of the World Trade Organization.

Although the concept of water security is relatively young, around a decade or so, in many ways it has been implicit in Spain's history from the onset. This is reflected in the development of both water infrastructure and institutions to deal with the inherent

Mediterranean climate variability, with Spain being the fourth country in the world in number of large dams, with over 1,200 reservoirs that can store over 55,000 million m³. This is also linked to a provision under water planning to include protection against extreme events, principally floods and droughts. In the past two decades, two new aspects of the concept of water security have been introduced, which refer, on the one hand, to the ability to *increase* domestic water security through the virtual water embedded in food imports (Allan, 2002; Guodong, 2003); and on the other hand a renewed emphasis on the importance of ensuring the good health of water-related ecosystems. These, in turn, would then secure all dependent activities and most water-related ecosystem services.

In this context, the concept of water security is strongly linked to the good status of aquatic systems, which is the basic goal of the European Water Framework Directive. Chapter 3 discusses how the WFD represents a *before and after*, a turning point for traditional Spanish water policy. The implementation of the WFD has marked a clear *road map* on the aim and the process to be followed, an aim fixed on the achievement of water good status as a primary planning and management objective; and a means due to a required and prescribed process of public participation in the elaboration of river basin plans. This has marked a quiet revolution for Spanish water policy planning and organizational *status quo*, catapulting environmental priorities to a more prominent role. This, however, has to be somehow integrated with the traditional planning framework and the *ethos* of meeting existing demands. A half-hearted attempt to introduce an as yet timid economic analysis of water services and a greater interest in water demand management are two additional implications. As explained in the book, the achievement of the WFD objectives faces several challenges and uncertainties that are technical, financial and political in nature. In this context, to a large extent, the successful implementation of the WFD may well lie in strengthening the link between land use and water management, and the institutional structures that facilitate co-responsibility and full cooperation between the central government and the regional administrations, which hold responsibilities for land use management and agriculture.

The book also identifies and discusses a series of key blocks for institutional reforms, namely, the legal and institutional framework, existing economic incentives, the organisational structure of the current water administration system, and procedures and paradigms for water policy and planning. Chapter 4 identifies areas where it is simply a case of making better use of existing policy instruments, by making these more efficient and effective. This includes strengthening the implementation of the current norms, increasing transparency in cost recovery, and reinforcing budgetary control and oversight. However, in some cases institutional efficiency and effectiveness simply will not suffice, and deeper structural reform might be needed. This is where there is currently a tension between what is politically feasible and issues of political will and skill. Politician will act if carried along or pushed by public support and at present there is a clear de-synchronization when in some ways *bad* water management is good politics. Thus a deeper *water democratization* is ultimately underpinned by a healthy and active civil society that has access to relevant information, and can act as a final check on the system to have a stronger overall accountability from all water institutions to achieve a transformative process of water policy and planning. Politicians will make good water management a priority when pushed by civil society.

The remainder of the book details illustrative examples that highlight how metrification can help shed new light on old problems, to provide a catalytic effect on facilitating change, both in terms of making existing institutions more responsive to change and also in being capable of introducing deeper institutional reforms, if necessary. In this context, political frameworks and institutions face both old and new challenges related to water and food security, which needs an effective and functional overall legal framework for water management and effective, accountable institutions. The *politics of water* should be part of a healthy democratic process where it is no longer the political rents associated with water that determine decisions, but rather the politics of water become part and parcel of a healthy process of water governance and negotiation.

3.2 Metrification of water uses

In this book the quantification of water consumption and its economic value has been undertaken mainly using the concept of the water footprint, which is an indicator of human appropriation of water resources that looks at both the direct and indirect water use of a consumer or producer (Hoekstra, 2003). It includes three components: blue water (referring to surface and ground water), green water (defined as the rainwater stored in the soil) and grey water (referring to pollution). The latter concept of grey water is an interesting, although complex and controversial concept. The water footprint combined with other socio-economic and environmental indicators is a tool that provides a transparent and multidisciplinary framework for informing and optimizing water policy decisions, and ultimately can help to achieve a more integrated water resource management.

In this book, the metrification of water uses, using the water footprint coupled with socio-economic and environmental data, has been applied at multiple scales: to the country as a whole, considering all types of water sources (Chapter 6); focusing specifically on groundwater (Chapter 7); at the river basin level (Guadiana and Guadalquivir, Chapters 8 and 9); and down to some particular products such as tomato and olive oil (Chapter 10).

Chapter 6 shows that, like in the rest of the world, agriculture is the main water user sector in Spain, while the urban and tourist uses represent a very small percentage of consumptive uses (Table 2). Nevertheless, the political weight of urban water supply is great due to its importance to public health and the general well-being of the population. The social and economic importance of urban and industrial sectors implies that drought cycles should not compromise the reliability of water supply,

Table 2 Consumptive blue and green water use and associated socio-economic aspects (year 2010).

Activity	Consumptive use (hm³)	GDP (M€)	Workforce (%)
Agriculture & livestock	38,990 (84%)	26,000 (2.5%)	4.2
Urban water supply	4,042 (9%)	4,000 (nil)	(nil)
Industry	3,251 (7%)	123,000 (11.8%)	14

Source: Garrido et al. (2010), INE (2012). [hm³ = cubic hectometre = million m³ = 10^6 m³].

because agricultural water use can be significantly curtailed at low or moderate costs to society, through e.g. virtual water imports.

As it is common in agriculture, green water holds the largest share of Spanish agricultural water consumption. This sector has experienced significant changes over the last 25 years, like changes in crop production that are giving blue water consumption an increasing importance, especially in the fastest growing sectors like vineyards and olive groves. On the contrary, the water footprint of the animal sector is turning *greener*, as the imports of rain fed-produced feedstuff have increased significantly. Nowadays, the intensive livestock sector in Spain relies primarily on imports from Brazil, Argentina, France, UK and USA, where livestock feeds are grown mainly under rain fed conditions.

Spain's agricultural trade has also grown significantly in volume and value during the last 15 years. The economic value of exports increased more than that of imports, but the total volume of traded virtual water followed an opposite path. The volume of virtual water[1] (VW) imported is mainly associated with cereals and oilseeds based on green water, and is significantly larger than the exported VW, composed increasingly of animal products and olive oil, in addition to the more traditional and stable exports of fruits and vegetables.

In the context of water management in Spain, groundwater plays a strategic role. As in most arid and semiarid countries, during the last half century the *silent revolution* of groundwater use has produced significant socio-economic benefits. Nonetheless, often limited attention has been paid to groundwater, due to a mix of professional bias, neglect and corruption (Llamas *et al.*, 2001). The EU WFD planning process has implied important advances in the knowledge of groundwater resources and their use in Spain. However, data on groundwater resources in some cases are still incomplete and a comprehensive official overview of groundwater resources (and their uses) is missing.

As a first estimation, Chapter 7 reports that Spain's groundwater demand is about 7,000 million m³/year, mainly for agriculture (73% of the total demand), followed by urban uses (21%). The development of intensive groundwater use, with little planning and control by the water authorities, has contributed to the degradation of this strategic resource. At present only 54% of the Spanish groundwater bodies are in good status[2], due to intensive aquifer exploitation or pollution. Full implementation of the WFD provides an opportunity to reduce this deterioration, and indeed water authorities expect to increase[3] to 80% the percentage of groundwater bodies that will be in good status by 2027, which marks the end of the WFD implementation process.

To understand the economic significance of groundwater uses, Chapter 7 also analyses the differences between the apparent economic value of crops irrigated with groundwater and surface water. Previous studies in Andalusia showed that groundwater irrigation was economically more productive than surface water irrigation. This

---

1 The virtual water of a product or service is the volume of freshwater used to produce the product or service, measured over its full production chain (Hoekstra, 2003).

2 41% of the groundwater bodies are in poor status. Currently, the status of 5% of the groundwater bodies is still under study.

3 Less stringent objectives have been established for 3% of the groundwater bodies, while for 17% of the the groundwater bodies there are no sufficient data to predict their status by 2027.

was attributed to a series of factors, chiefly its resilience to long dry spells, and it was thought this would be applicable also to other regions in Spain. Data obtained for the whole country and presented in this book seem to question this idea, with no clear correlation found between the source of water and the apparent water productivity in irrigated agriculture. This is an issue that merits further analysis, combining local and country-wide data to refine the calculations.

The calculation of the extended water footprint of specific river basins (Guadiana and Guadalquivir) yields closer insights into water uses. In Garrido et al. (2010), the authors looked in detail at the Guadiana basin. In Chapter 8 a similar approach is carried out for the Guadalquivir basin. The innovative aspect of the water footprint analyses in these two river basins is the separate analysis of both green and blue water consumption and the associated economic value, using the economic productivity indicator at the basin and sub-basin scales. In the case of the Guadalquivir river basin, blue and green water consumption for direct human use (agriculture, urban and industrial supply), and the green water consumption by forest uses (forests, shrublands, grassland, etc.) have also been considered. The analysis of the water demand by forest ecosystems, closely related to land use changes, is an issue of growing importance.

In both the Guadiana and Guadalquivir river basins, agriculture is the main water consumption sector, accounting for 95% and 93% of the total blue and green human water footprint, respectively. The economic productivity analysis shows that better water management could be reached through reallocation of water resources between the different uses. This reallocation may occur without social conflict with farmers since the quantities of blue water required for specific high-value uses (for example thermo-solar energy production, vegetable production, tourism) constitute a small amount of the current total blue water use. At the same time, the competent authorities should promote a win-win solution, facilitating farmers a change towards more productive and less polluting production techniques.

The tomato and olive oil product-level analysis looks not only at green and blue water consumed, but also at pollution aspects and the associated economic value of water uses. The study confirms the importance of a detailed water footprint supply chain assessment in the case of agricultural-based products, since these represent most of the green, blue, and grey water footprint, thus emphasizing improved agricultural management as fundamental to improving water management. Water accounting in the agricultural sector shows significant differences in results depending on the database, assumptions (e.g. spatial-temporal), and methods used. The main lessons learnt from these discrepancies is that the methods for estimating the water footprint of crops and products are prone to a cascade of errors, mainly when applied at a larger regional scale. Therefore, results from water footprint calculations have moderate accuracy, in line with most hydrological data.

3.3   Looking at the environment and sector uses

Administering an increasing water demand under changing environmental conditions represents a major challenge for managing water resources in a semi-arid country like Spain. Recent water planning documents in several river basins have reported a reduction of 10–20% of available water resources during the second half of the 20th century (Chapter 15). In some of these basins (Duero, Guadalquivir, Guadiana and

Jucar) mean annual precipitation has also decreased in the last six decades between 2–8%, while in other basins, like the Ebro and the Inner Basins of Catalonia, precipitation reduction has not been as significant. Despite the natural variability of the Mediterranean climate, these trends seem to be pointing to a potential transition towards a new climatic situation – as a possible signal of climate change – where annual rainfall is decreasing, thus decreasing naturally available water resources. Nevertheless, the observed runoff reduction cannot be explained by the reduction in precipitation alone. Ongoing changes in land use and land cover, and the increasing water abstraction rates, are also important drivers for the observed reduction in available blue water resources (Lorenzo-Lacruz *et al.*, 2012).

Looking at the future, a recent official report (CEDEX, 2011) forecasts that the observed trend in runoff reduction is likely to continue during the 21st century, especially in the southern, most water-stressed basins. This book argues that, whilst acknowledging the need for water authorities to adopt a precautionary approach to consider the potential intensification of current hydrological regimes, predictions have to be studied carefully because uncertainties remain high as forecasts have considerable error intervals.

Although this book does not address these phenomena in detail, Chapter 15 discusses how rates of droughts and extreme rainfall events have increased since the 1950s in different regions across Spain, although no significant trend has been observed at national scale. In the mean- and long-term scenarios an increase in droughts and in extreme rainfall events in much of Spain is forecast, especially in the central and eastern basins (CEDEX, 2011). However, uncertainty is high.

A reduction in water availability, coupled with an increase in temperature and extreme events, is likely to have major consequences for water-dependent sectors. Efforts are needed in the implementation of adaptation measures such as the development of drought resistant crops, improved forest management to enhance forest's resilience, or the establishment of institutional mechanisms such as a review of existing water rights and allocation, while others focus more on *hard infrastructure*, like improved drainage. Nevertheless, if climate change is considered as an effect rather than a cause, more efforts need to be made in searching for mitigation measures, and in this respect agriculture and urban areas in Spain will have a major role to play.

An important adaptation measure in Spain to climate change not yet explored in detail is how forest management can help to optimize the water cycle and the provision of runoff downstream. In fact, as mentioned earlier, the reduction in blue water resources observed across different basins in Spain cannot be explained only by a decrease in the precipitation regime. One potential driver of the reduction in runoff could be the observed increase in forest cover across large areas of the Spanish territory. According to official data (MAAM, 1996; 2006), forest area in Spain has increased by approximately 1.5 million ha. In this respect, Chapter 11 provides a first estimation of the major green water requirements of main forest uses, including forests, shrubs, agro-ecosystems and pastures. The preliminary results reveal phenomena seldom considered in the WFD and with great potential importance for water planning in Spain: forests consume four times more water than the whole of the Spanish agricultural sector (not including livestock) and use 39% of the mean annual Spanish rainfall. This proportion can increase up to 42% during droughts. Land use changes, and to a large extent an increase of forest cover linked to land abandonment and afforestation programs, seem to have a great impact on water resources in Spain, as these appear to have reduced national blue water resources by up to 4% since the 1980s. The magnitude of these reductions is especially relevant in those regions which are most water-stressed. Nevertheless, these results need to be assessed more in detail, foremost because current official sources of land use and land cover data (e.g. CORINE Land Cover project, Crop and Land Use Map) report different trends of forest evolution in Spain over the last decades, making it difficult to ascertain the real impacts changing forest area has had over the last years on national water resources. Overall, integrating land and forest management into water planning might be a cost-effective solution to optimize the supply of water in catchments, even though efforts still need to be placed on how to use water more efficiently and for a better management of water demand. Such integration would also highlight the possible negative *co-effects* that climate change mitigation afforestation programs could have from a water perspective.

A major environmental concern for Spain and most European countries is the problem of diffuse pollution caused by agriculture (Chapter 12). Nitrate pollution is a major issue in areas where intensive cropping and livestock (mostly from pig farms) occur. The use of chemicals in agriculture and livestock rearing does not differ from other European countries, although the inherent climate and soil conditions of Spain explain the differences in terms of application rate and types of chemicals used. Fulfilling the mandates on water quality from European Directives poses specific technical and also administrative coordination problems due to the different authorities responsible, on the one hand for water resources, and on the other for the environment at national and regional levels. Many aquifers and the related springs and surface water bodies will not reach good status by 2015, as required by the WFD. This means that new terms will have to be negotiated and new strategies will need to be carefully defined, taking into account technical, economic, social and administrative elements of intervention. Most work done so far deals with nitrate pollution, vulnerable zones and good agricultural practices, but there is also the issue of salinity pollution due to the evapoconcentration effect, when irrigation water is originally of poor chemical quality, something not rare in Spain. Pesticides and emergent pollutants in groundwater and the related surface water are being studied, since they may impact the quality of drinking water, but applied results are still in the early stages.

The water-energy nexus (Chapter 14) has become a relevant issue for Spain's sustainability standing and prospects. A key aspect stemming from the analysis of the water-energy nexus is the importance of the increasing use of energy in the irrigation sector. Since 2002, Spain is deeply involved in a process of modernization of its irrigation systems, with the aim of saving water, which leads to the unintended consequence of increasing energy consumption in agriculture. This came along in parallel with the reform of the electricity sector, which resulted in an increase in electricity costs to farmers by 80%. Moreover, from a water-energy perspective, unless there is a low initial water use efficiency (around 50%), modernization may not be the best option for increasing water availability. Desalination or reclaimed water may be a better suited alternative from the point of view of efficient resource use.

On the side of water for energy, the book highlights that the energy sector is mostly a non-consumptive one. The analysis of seven scenarios of electricity production for year 2030 with different technology mix suggests that the largest decrease in (non-consumptive) water use is achieved when more renewable energy systems and less nuclear energy are present in the technology mix.

The analysis of the water-energy nexus also links with urban and industrial water demand and the associated reclaimed water production, which is an important challenge in the European Union. The careful use of the potential efficiencies in the water-energy nexus could become an economic option, with the added advantage of providing an economic use for a raw material like wastewater. Extending the wastewater treatment system to include tertiary treatment all over Spain would account for 3% of the water-related electricity consumption or 0.2% of the total Spanish electricity demand.

Chapter 13 delves into challenges of urban and industrial water uses beyond electricity demand for water supply and treatment. The spectacular urbanization and population growth occurred over the past six decades affects the hydrological cycle and increases the complexity of sustainable water management. In this context, the steady increase in the deterioration of water quality due to urban, industrial and agricultural pollutants represents a huge challenge for urban supply, since it requires increasingly more expensive and sophisticated treatments to make water potable. Overall, the response to water pollution has been satisfactory although with three weak points: most of the investment in wastewater treatment was paid by EU funds that have now ended; most small cities and rural settlements still do not treat their grey water, and water prices are subsidized. Spain has some 200,000 km of water supply pipes, around 70,000 km of sewers and about 5,000 water treatment plants. An important part of these assets is or will soon become old, and revenues from current urban water treatment tariffs do not fully allow maintaining or replacing them. Thus, the decrease in the flow of EU funds and the current economic crisis are likely to mark a turning point from subsidies to full urban water cost recovery.

To face the upcoming challenges related to urban and industrial water supply a cultural change is needed, in particular for the major players – ranging from decision-makers to all stakeholders – so that they truly have an integrated approach to water problems. For this to happen these actors have to be trained to address problems from a broader perspective, and the general public has to be educated to value long-term policies, even when sometimes these require unpopular decisions in the short term. Other key elements to face urban and industrial water use challenges are the integration of water policies with other environmental policies (e.g. land use, energy), passing all the costs of water services to users (and reinvest the resulting revenues in the water system), progressively renewing water infrastructures, maintaining the commitment to provide a good quality supply, and improving knowledge and research to continue making progress in technology.

3.4   Possible mechanisms and enabling conditions

Based on the book's multiple angles of metrification, the adaptive reallocation of water rights among existing users or new emerging users is key to finding solutions to current tensions within the water sector and between human water uses – mainly agriculture – and the environment. Water trade and market mechanisms are often discussed as a possible way of facilitating that reallocation. In 1999, following the example of California and other countries, the Spanish Government decided to relax the Water Act of 1985 so that public water concessions could legally be traded to other users. Chapter 16 reviews the Spanish water market regulation established in the 1999 Water Law reform and overlooks on the type of exchanges that took place between 2004 and 2008, when market exchanges were more frequent due to a dry spell. While exchanged amounts were not very significant in absolute terms, those that involved inter-basin transfers raised most of the concerns. The main concerns are related to the effects of altering the timing and location of water abstraction, as well as changing the place and volume of the return flows of the traded water. For example, trading water belonging to downstream users from the head of a basin to another basin implies that the traded resource will not be available for other users and for in-stream flows on its way to the sea. Moreover, often much of the water associated with a water right is not abstracted or, if it is abstracted, it is not really consumed and returns to the basin, to be used by other users. In terms of implementation, this book outlines legal, political and economic constraints that hinder the growth of water trading in Spain, and lists a number of desirable elements to make them more efficient, environmentally friendly and legitimate.

Groundwater is a strategic resource during dry spells and in areas where no or limited surface water resources are available. Groundwater management may also play a significant role in adaptation to potential impacts of climate change. In Spain, as in most arid and semiarid countries, during the last half century an intensive use of groundwater for irrigation has occurred, providing stupendous socio-economic benefits (Shah, 2009). Nevertheless, due mainly to the fact that this intensive development has been done with little planning and control by water authorities, some problems, mainly environmental ones, have occurred. Since 1985 the Spanish water law developed measures to regulate and control abstractions by declaring an aquifer overexploited, yet these measures in most cases have failed to achieve control of groundwater uses and ultimately to improve the quantitative and qualitative status of the resource, as required under the WFD. As an answer to this situation of mismanagement, there have been spontaneous user-led initiatives, to develop a range of collective action institutions. These young groundwater collective institutions have increasingly focused on reducing water users' risk through the development of a portfolio of water resources, yet leaving some questions unanswered on the overall resilience of the system to intensive groundwater use. Nevertheless, the Spanish experience on these collective institutions to manage groundwater may be very useful for many other countries which have also developed an intensive use of groundwater.

The improvement of water efficiency is one of the pillars of the *Green Economy* and the modernization of irrigation systems represents a facet of this move towards a more efficient use of water. The irrigation modernization programme implemented in Spain during the past decade is possibly the largest in terms of surface area and investment in the whole of Europe and one of the largest in the world (Chapter 19). It is often presented as the flagship for the achievement of sustainable development in the water sector. This plan was a state-led effort to increase water efficiency in irrigation and generate water savings at plot and basin levels, particularly to reduce water stress during drought periods. Chapter 19 shows some evidence that estimated water savings have not been achieved, and that in some cases the increase in water efficiency application may have entailed an expansion of active irrigated land and/or some crop changes, leading potentially to a higher overall local irrigation water consumption. However, there are other unintended consequences and in some cases co-benefits in terms of reduced used of fertilisers due to fertirrigation, and better traceability and control of water use due to technological improvements. Yet the lack of reliable and consistent information on the actual aggregated consequences of this large public investment programme highlights the need for a detailed assessment of the outcomes and logic of the modernization process.

All the mechanisms presented in this subsection underscore that public access to data about water management and a stronger involvement of stakeholders in decision-making and implementation of policies is needed. Although the implementation of the WFD requirements is contributing to improving transparency and participation, in Spain the tradition of public accessibility to data and public participation in management decisions is still rather poor. Chapter 17 provides an assessment of these two *enabling conditions* for good water governance and presents results of an assessment of information transparency by the Spanish river basin authorities. In general, there is evidence that the transparency of the Spanish water agencies is slowly improving. Nonetheless, there is still a long path ahead for having all the relevant information available, especially in relation to water use and management, and water-related contracts and tenders. The most compelling challenge is possibly ensuring the reliability and the consistency of the information made available by the public administrations. Another key issue is making the information accessible to different target audiences by adapting information to different levels of interest and technical capacity.

3.5 Case studies

Some of the mentioned dimensions of water governance and the daunting task of balancing agricultural water uses and the maintenance of ecosystems are presented in different case studies, which show the difficulties and opportunities of working on specific contexts: Daimiel, Doñana, and the Canary Islands. The three case studies focus on groundwater resources in different regions of Spain, analysing the challenges and opportunities for improved water management and the implications of water use for the economy and the environment. This includes the pressures and impacts that human activity has caused on some unique areas close to important wetlands of the Guadiana basin (Las Tablas de Daimiel wetlands, Chapter 20) and Guadalquivir basin (Doñana wetlands, Chapter 21). The analysis of the situation in the Canary Islands (Chapter 22) gives an interesting view of the strategies adopted in a water scarce region.

In these three Spanish cases a spectacular increase in groundwater development for irrigation has taken place during the last half-century, carried out mostly by the private initiative of thousands of modest farmers in the pursuit of significant economic benefits. However, the chaotic nature of groundwater development, coupled with the attitude of many water policy-makers towards groundwater, has sometimes resulted in a series of unwanted environmental effects, such as water table depletion, groundwater quality degradation, land subsidence, or ecological impacts on aquatic ecosystems.

Since half a century ago, wetlands have changed from wastelands to almost sacred ecosystems all over the world. The concern for the topic and the analyses of many cases of wetland degradation by human actions, mainly by groundwater abstraction, has increased during the last decades in Spain. Two emblematic cases are presented in this book: the development of irrigated agriculture close to the National Park of Las Tablas de Daimiel, a UNESCO Reserve of the Biosphere, and the appearance of an intensively irrigated strawberry-producing area in the north-west of the Doñana National Park, a Ramsar site.

In the case of Las Tablas de Daimiel, the strain between nature and socio-economic development is far from being resolved. The consequences of intensive groundwater abstraction have been particularly severe in the groundwater dependent wetlands of Las Tablas de Daimiel National Park. In fact, in 1994, the Guadiana River Basin Authority declared the aquifer feeding these wetlands legally over-exploited. The last attempt of regulation, the Special Plan for the Upper Guadiana, aimed at reorganizing the water rights structure and reducing extractions to obtain water for ecological flows to Las Tablas de Daimiel. However, the financial cost has conditioned its full implementation. The challenge is to change the current agricultural model towards a more balanced model which aims to allocate water more equitably between all users, including the Biosphere Reserve. In the case of water in arid environments, which face strong competition between users, moving away from current trade-offs (and stand-offs between sectors) requires identifying win-win solutions.

In the strawberry-producing area of Doñana, problems are more recent and the consequences of intensive irrigation, while less spectacular (relative to the Daimiel area) in terms of water table reduction, can also be clearly traced in the transformation of land use and the degradation of groundwater quality. During the past decade, the increase in water abstractions has to some extent been limited through improvements in water efficiency, using improved irrigation technology and the enforcement of administrative constraints to the expansion of irrigated lands and tourist resorts. However, the complex nature of the resource system, the presence of unlicensed groundwater extractions, unsolved issues of land ownership, and a fragmented institutional structure hinder the achievement of long-lasting solutions.

Chapter 22 presents a summary of the characteristics of the Canary Islands archipelago. In a certain way, its hydrological and socio-economic characteristics suggest the need for a reconsideration of the prevailing ideas on the evaluation of water scarcity in the region. Indeed, this archipelago, with a water availability ratio around 300 m³/year *per capita* should be in a catastrophic situation if it is analysed through the prism of traditional water scarcity indicators. Indeed this water availability ratio is well below the 500 m³/year *per capita* threshold considered as an extreme water scarcity situation according to these types of indicators. The Canary Islands are in the lowest quartile among the Spanish regions in per capita terms, and yet the islands have not suffered in the last half century any serious economic and social problems. This has been possible mainly thanks to two factors: a) the intensive use of groundwater and the existence of privately managed water markets, and b) the desalination of sea water. Nevertheless intensive use has had a strong impact on the hydrology and social consideration of water in the archipelago, which varies from island to island, according to their specific characteristics. This leads once again to consider a broader perspective to analyse what is meant by sustainable development.

4 ISSUES THAT ARE NOT INCLUDED IN THIS BOOK

This book does not address a number of important topics related to water in Spain. Some relevant aspects have not been covered or are only considered at a preliminary level. This was due either to data not being available at the time this book was completed, or that their collection and study needed time, additional effort or an expertise that was not at hand. Some of these issues are:

- How food security is considered in Spain by the different Autonomous Communities that form the State, and how this agrees with the European Union policies, from the point of view of water resources.
- How droughts influence water policy and food security in Spain and the European Union. Spain is subject to recurrent, severe meteorological and hydrological droughts that affect people, economy and political life. These conditions vary across the territory, with differences between the Mediterranean area and the rest of the country, between the north and the south, including the archipelagos, one Atlantic and the other Mediterranean. The subject has been considered in a recent paper (Estrela & Vargas, 2012).
- How the Common Agricultural Policy (CAP) of the European Union affects water resources use in Spain, and what the future prospects are. This has an effect on the world's food trade and virtual water transfers. The CAP has driven agricultural water uses, sometimes promoting water demand, other times indirectly curtailing it. There has been no comprehensive study to estimate the net impact of the CAP on water use and water pollution. However, based on the available literature one could claim that hitherto the CAP has not had a positive impact on agricultural water use in Spain. Not including this major driver in this book is justified by the fact that a lot of uncertainty still exists about the details of the CAP for 2014–2020, especially about those issues related to irrigated agriculture.
- How the political divisions in Spain, where autonomous regions have responsibility for territorial resources and environment conditions, affect water use, including internal virtual water trade. This requires putting together regional goals and responsibilities, and central government policies, including energy use, environmental objectives and land-use planning.
- The costs and the economics related to water quality in agriculture, human supply and industrial production. These have to be considered taking into account actions (from no action to full commitment) to manage natural water quality (which is poor in some cases), contamination due to hydrodynamic changes induced by water use, and pollution, with special emphasis on that due to agriculture and feedstock, but also due to mining. This evolves over time, especially in groundwater, where the appearance of changes is delayed after some pressure is introduced.
- A wider set of case studies. The Guadiana and Doñana case studies are broadly studied from an agricultural point of view, yet water quality aspects are not fully addressed. A few study cases could be added, such as the Plana de Castelló, the lower Júcar and the Río Verde, but these would need additional time and effort. One of the best studied cases is the Llobregat basin, and especially its lower part,

a main source of water to Barcelona's metropolitan area. This basin is dominated by urban and industrial water supply and involves diverse water sources and activities, which include water transfers from other basins, artificial recharge of the aquifers, advanced river water treatment, seawater desalination, waste water reclamation and the integrated use of aquifers in the framework of all water resources. There, water quality is an important issue (Custodio, 2012).

–   Food production and its security not only refer to agriculture and feedstock, but to fisheries and other production in open water. Considering open sea fisheries is out of the scope of this book, but other aspects have a close relationship, such as fish and algae production in rivers and lakes, estuaries and deltas, and in the littoral sea affected by continental surface and groundwater discharge. Water policy and food production have an important influence, but are poorly known components.

–   A more detailed, basin-scale assessment of the role of land use and land cover changes in the observed reduction of runoff and the analysis of the potential trade-offs of carbon mitigation measures like afforestation from a water management perspective.

–   The assessment and valuation of the multiple ecosystem services supplied by Spain's water-dependent ecosystems to reveal the socio-economic importance of securing water bodies and the need to recognize nature as a full water user.

–   The analysis and further development of the different methodologies used in the water footprint assessment at river basin level in Spain.

–   A more precise picture of groundwater in Spain (e.g. based on the 2009 Agrarian Census), which is necessary to: a) achieve a good status of surface and groundwater bodies; b) to be able to enforce the Water Code; c) to know the manpower and economic means that require the Water Authorities to cope with these issues, and to negotiate with farmers' lobbies.

5 CONCLUSIONS AND RECOMMENDATIONS

The results of this book can be summarized stating that currently in Spain, as in other semi-arid countries, water problems are not due to physical water scarcity but rather to poor water governance and significant inertia and resistance to evolving and adapting to the present challenges. Based on the results obtained at different scales it seems clear that there is room to improve the allocation of water resources at a moderate cost. This cost comes in the form of foregone agricultural production, primarily products with large blue water footprint and low economic value per cubic metre. This might be achieved without producing social disturbances or triggering scarcity of staple foods. However, it is necessary to take into account that agriculture plays a very important role from the standpoint of landscape, biodiversity and rural life, and that any changes in land use affect the whole human and natural systems. Thus, any transition needs to be undertaken keeping in mind indirect impacts and intangible values associated with land and water uses.

Spain has experienced drought cycles for centuries. Its institutions have evolved and become more complex, flexible and diverse. Some flexibility was introduced with water markets since the Water Law was reformed in 1999, but exchanged volumes so far have been small, highly concentrated in some areas and far from the volumes that would perhaps need to be reallocated in the future. Current environmental degradation shows that ecosystems must be better preserved and ecosystem services boosted. An important opportunity will be to release significant amounts of blue water currently used in producing low-economic value products back to nature and to more competitive sectors, especially if climate change predictions are confirmed.

In Spain, 90% or more of consumptive uses of water for human activities are due to agriculture. International experience shows that farmers' lobbies are extremely powerful and it is almost impossible to achieve sound water and agriculture policies without their collaboration: it is necessary to look for win-win solutions. Farmers have to increase their economic productivity (*more cash per drop*) and also ensure that farming practices become less polluting and compatible with natural ecosystems, that is, the maintenance of *more care for nature*. Today it is a fairly widespread opinion that the former can be achieved with relative ease thanks to continuing advances in agricultural technology. Achieving the second objective does not seem so easy, although the future CAP for 2014–2020 is exploring options to promote greener payments. The current movement driven by the United Nations toward a *Green Growth* may also help.

Water quality related aspects are less developed since information sources in Spain are scarcer and less comprehensive studies are available, shadowed by the high pressure on water quantity that is common in arid and semi-arid countries. This has to be further developed in future publications since it is becoming increasingly relevant, particularly for the agricultural sector, which plays an important role as a polluter and also as a sector susceptible to significant improvements.

Technological advances during the last decades clearly show that most current problems can be solved with means that were unthinkable only twenty years ago. These means are mainly related to the international food (virtual water) trade made possible by modern food storage and transportation systems, the membrane technology, the intensive use of groundwater (but better planned and controlled), and remote sensing techniques as a tool for water and land use monitoring. Moreover, advances in communication technology (mobile phones, Internet) can contribute to increasing the general information transparency and stakeholders' participation in decision-making processes. These are in our view the main findings of the book, and those that have received most attention in many of its chapters.

The rational allocation of water to uses that are considered to be of high value to society – be it economic, social or environmental – would help in reframing the idea of physical water scarcity. Nonetheless, water reallocation among users or from users to nature is far from simple. Similarly, initiatives portrayed as the solution to the water governance *jigsaw*, such as water trade, improved water use efficiency, users' collective action and public participation, are not free from difficulties and shortcomings. However, knowing the economic value of water uses helps dilute the idea that having water is automatically followed by thriving societies and opens the possibility to shift water debates from rights ("water is mine for historical reasons") to needs ("I need water for covering these needs") and from needs to sharing benefits ("how can I cover these needs").

In Spain water has often been used as a political weapon (Llamas, 2005; López-Gunn, 2009) and, as matter of fact, during the past decade water policy issues have significantly influenced the vote in several regions and have led to tensions between regions, between regions and the State government, and among stakeholders. In the last five years several influential people have advocated that, to temper this political factor in Spanish water policy, it might be necessary to have a political agreement – a Water Pact – between the main political parties based on a shared vision of the future of water in Spain. Having transparent data on available water resources and their uses seems a sensible first step to create awareness of the great changes in the water paradigms that have occurred in the Spanish and global water context. The following step is moving away from nested positions on water, towards the achievement of this Water Pact.

Our work is motivated precisely to provide results and findings that always come with some qualification as to their reliability and trustfulness. Some of our findings are robust, some others subject to further scrutiny. And yet, with all the caveats that our own results and those of others may have, there are reasons to be optimistic about water policy in Spain. We believe this book offers enough evidence to support the contention that a number of crucial measures could make a difference to water policy in Spain. The social and economic costs require a careful and sensible implementation process, yet in our view are affordable and possible.

REFERENCES

Allan, A. (2002). *Water Security in the Middle East: The Hydro-Politics of Global Solutions.* Columbia International Affairs Online (CIAO), New York, USA.

CEDEX (2011). *Evaluación del impacto del cambio climático en los recursos hídricos en régimen natural* [Evaluation of climatic change impact on water resources under natural conditions]. CEDEX, Madrid, Spain. Available from: http://www.magrama.es/es/agua/temas/planificacion-hidrologica/planificacion-hidrologica/EGest_CC_RH.aspx [Accessed 15th January 2012].

Custodio, E. (2012). The Llobregat aquifers: intensive development, salinization, contamination and management. In: Sabater, S.; Ginebreda, A. & Barceló, D. (eds.) *The Llobregat: the story of a polluted Mediterranean River.* Springer, Dordrecht, the Netherlands.

Estrela, T. & Vargas, E. (2012). Drought management plans in the European Union. The case of Spain. *Water Resources Management,* 26(6): 1537–1553. doi: 10.1007/s11269-011-9971-2.

Garrido, A.; Llamas, M.R.; Varela-Ortega, C.; Novo, P.; Rodríguez-Casado, R. & Aldaya, M.M. (2010). *Water Footprint and Virtual Water Trade in Spain: Policy Implications.* Springer, New York, USA.

Guodong, C. (2003). Virtual Water. A strategic instrument to achieve water security. *Bulletin of the Chinese Academy of Sciences,* Issue 4. Available from: http://en.cnki.com.cn/Article_en/CJFDTOTAL-KYYX200304005.htm [Accessed: November 14th, 2011].

Hoekstra, A.Y. (ed.) (2003). Virtual water trade: Proceedings of the International Expert Meeting on Virtual Water Trade. Value of Water Research Report Series, 12. UNESCO-IHE.

INE (2012). National Statistics Institute (Spain). Available from: http://www.ine.es [Accessed 3rd December 2011].

Llamas, M.R.; Fornés, J.; Hernández-Mora, N. & Martínez Cortina, L. (2001). *Aguas subterráneas: retos y oportunidades* [Groundwater: challenges and opportunities]. Marcelino Botín Foundation and Mundi-Prensa, Madrid, Spain.

Llamas, M.R. (2005). Los colores del agua, el agua virtual y los conflictos hídricos [Water colours, virtual water and water conflicts]. *Revista de la Real Academia de Ciencias Exactas, Físicas y Naturales*, 99(2): 369–389.

López-Gunn, E. (2009). *Agua para todos*: A new regionalist hydraulic paradigm in Spain. *Water Alternatives*, 2(3): 370–394.

Lorenzo-Lacruz, J.; Vicente-Serrano, S.M.; López-Moreno, J.I.; Morán-Tejeda, E. & Zabalza, J. (2012). Recent trends in Iberian streamflows (1945–2005). *Journal of Hydrology*, 414–415: 463–475.

MAAM (Ministerio de Agricultura, Alimentación y Medio Ambiente) (1996). Datos de superficies del II Inventario Forestal Nacional. Available from: http://www.marm.es/es/biodiversidad/temas/inventarios-nacionales/inventario-forestal-nacional/ [Accessed on 3rd September 2011].

MAAM (Ministerio de Agricultura, Alimentación y Medio Ambiente) (2006). Datos de superficies del III Inventario Forestal Nacional. Available from: http://www.marm.es/es/biodiversidad/temas/inventarios-nacionales/inventario-forestal-nacional/ [Accessed 3rd September 2011].

Shah, T. (2009). *Taming the Anarchy: Groundwater Governance in South Asia*. Resources for the Future, Washington, D.C., USA; and International Water Management Institute, Colombo, Sri Lanka.

Part I

Political framework and institutions

Chapter 2

The concept of water and food security in Spain

Elena López-Gunn[1], Bárbara Willaarts[2],
Aurélien Dumont[1], Insa Niemeyer[2] &
Pedro Martínez-Santos[3]

[1] *Water Observatory of the Botín Foundation; Department of*
Geodynamics, Complutense University of Madrid, Madrid, Spain
[2] *Water Observatory of the Botín Foundation;*
CEIGRAM, Technical University of Madrid, Madrid, Spain
[3] *Department of Geodynamics, Complutense University of Madrid,*
Madrid, Spain

ABSTRACT: Water and food security are tags used widely yet hiding very different meanings depending on the context in which they are used. This chapter looks at what these concepts mean for Spain and across scale linkages due to globalisation. Since food production and access is largely guaranteed in Spain, food security here is linked to the idea of guarantying food safety and food health. As in other European countries, there has been a substantial shift in the dietary habits of Spanish consumers with changes to the recommended Mediterranean diet, with higher meat and processed food consumption, and a drop by half in the intake of cereals, legumes, fruits and vegetables. This chapter argues that dietary shifts have increased the water footprint of an average Spanish diet by 8%, which has been possible thanks the imports of green virtual water from third countries, without compromising Spain's water security. The chapter also reflects on the different dimensions of water security in Spain, and whether some aspects of water security (like protection from hazards or water availability) have been secured others represent important – sometimes contradictory – challenges like securing water for food or the environment. These links can be understood when framed by a global system with feedbacks between food production and consumption, impacting on agricultural production and water resources, food supply capacities, and environmental security.

Keywords: Mediterranean diet, nutrition, globalization, virtual water, environmental trade-offs

1 INTRODUCTION

Water and food security are tags widely used nowadays although these can hide very different meanings depending on the socio-economic context. This chapter sets the frame of what these concepts mean particularly in the context of Spain and its linkages across scales due to globalisation. Also, it emphasises that, in the long term, the overarching goal for Spain and other countries is how to meet water and food security without compromising national and international environmental security.

2   WHAT DOES FOOD SECURITY MEAN GLOBALLY AND FOR THE PARTICULAR CASE OF SPAIN?

The food price crisis of 2008 evidenced the vulnerability to hunger and food shortages of many countries throughout the world, and the need to place global food security as a high priority on the international agenda. In poor countries the effects of this price spike on staple food put at risk sufficient food access, exacerbating the hunger problem. In developed countries the price spike did not compromise food access of households for the most part, although it impacted on consumer's food purchasing power. For example, the average food consumer price index in Spain increased by about 7% compared to only 3% before (CEC, 2008). This differentiated set of consequences from price changes and volatility supports the idea that challenges to increase food security worldwide will vary depending on the socio-political context of countries.

The UN Food and Agriculture Organization (FAO) defines a country as food secure when "all people at all times have physical, social and economic access to sufficient, safe and nutritious food to meet their dietary needs and food preferences for an active and healthy life" (FAO, 1996). Hence, the overall goal is not only about ensuring sufficient food production, but also assure a stable access, with the added guarantee of quality at affordable prices. Yu et al. (2010) used a set of four different indicators (food intake, food production, trade security and agricultural potential) to quantify the degree of a nation's food security (see Figure 1) after Diaz-Bonilla et al. (2000). The dichotomy North-South is once again evidenced, with most Northern developed countries classified as food secure and most African and Andean countries being food insecure due to their insufficient production capacity and unstable market access, despite having a large agricultural potential. This evidences that in developing countries efforts could be placed on increasing agricultural productivity to satisfy the food demand from an increasing population. Ensuring sufficient production but above all a fair and stable access to food are two fundamental prerequisites to bridge the food gap. Also, a greater participation of these countries in world markets could increase their food security, as it will mean larger capacity to buy or sell food and to adjust their production to global price signals, which will generate larger government revenues and overall economic growth, all of which have direct or indirect impacts on the nutritional status of people in the country (Nouve, 2004). However, to reach this target more just and equitable international food trade regulations need to be formulated in the context of the World Trade Organization (Von Braun, 2008). In developed countries like those of EU-27 (and Spain as a full member of the EU), advances in food security are linked to maintaining both production and market access, ensuring the exporting capacity, guaranteeing the quality of products and reducing the environmental impacts associated with production processes (EC, 2010).

Yet the food security debate at the global level is very much focused on the quantitative side, although more recently the FAO is turning increased attention to the importance and potential for quicker progress by considering the qualitative aspect of "safe and nutritious food" and "food preferences for an active and healthy life". In countries like Spain production and access are largely guaranteed, and food security is linked to the idea of guarantying food safety and food nutrition. Figure 2 offers a

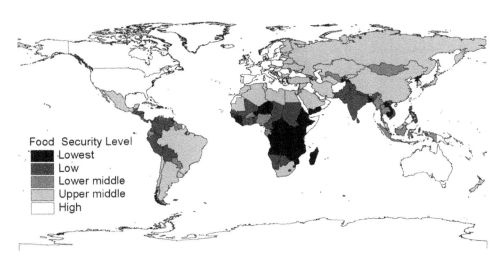

Figure 1 Global Food Security. (Source: Yu *et al.* (2010); calculation based on FAO (2009)).

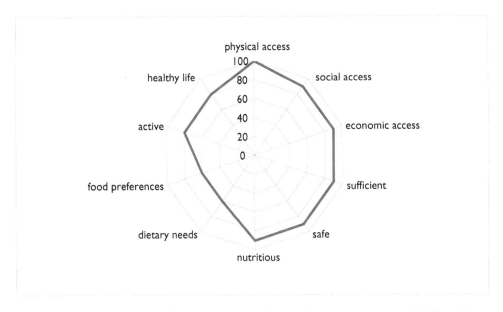

Figure 2 Different dimensions of food security applied to Spain on the basis of FAO definition. (Source: Own elaboration).

conceptual diagram to consider all the different qualitative aspects of food security definition applied to Spain.

The main initiatives taken on food safety in Spain so far refer to issues related to food labelling, and risks associated with the whole food chain. Thus, improving food safety in Spain is very much focused on increasing consumer's confidence after scares related to e.g. BSE *Bovine spongiform encephalopathy*, colza oil, and

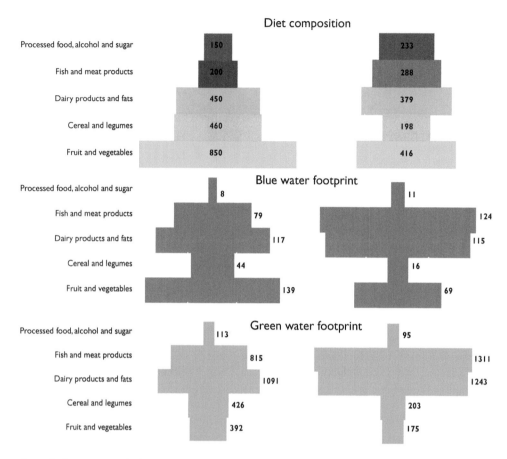

Figure 3 Composition (gr/person/day), blue and green water footprint (L/day) of the recommended Mediterranean diet (left) and the current average diet (right) of a Spanish adult consumer. (Source: Own elaboration based on data from SENC (2005) and AESAN (2006)).

more recently E-coli. To this end Spain set up in 2002 a National Food Security and Nutrition Agency (Agencia Española de Seguridad Alimentaria y Nutrición, AESAN). This agency is in charge of promoting food safety and the health of citizens, by adopting precautionary principles in food consumption and production, e.g. in relation with Genetically modified organisms or GMOs and food traceability to minimize food risk. Despite the efforts, Spain is confronting a nutritional problem, as 12.9% of the Spanish population is now obese, compared to e.g. 20% in the USA and 5% in countries like China or Japan (Aranceta-Bartrina *et al.*, 2005). As in other European countries, there has been a substantial shift in the dietary habits of Spanish consumers in the last decades (see Figure 2). Today Spanish consumers eat 30% more animal proteins and 60% more processed food (sugar, bakery, fast food) and alcohol compared to the gold standard Mediterranean diet, a diet officially recognized by the World Health Organization for its health benefits and its equilibrated nutritional balance (Padilla *et al.*, 2006; 2010), and also considered part of Humanity's immaterial cultural heritage by UNESCO in November 2010. Meanwhile, the intake of cereals, legumes, fruits and vegetables has dropped by half. This has been possible to a large extent because of Spain's animal feed imports from third countries (mostly from Latin America), which have allowed the development of a growing livestock sector. From a water-requirement perspective, Spain can maintain its rate of livestock production thanks the imports of virtual water embedded in the soybean produced in countries like Brazil or Argentina. According to Fereres *et al.* (2012) the changes in the Spanish diet in the period 1964 to 1991 were about 20% of the water footprint in the average diet, whereas from 1991 the increase in the water footprint has been smaller (less than 5%). This is mainly attributed (as can be seen in Figure 2) to an increased consumption in animal proteins and fats (Fereres *et al.*, 2012). Comparing the water footprint of the current Spanish average diet with the recommended Mediterranean diet, our preliminary results show that following the recommended diet would mean a reduction of 8% of the total water footprint (WF) (green and blue) (from 1.1 to 1.0 hm^3/person/year). This decrease would be largely driven by a decrease in the green WF embedded in the current consumption of a greater quantity of animal proteins. However, the blue WF would rise (around 12%) as a result of a higher intake of fruits and vegetables, which is nowadays largely produced with blue water resources. The link can be interpreted as a feedback system between food consumption, production and linkages with global agricultural production and water resources, which affect food supply capacities and food consumption.

3   WHAT DOES WATER SECURITY MEAN GLOBALLY AND FOR THE PARTICULAR CASE OF SPAIN?

The concept of water security is relatively young, and has evolved since it first appeared in the literature at around the turn of the 21st century. Originally, the water security concept was approached from a physical perspective linked to national security, a concern over potential *green wars* and to cases where water could be a potential cause of conflict between neighbouring countries. Therefore, securing sufficient water resources and guaranteeing access was seen as strategic for national security (Houdret, 2004). More recently, attention has shifted to domestic conflicts, between groups or users. In Spain this in some ways echoes domestic trade-offs and regional conflicts that will have to be faced in the re-allocation of water between sectors and regions (Fereres *et al.*, 2012) and the potential securitization of water. This highlights the inherent scale specific nature of water security as discussed by Cook & Bakker (2012). Lately, the term has also been associated to virtual water as a means for a State to ensure *water security* through food imports (Allan, 2002; Guodong, 2003).

A more modern acceptation refers to the ability to assure the good functioning of the water cycle and all the functions associated to water, to allow the perennial use of water for all dependent activities and particularly the production of food and services associated to the good health of water-related ecosystems. In this sense ensuring *water security*, is seen as synonymous to providing resilient systems. Thus

Table 1 Definitions on the concept of water security.

	Water security: definitions
UNEP (2009)	"[...] water security represents a unifying element supplying humanity with drinking water, hygiene and sanitation, food and fish, industrial resources, energy, transportation and natural amenities, all dependent upon maintaining ecosystem health and productivity."
UNESCO-IHE (2010)	"Water security involves protection of vulnerable water systems, protection against water related hazards such as floods and droughts, sustainable development of water resources and safeguarding access to water functions and services."
Global Water Partnership (2010)	A water secure world harnesses water's productive power and minimizes its destructive force. It is a world where every person has enough safe, affordable water to lead a clean, healthy and productive life. It is a world where communities are protected from floods, droughts, landslides, erosion and water-borne diseases. Water security also means addressing environmental protection and the negative effects of poor management. A water secure world means ending fragmented responsibility for water and integrating water resources management across all sectors – finance, planning, agriculture, energy, tourism, industry, education and health. ... A water secure world reduces poverty, advances education, and increases living standards. It is a world where there is an improved quality of life for all, especially for the most vulnerable – usually women and children – who benefit most from good water governance.

this coincides with the definition of Grey & Sadoff (2007) which defined water security as: "the availability of an acceptable quantity and quality of water for health, livelihoods, ecosystems and production, coupled with an acceptable level of water-related risks to people, environments and economies". Throughout history, the improvement of water security has been associated to the development of storage capacity, with Spain as a paradigmatic case. However, an integrated view of the way to achieve water security has emerged more recently, particularly associated to the *ecosystems services* concept, where the *good functioning* of the hydrological cycle and associated ecosystems can bring resilience and security (Willaarts *et al.*, 2012). Thus in the most recent acceptations water security (and food security as was previously shown) are now encompassed within the wider concepts of human security, in turn ultimately sustained by environmental security (López-Gunn *et al.*, 2012) (See Table 1).

As Cook & Bakker (2012) summarize, there are four interrelated themes that dominate water security: water availability, human vulnerability to hazards, human needs, with a special emphasis on food security and sustainability. All these threads are present in the kaleidoscope that is water security in Spain. First, in terms of water availability, from a historical standpoint water security has been a concern for its inhabitants from early times. This is only to be expected in a region subject to a semi-arid climate. Even today, Roman aqueducts and Arabic waterwheels scattered across Spain bear witness to the country's long-standing water management traditions. So do relatively sophisticated water supply networks such as Madrid's *qanats*. In other words, the *official* strategy was to alleviate scarcity by increasing the supply through dams and canals. This explains why Spain was one of the pioneering countries to institute river basin authorities. It also explains why Spain ranks fourth, one of the world's leading countries in number of dams per million inhabitants. Second, from the perspective of human vulnerability to hazards, these are included into Spanish planning as a country vulnerable to climate variability and extreme events. For example, according to Estrela (pers. comm.), ex deputy director for water planning in Spain, the EU Directive 2007/60/EC on the assessment and management of flood risks, requires Member States to assess if all water courses and coast lines are at risk from flooding to map the flood extent, assets and humans at risk in these areas, and to take adequate and coordinated measures to reduce this flood risk. Its aim is to reduce and manage the risks that floods pose to human health, the environment, cultural heritage and economic activity. In a new concern over climate change, there is now added interest on the intensification of the water cycle and protection against extreme events, principally floods and droughts (see Chapter 15). Third, in relation to human needs, with a special emphasis on food security, there has also been an on-going concern in Spain for *crop water security* marked by the importance of irrigation and securing water for the dominant water consumptive sector. However, the most predominant and on-going challenge for Spanish water security, as highlighted in Chapters 3 and 12, is environmental sustainability and the ecological functions aspects of water security.

The transposition of the EU Water Framework Directive (WFD) into Spanish law brought about certain paradigm shifts in water security, including the emphasis on the need to enforce environmental sustainability in water management practices. In this regard, the first step consists in establishing the baseline conditions of water bodies. This implies delineating the water bodies and establishing the uses and pressures these are subject to, as well as the economic value of water and a series of environmental objectives and measures to attain them. The WFD also establishes the mandate to maintain the *good status* of water bodies, and demands the restoration by 2027 of those that have been subject to severe modifications by human beings (see Chapter 3 on the WFD).

As explained earlier, the concept of water security is manifold. If water planning is broadly described as the process whereby water security is attained, then the nature of the water planning process is both technical and political. It is technical because it implies a series of objective studies to establish the status of water bodies and monitor their evolution, and it is political because it implies a process of negotiation on what the priorities are and whether exceptions need to be made. Regarding water security, the implementation of the WFD brings about changes that are far from subtle. This is because water security is no longer understood as *guaranteeing water supply for all human activities*. Rather, water security is now described as the simultaneous achievement of three general objectives. The first one relates to maintaining the *good status* of water bodies, the rationale being that water security begins by protecting the resource base on ecological functions. In second place, water planning should aim at meeting demands as identified by basin plans; finally, the law establishes that water management practices should strive to harmonize both principles by making a rational use of water. In other words, the water security concept has become richer at the cost of losing some focus. While this can be perceived as adequate in some ways, it also implies that water security is now subject to interpretations as to what *a [sufficiently] good status* means.

4　CONCEPT OF FOOD SECURITY AND VIRTUAL WATER TRADE IN SPAIN IN A GLOBALIZED WORLD

Food security and water security, as discussed earlier, are highly interconnected concepts. From a production perspective, one question is how to fill this water gap in order to achieve global food security goals. One important solution to close the supply and demand gap lies in technological innovation to increase yield (the yield gap), e.g. via agronomy and plant breeding (Fereres *et al.*, 2012), and also by improving water productivity. In addition, fostering international food trade is another piece of the puzzle in global food security. Trade – if adequately regulated – has the potential to bridge the mismatch between areas with the largest production potential and those with large populations without the capacity to increase production (Rockström *et al.*, 2009). Generally speaking, food trade improves physical and economic access to food by increasing food availability and lowering food prices for domestic consumers. From a water perspective, food trade can be described as the virtual flow of water from producing and exporting countries to importing and consuming countries. Several authors (Allan, 1993; Niemeyer & Garrido, 2011) have described how water short countries can enhance their food security by substituting national production by imports of water intensive food crops. However, from a natural resource point of view, the increased reliance on farm trade raises some crucial issues as exporting countries might not have the capacity or political willingness to curtail powerful exporting sectors on the basis of environmental constraints, compromising their natural capital or increasing the pressure on already stressed water ecosystems. Therefore, the consequences of globalisation on water and food security have to be examined on a case-by-case basis and cannot be generalised.

The study by Garrido *et al.* (2010) on the Spanish water footprint and nationwide virtual water trade showed that the Spanish water footprint amounts to about 45 km^3, of which 85% is from agriculture. From a water management and food security point of view, in absolute terms, Spain is a net importer of virtual water (Garrido *et al.*, 2010). A high proportion of these virtual water imports result from feed stuff produced in Brazil and Argentina. This is in line with the changing dietary habits discussed earlier. The shift towards a more meat oriented diet leads to higher water (and land) resource requirements that cannot be fulfilled through self-sufficiency. Imports of animal feedstuff with low economic value and high water content are used as feed for high economic value livestock, which are then exported to large parts of Europe. In this regard international trade has contributed to national food security in Spain, at least in quantitative terms, but it has also brought other important issues that cannot be obviated: a growing health problem and a greater pressure on the natural resources of production countries (Willaarts *et al.*, 2011).

5　CONCLUSION

According to FAO/OECD (2011), the rise in price commodities experienced in 2007/08 is predicted to continue to increase over the next decade by 20–30% compared to the period 2001–2010. The FAO/OECD attribute these predictions to several causes, such as a slower annual growth in agricultural stock levels and other compounding factors like climate variability, energy prices, exchange rates, growth in demand, and potential trade restrictions and speculation. Current drivers of change include globalization, the rise of new powers in the emergent economies, the erosion of international institutions, and the rising geopolitics of energy, increasingly tied up with water through the water/energy nexus (see Chapter 14). This new emerging global system is multi-polar, with potential future rivalries predicted over trade, investments and technological innovation. The transfer of wealth, from old economies to new economies is happening at a speed and size (an *accelerated* phase) never witnessed before, in the so called era of *the anthropocene* (Steffen *et al.*, 2011). All these changes in the world balance have deep ethical implications for global water and food security (Lopez-Gunn *et al.*, 2012). In this context Spain is, in many ways, in a relative solid position thanks to a relatively strong water and food security. The main challenge lies on how to make the long term viability of water and food security, fit in with national and international environmental security, understood as securing that water and food security are not at the expense of environmental capital, either in Spain or as an external impact, through the import of food (virtual water) which could affect the achievement of water and food security in other nations now linked through international trade. As the FAO (2011) states, in many ways emergent and developing countries are, and will become, the engines for growth, while developed countries have a lot to offer in terms of food safety and environmental standards. The challenges are global regulations to prevent environmental and social dumping at a global scale.

REFERENCES

AESAN (*Agencia Española de Seguridad Alimentaria*) (2006). *Modelo de Dieta Española para la determinación de la exposición del consumidor a sustancias químicas* [Model of Spanish diet to determine consumer exposure to chemicals] Available from: http://www.aesan.msps.es/AESAN/docs/docs/notas_prensa/modelo_dieta_espanola.pdf [Accessed 12th January 2012].

Allan, J.A. (1993). Fortunately there are substitutes for water; otherwise our hydro-political futures would be impossible. In: *Priorities for water resources allocation and management*, ODA, London: 13–26.

Allan, J.A. (2002). Water Security in the Middle East. The Hydro-Politics of Global Solutions. *Columbia International Affairs* Online (CIAO), New York, USA.

Aranceta-Bartrina, J.; Serra-Majem, L.; Foz-Sala, M.; Moreno-Esteban, B. & Seedo, G. (2005). Prevalencia de obesidad en España [Prevalence of obesity in Spain]. *Med. Clin (Barc)*125 (12): 460–466.

CEC (Commission of the European Communities) (2008). *Food prices in Europe*. Available from: http://ec.europa.eu/economy_finance/publications/publication13571_en.pdf [Accessed 2nd March 2012].

Cook, C. & Bakker, K. (2012). Water security: debating an emerging paradigm. *Global Environmental Change*, 22: 94–102.

Diaz-Bonilla, E.; Thomas, M.; Robinson, S. & Cattaneo, A. (2000) Food security and trade negotiations in the world trade organization: a cluster analysis of country groups *TMD Discussion Paper NO. 59*; IFPRI, Washington DC, USA.

EC (2010). The CAP towards 2020: *Meeting the food, natural resources and territorial challenges of the future*. Communication from the Commission to the European Parliament, the

Council, the European Economic and Social Committee and the Committee of the regions, Brussels. Available from: http://ec.europa.eu/agriculture/cap-post-2013/communication/com2010-672_en.pdf [Accessed 8th February 2012].

FAO (1996). Rome Declaration on world food security and world food summit plan of action. In: *World Food Summit* 13–17 November 1996, Rome, Italy.

FAO (2009). *Assessment of the world food security and nutrition situation*. Committee on World Food Security, 34th Session. Rome, Italy.

FAO (2011) *World Livestock 2011. Livestock in food security*, Rome, Italy.

FAO/OECD (2011). Agricultural Outlook 2011–2020. FAO/OECD, Rome, Italy.

Fereres, E.; Orgaz, F. & González-Dugo, V. (2012). Reflections on food security under water scarcity. *Journal of Experimental Botany* 62: 4079–4086.

Garrido, A.; Llamas, M.R.; Varela-Ortega, C.; Novo, P.; Rodríguez-Casado, R. & Aldaya, M.M. (2010). *Water Footprint and Virtual Water Trade in Spain*, Springer, New York.

Global Water Partnership (2010). *What is water security?* Available from: http://www.gwp.org/The-Challenge/What-is-water-security/ [Accessed 14th November 2011].

Grey, D. & Sadoff, C. (2007). Sink or Swim? Water security for growth and development. *Water Policy*, 9(6): 545–571.

Guodong, C. (2003). *Virtual Water – A Strategic Instrument to Achieve Water Security*. Bulletin of the Chinese Academy of Sciences. Issue 4. Available from: http://en.cnki.com.cn/Article_en/CJFDTOTAL-KYYX200304005.htm [Accessed 14th November 2011].

Houdret, A. (2004). *Water as a security concern – conflict or cooperation?* 5th Pan-European Conference of International Relations. September 8–11, 2004. The Hague, Netherlands. Available from: http://www.afes-press.de/pdf/Hague/Houdret_Water_Security.pdf [Accessed 14th November 2011].

López-Gunn, E.; De Stefano, L. & Llamas, M.R. (2012). The role of ethics in water and food security: Balancing utilitarian and intangible values. *Water Policy*, 14, Supplement 1: 89–105. Available from www.fundacionbotin.org/agua.htm [Accessed 14th July 2012].

Niemeyer, I. & Garrido, A. (2011). International Farm Trade in Latin America: Does it Favour Sustainable Water Use Globally? *Value of Water Research*, Report Series, 54, UNESCO-IHE.

Nouve, K. (2004). *Impacts of global agricultural trade reforms and world market conditions on welfare and food security in Mali: a CGE assessment*. Dissertation, Department of Agricultural Economics, Michigan State University, USA.

Padilla, M.; Hamimaz, R.; El Dahr, H.; Zurayk, R. & Moubarak, F. (2006). *The development of products protecting the health and the environment in the Mediterranean region*. CIHEAM Analytic No. 5, March 2006. Available from: http://portail2.reseau-concept.net/Upload/ciheam/fichiers/ANP5.pdf [Accessed December 2012].

Padilla, M. (2010). *Is the Mediterranean diet, world paragon, sustainable from field to plate?* Presentation at Food day/World food week 2010 International Scientific Symposium on "Biodiversity and sustainable diets" United against hunger, 3–5 November 2010; Organised by FAO and Biodiversity international.

Padilla, M. (2001). Evolution of Mediterranean Diet: Facts, Causes, Effects. In: Amado, R.; Lairon, D.; Gerber, M.; Maiani, G. & Abt, B. (eds.) *Bioactive micro nutrients in the Mediterranean diet and health European communities* pp. 263–272; Luxembourg.

Rockström, J.; Steffen, W.; Noone, K.; Persson, Å.; Chapin, III, F.S.; Lambin, E.; Lenton, T.M.; Scheffer, M.; Folke, C.; Schellnhuber, H.; Nykvist, B.; De Wit, C.A.; Hughes, T.; van der Leeuw, S.; Rodhe, H.; Sörlin, S.; Snyder, P.K.; Costanza, R.; Svedin, U.; Falkenmark, M.; Karlberg, L.; Corell, R.W.; Fabry, V.J.; Hansen, J.; Walker, B.; Liverman, D.; Richardson, K.; Crutzen, P. & Foley, J. (2009). Planetary boundaries: exploring the safe operating space for humanity. *Ecology and Society* 14(2): 32. Available from: http://www.ecologyandsociety.org/vol14/iss2/art32/ [Accessed 3rd April 2012].

SENC (*Sociedad Española de Nutrición Comunitaria*) (2005). *Guía de la alimentación saludable* [Guide to Healthy Eating] Ed. Everest: 21–22. ISBN: 9788424108519.

Steffen, W.; Grinevald, J.; Crutzen, P. &McNeill, J. (2011) The Anthropocene: conceptual and historical perspectives. *Philosophical Transactions of the Royal Society A: Mathematical Physical and Engineering Sciences*, 369: 842–867.

UNESCO-IHE (2010). *Water security*. Available from: http://www.unesco-ihe.org/Research/Research-Themes/Water-security [Accessed 16th November 2011].

UNEP (United Nations Environment Programme) (2009). *Water security and ecosystem services: the critical connection*. Available from: http://www.unep.org/themes/freshwater/pdf/the_critical_connection.pdf [Accessed 16th November 2011].

Von Braun, J. (2008). *High Food Prices: The What, Who and How of Proposed Policy Actions*. International Food Policy Research Institute, Washington, DC, USA.

Yu, B.; You, L. & Fan, S. (2010). *Toward a typology of food security in developing countries*. IFPRI Discussion Paper 00945, Development and Strategy Governance Division. January.

Willaarts, B.; Niemeyer, I. & Garrido, A. (2011). Land and Water requirements for soybean cultivation in Brazil: Environmental consequences of food production and trade. *Conference Proceedings prepared for the International Water Resources Association* (IWRA), Brazil.

Willaarts, B.; Volk, M. & Aguilera, P.A. (2012). Assessing ecosystem services supplied by freshwater flows in Mediterranean agroecosystems. *Agricultural Water Management*, 105: 21–31.

Chapter 3

Water planning and management after the EU Water Framework Directive

Lucia De Stefano[1] *& Nuria Hernández-Mora*[2]

[1] Water Observatory of the Botín Foundation; Department of
  Geodynamics, Complutense University of Madrid, Madrid, Spain
[2] Founding Member, New Water Culture Foundation, Madrid, Spain

ABSTRACT: This chapter provides an overview of the different legal, administrative and economic factors that provide the institutional context for water management in Spain, focusing on the effects of the 2000 European Water Framework Directive (WFD). At present and partially due to the WFD implementation process, the Spanish water sector is experiencing a slow transition from old to new water paradigms. Highlights in this sense are the consideration of the achievement of ecological quality as a primary planning and management objective; an increase in public participation and transparency in water-related decision processes; the economic analysis of water services; and an increased emphasis on water demand management. The achievement of the WFD objectives faces several challenges and uncertainties that are of technical, financial and political nature. However, possibly the key to a successful implementation of the WFD and a real shift of paradigm lays in strengthening the link between land use and water management and in creating institutional structures that facilitate co-responsibility and full cooperation between the central state and the regions, who hold most of the responsibilities on land use management.

Keywords: water policy, Water Framework Directive, water paradigm, IWRM, assessment

1 INTRODUCTION

The start of the 21st century has brought about significant changes in water policy and management in Spain. These include the approval of several new pieces of water-related legislation, the resulting reform of river basin planning procedures, and a progressive decentralization of water management. The latter in turn has led to the creation of new regional river basin authorities and changes in the distribution of functions and authority on water planning and management between the central and regional governments. This chapter provides an overview of the different factors that provide the institutional context for water management in Spain, focusing on the effects of the European Union's Water Framework Directive (WFD, *Directive 2000/60/EC*) approved in 2000 and transposed into Spanish legislation in December 2003.

2   LEGAL, ADMINISTRATIVE AND ECONOMIC FACTORS

While it is not the objective of this chapter to make a comprehensive analysis of all the elements that influence current Spanish water policy, it is worth highlighting some of the main legal, administrative and economic changes that have taken place during the past decade and define Spain's water management framework today.

In the legal sphere, the European Union (EU) has approved a number of new Directives that have important implications. First among them is the WFD, which represents the main focus of this chapter and will be further discussed in the next pages. Other relevant EU Directives are the Directive on the Protection of Groundwater Against Pollution and Deterioration (2006/118/EC), the Dangerous Substances Directive (2006/11/EC), the Environmental Quality Standards Directive (2008/105/EC) and the Directive on Integrated Pollution Prevention and Control (2008/1/EC), all of which set strict qualitative requirements and management standards for both ground and surface waters. The Directive on Public Access to Environmental Information (Directive 2003/4/EC) adapts European legislation to the requirements of the Aarhus Convention (Convention on access to information, public participation in decision-making and access to justice in environmental matters) and reinforces the WFD requirements in relation to public participation (see Chapter 17). Additionally, during the past decade, Spain has made significant progress in compliance with two Directives approved in the 1990s that have great influence on the status of water bodies: the Wastewater Treatment Directive (91/271/EEC) and the Directive for Protection of Waters against Pollution caused by Nitrates from Agricultural Sources (91/676/EEC).

At the national level, the Spanish National Hydrological Plan (NHP) (Law 10/2001) – a crucial element in pre-WFD Spanish water law – and its subsequent amendments (Royal Decree 2/2004 and Law 11/2005), dominated Spanish water policy debates in the first half of the decade. The key element of this plan was the highly disputed Ebro water transfer scheme, a 914 km long transfer of 1,000 hm³/year [hm³ = cubic hectometre = million m³ = 10^6 m³] from the lower Ebro river in the northeast of Spain to the different provinces along the Mediterranean coast. The abrogation of this project by Royal Decree (RD 2/2004) marked a milestone in Spanish water policy for several reasons. First, it meant a rupture with the long-standing discourse of *hydro-solidarity* among *water abundant* and *water scarce* regions, and a clear example of the use of water as a political weapon.[1] Second, in the Mediterranean regions it implied a shift of the emphasis from reservoirs and transfer schemes to desalination plants as a means of augmenting water supply.[2] Third, it emphasized (at least on paper) demand

---

1 The donor and recipient areas of the transfer were of opposite political colour and the abrogation or continuation of the transfer was a political flag during the 2004 national elections campaign. The abrogation of the project was the second major act by the newly elected Government, after the withdrawal of Spanish troops from Iraq.

2 The AGUA Program (Actions for Water Management and Use), approved in 2005, emphasized the construction of desalination plants along the Mediterranean coast to substitute the water that the Ebro transfer would have contributed. The planned desalination capacity amounted to over 700 hm³/year. Today many of the planned plants are operational, and desalinated water is heavily subsidized to make it affordable for farmers. However, plants are dramatically underused (only operational at 20% capacity for the most part), and farmers continue to rely on regulated or unregulated groundwater resources, aggravating the situation of many intensively used aquifers along the coast.

management approaches, boosted wastewater, recycling and reuse and, although timidly, increased the focus on water governance issues.

The NHP Law also improved drought management practices requiring the elaboration of normative Drought Management Plans at the river basin level and Drought Emergency Plans for cities of over 20,000 inhabitants. These plans imply a shift from reactive responses to cyclical droughts to proactive management approaches and are in line with the recommendations set out in the 2007 European Commission Communication on Water Scarcity and Droughts. Drought Management Plans were developed and approved in 2007 for all river basins. The NHP also included a specific requirement for the development of a Special Management Plan for the Upper Guadiana basin, a region where intensive (and often uncontrolled) groundwater development starting in the 1970s contributed to the degradation of protected groundwater-dependent wetland ecosystems. After much negotiation and debate, the Upper Guadiana Special Plan was approved by Royal Decree in 2008, representing a first attempt to tackle the problem of uncontrolled groundwater use in that region in a comprehensive and participatory way. It acknowledged that any viable answer to groundwater mismanagement required devising broadly acceptable solutions with the participation of the main water users (see Chapter 20).

In the administrative sphere, the advent of democracy in Spain in the 1970s brought with it political decentralization, with the transfer of authority over an increasing number of policy areas (education, healthcare, agriculture, environmental and land use policy, among many others) from the central government to autonomous regions.[3] This process also affects the management of water resources. The Spanish Constitution clearly establishes that inter-regional river basins, that is, those that cross more than one autonomous region, should be managed by the central government (through the River Basin Authorities or *Confederaciones Hidrográficas*), whereas river basins that flow entirely within an autonomous region should be managed by its autonomous government. However, the Spanish political landscape has significantly changed since the definition of that constitutional rule, and regions have looked for legal means to increase their role in water management in inter-regional basins. The highly emotional nature of decisions surrounding water and the use of water policy debates for political gain have made it difficult to reach satisfactory solutions in many cases (López-Gunn, 2009). The jurisdictional uncertainty originated by these territorial tensions has hampered river management and planning processes in some cases. For example, in 2007 management authority for the Guadalquivir river (90% of whose area is within the autonomous region of Andalusia), was transferred from the national to the Andalusian regional Government through emergency legislation. A constitutional challenge by the region of Extremadura (which has 6% of the basin within its borders) resulted in a 2011 ruling by the Spanish Constitutional Court to return management responsibilities to the central government. This has led to a prolonged transition period and the associated confusion among water users and River Basin Authority officers. In the Júcar river basin, a legal dispute among the two primary regions sharing the basin (Castilla-La Mancha and Valencia) over

3 The 1978 democratic constitution divided the country into 17 autonomous regions and 2 autonomous cities with varying, and increasing, degrees of autonomy.

the delimitation of the Júcar River Basin District, is preventing the publication of the draft basin management plan and therefore blocking the water planning process.

Given the preponderance of agriculture in the consumption of Spanish water resources, the evolution of agricultural policies has a significant impact on water management and use. Over the past decade there have been significant changes in the EU Common Agricultural Policy (CAP) with an increased emphasis on environmental conditionality and the progressive decoupling of subsidies from production. These changes have brought about a decrease in the production of water-intensive crops that benefited from subsidies in the pre-2003 CAP scheme. For instance, between 2004 and 2006 the surface dedicated to irrigated corn in Spain has decreased by 22%, and for legumes has decreased by 33%, while the surface of less water-intensive crops like winter cereals, vineyards and olive trees has increased between 13 and 18% (Garrido & Varela-Ortega, 2008).

Agricultural policies have also experienced a significant evolution in Spain over the past decade. An increasing emphasis on improved efficiency of irrigation systems materialized in several regional agricultural modernization initiatives in the late 1990s and the early 21st century, and the approval in 2006 of a National Plan for Irrigation Modernization, funded with over 2,000 M€. Although the allocation of saved water to new irrigation uses is casting doubts about the effectiveness of these modernization plans in achieving their stated objectives (see Chapter 19), they have contributed to modify the irrigation dynamics in several areas of Spain.

The Spanish agricultural sector was also affected by the rise of cereal prices in 2008 and the price volatility that followed, reframing the concept of both EU and national food security and the strategic role of staple crop production (EC, 2010).

Finally, the context for water management in Spain should be analysed in light of the global economic recession triggered by the USA subprime mortgage crisis in 2008. The recession has deeply affected the Spanish economy, with significant implications for water management. Budgetary restrictions have dramatically reduced the public administrations' investment capacities. As a result, existing water-related plans and programmes are being revised and reduced. The draft Basin Management Plans' Programmes of Measures will have to be thoroughly reviewed to make them financially viable and will likely have a far more modest emphasis on new infrastructures for water resources development (dams, desalination plants, water transfers) and a greater emphasis on non-structural measures. Moreover, budgetary restrictions could imply a reduction in water-related investments such as the construction of urban wastewater treatment plants, which soon will no longer be backed by EU funds (see Chapter 13). On the other hand, the collapse of the economic growth model based on real estate development[4] is likely to decrease the pressure on water resources, especially in water-stressed Mediterranean regions where much of the tourism-related second home development was concentrated. While this pressure is not large in terms of overall quantity (relative to agricultural uses), it is significant because it is concentrated in space and time, placing a significant burden on municipalities and local environmental conditions.

---

4 Between 2007 and 2010 the number of new homes whose construction was started in Spain decreased by 80% (INE, 2011).

3   THE WATER FRAMEWORK DIRECTIVE: OPPORTUNITIES AND CHALLENGES

The Water Framework Directive set ambitious environmental objectives to EU Member States, who are required to achieve and maintain a good status of all their waters (surface and groundwater; transitional and coastal waters) by 2015, as well as to prevent any further deterioration of that status. According to the WFD, the achievement of these objectives will contribute to the "provision of the sufficient supply of good quality surface water and groundwater as needed for sustainable, balanced and equitable water use, a significant reduction in pollution of groundwater, the protection of territorial and marine waters, and achieving the objectives of relevant international agreements ..." (WFD, Article 1).

A critical milestone in the implementation of the WFD is the approval of a River Basin Management Plan (RBMP) for each River Basin District (RBD). According to the official WFD calendar this should have occurred in December 2009, to allow for a 6-year implementation cycle until 2015. In Spain, work on the WFD implementation started in earnest in 2004 (Hernández-Mora et al., 2011). The resulting RBMPs have to be approved by a governmental Royal Decree, thus having a normative value. However, by the spring of 2012 only the RBMP of the District of the Internal Catalan Basins had completed the approval process, while plans in most of the other RBDs were pending legal approval by the RBD's consultation boards (*Consejos del Agua*) or by the Central Government. In a few cases – Tagus, Segura, and Júcar – the strenuous negotiation over some substantial aspects of the RBMP between the Central Spanish Government and the affected regions was still blocking the publication of the draft plans in May 2012. Consequences of this generalized delay are not only legal actions initiated by the EC against Spain (in 2010 and 2011), but also and more importantly that in most of the river basins the 6-year implementation process will be significantly reduced, thus curtailing the effectiveness of the RBMPs.

The adoption of the WFD implies moving past the water development and supply augmentation paradigms that were forged and successfully applied throughout the 20th century, when Spain needed to harness and use water in order boost its socioeconomic development. The WFD has added new dimensions and challenges to Spain's water policy, requiring that all water bodies achieve good status as a primary management goal. However, some authors have argued that a more substantial reform of water legislation goals and priorities would have been necessary to truly comply with WFD requirements (La Calle, 2008; Hernández-Mora et al., 2011). The transposition of the WFD to Spanish legislation has resulted in the new goals and priorities being superposed to existing demand satisfaction objectives, without truly transforming water policy goals. The European Commission has supported this view and, in 2010, issued a reasoned opinion to the Kingdom of Spain for poor transposition of the WFD into the national legislation.

The WFD sets ambitious public participation requirements as an integral part of the water planning and management process. Prior to the WFD, Spain had a long tradition of user participation in water decision making through the representation of permitted users in the formal consultation bodies of River Basin Organisations (RBOs). Non-economic users or other interested parties, as well as the wider public, had limited access to decision-making processes and had difficulty making their voices heard. The WFD requires RBOs to incorporate all stakeholders, including users, and the wider public in decisions over water, thus widening the circle of participation. In Spain, with a limited tradition in this wider concept of participation, this requirement is hampered by lack of expertise and means, as well as the need to overcome the institutional inertia of the established system (see Chapter 17).

The economic requirements set by the WFD have also posed a challenge. The Directive requires Member States to estimate the economic value of water uses, the cost of the associated water services, and how much of that cost is recovered from users, encouraging the use of water pricing as a tool to achieve an efficient use of water. When designing the Programme of Measures of the RBMPs, RBOs should apply cost-effectiveness analysis to the selection of measures to be implemented. In Spain, the debate on a new water pricing policy was addressed in the failed 2007 Water Act reform effort. The draft Water Act proposed reforming the existing economic and financial regime that applies to water uses. It set a groundwater use fee to cover groundwater management costs. In terms of surface water users, official reports stated that existing fees covered between 85 and 98% of all water service costs (MIMAM, 2007a). These figures, however, were questioned by some authors (Arrojo, 2008), who argued that existing water legislation limits the ability of the water administration to recover even the full financial cost of hydraulic infrastructures. In the case of groundwater, users cover well construction, maintenance and operational costs. But public management and control costs, environmental externalities, or the cost of corrective measures to mitigate the impact of these externalities are neither calculated nor, by consequence, recovered. The pricing reform met frontal resistance from irrigator associations and other stakeholder groups. The Water Act reform proposal did not make it out of congressional debates and was eventually abandoned.

Since the creation of the *Confederaciones Hidrográficas* or RBOs in Spain beginning in the 1920s, water has been managed using the river basin as the basic administrative unit. Hence, the WFD requirement that water should be managed according to River Basin District has not implied a radical transformation in the water management institutional setting as has occurred in other countries (see for instance Hedelin & Lindh, 2008). Nonetheless, the WFD has contributed to unsettling a model that from the outside seemed robust but which internally was strained, since it was designed to deal with objectives, challenges and social demands that greatly differ from current ones. The WFD requires a better integration of sectoral and water policies, and of continental and coastal water management. To facilitate this integration, the Spanish law that transposed the WFD created a coordinating body, the Committee of Competent Authorities, that includes representatives from different administrative levels (national, regional and local), as well as ports and coastal management. In a country where autonomous regions have powers over an ever increasing range of issues, their role in water policy and decision needs to be redesigned. Regional governments have claimed authority over water management with increasing vehemence over the past decade, thus increasing the strain on the current institutional model (for more on this point see Chapter 4). The reform of the current administrative model is complex and full of difficulties but clearly necessary, since it represents a bottleneck in the achievement of WFD objectives.

The WFD has also implied a tremendous effort in terms of collecting, organizing and analysing water-related data. Given the shift in policy and management priorities from water development to ecological protection, information and control systems had to be adapted. Biological monitoring networks have been created where none existed before, and other monitoring and control networks have been reinforced and expanded to comply with WFD requirements. Just as an example, in the nine RBDs managed by the central government, the number of piezometers increased from 831 in 2002 to over 2,200 in 2009, with a total investment of over 65 M€ (Carceller, 2011).

The planning process has provided a vast amount of highly technical information and data, which has been made available to the public at the different planning stages. However this has not necessarily led to better understanding of planning goals or the water management process by stakeholders and laypeople. The scarcity of summary documents easily understandable for non-technical audiences, the sheer volume of information to be processed, and the difficulty in obtaining the background studies from which plans and proposals are developed, have made it hard for stakeholders and the general public to actively participate.

The evaluation of the economic value of water uses, the analysis of biological and hydrogeomorphological parameters, the identification and characterization of water bodies according to their ecological status, are all new planning requirements that have given rise to an intense methodological debate and have absorbed a good amount of resources. Moreover, the WFD requires working with new geographical management units (water bodies) different in both scale and definition criteria from those used in former water planning efforts in Spain.

4   THE RESULTING SCENE: A SLOW SHIFT OF PARADIGMS?

Twelve years after the approval of the WFD, the Spanish water sector is experiencing a slow transition from old to new water paradigms. Progress towards the water management model prompted by the WFD and the so-called *New Water Culture* (Martínez Gil, 1997) is slowed by inertia (or active opposition) to change of traditional systems, by constraints in the human and social capital, and by the limited technical and financial resources to fully adapt to the WFD mandate (for a broader debate see Hernández-Mora *et al.*, 2011).

The new RBMPs drafted under the WFD guidelines, despite having weaknesses and gaps, represent a substantial progress relative to the RBMPs developed in the 1990s. Highlights in this regard are the consideration of the achievement of ecological quality as a primary planning and management objective; the increase in both public participation and transparency throughout the planning process; the economic analysis of water services; and the emphasis on water demand management as an effective measure to deal with situations of scarcity. Nonetheless, several unresolved issues remain, some of which are summarized in Table 1.

The achievement of good status for all waters within the established deadlines is surrounded by a high degree of uncertainty, due to a number of reasons. The first one is of a technical nature and is inherent to any programme aimed at improving the ecological status of a natural resource, as there is limited understanding of an ecosystem's response to the implementation of management measures. A second limitation derives from the fact that surface and groundwater bodies are physically interconnected.

Table 1 Elements of the transition from the old to the new water paradigm.

Progress towards the new water paradigm	Unresolved issues
More and better data on water availability, ecological status and water uses.	Lack of data overview, remaining information gaps for water bodies and certain types of data (e.g. hydro-geomorphological elements).
Inclusion of demand management measures (water markets, increased water efficiency).	Little debate on cost recovery and water pricing to achieve economic sustainability.
Acknowledgement of the need to set limits to the expansion of irrigated agriculture.	Need for a transparent and open debate on the role of agriculture in Spanish society and economy.
Boost to public participation.	The composition of formal participatory bodies does not reflect the WFD mandate (wider public, non-consumptive users).
Consideration of environmental objectives in the planning process.	Incomplete transposition of WFD into Spanish legislation (superposition of objectives and goals).
Acknowledgment of the need to act on land use drivers to improve water quality.	Difficult inter-administrative cooperation (between regional and national authorities, between different sectors).
Acknowledgement of the clear link between chemical and quantitative status.	Diffuse water pollution.
Definition of a standard approach to the establishment of in-stream flow regimes.	Opaque and limited in-stream flow regime negotiation processes further hampered by overallocation of existing resources in some river basins.

Source: Own elaboration.

Moreover, the good quantitative status of a water body is crucial to achieve its good qualitative status. As a result, the achievement of the WFD objectives requires a coordinated implementation of the measures on all the interconnected water bodies and addressing both quantitative and qualitative water problems. To this uncertainty, one must add that diffuse water pollution arising from agriculture is a complex and severe problem that cannot be tackled without changes in current farming practices (see Chapter 12). Thus, it also requires a strategic review of Spanish agricultural policy and cannot be solved exclusively through the RBMPs. In other words, a major WFD challenge is the achievement of an effective integrated land and water management and the high inter-administrative and inter-sectoral cooperation required for it.

A second level of uncertainty is related to the actual availability of financial resources to fund the measures defined in the RBMPs. In light of the economic recession and the resulting public spending cuts, it is very likely that each RBO will have to revise the Programmes of Measures to adapt them to the funds actually available. In this situation of uncertainty, it seems crucial to prioritize cost-effective measures that make the best use of limited resources. For instance, it may be worth critically evaluating whether the large amount of funds that are being allocated to the modernization of irrigation systems could be applied to alternative measures that would be potentially more effective in achieving WFD goals. Furthermore, in some RBDs the water monitoring and control networks are being significantly curtailed as a result of budgetary restrictions (Iglesias, 2011), thus limiting the RBOs' capacity to adequately monitor the effectiveness of the WFD implementation.

A third level of uncertainty is related to the political will to implement non-technical measures that may encounter some degree of social opposition, such as strict control on water abstractions, the restriction or cancellation of water use permits in over-allocated basins, or the application of more ambitious cost recovery measures. In a context of economic crisis, the latter is closely related to the sustainability of the water management system itself. Low water prices that do not fully capture investment and management costs can lead to high deficit levels that cannot be sustained in the long term.

In several Spanish RBDs agriculture is the largest user of water and must there-fore play a key role in achieving WFD objectives. The low profitability of much of Spanish continental agriculture (MIMAM, 2007b) together with the lack of gen-erational replacement in the agricultural sector in many regions, pose a significant challenge to the future of agriculture in the context of global markets and increas-ing deregulation. Setting aside broader social and territorial considerations, these challenges may lead to a self-regulation of agricultural water use in some regions. It is strategically necessary to review the role of agriculture in the economy, in the conservation of biodiversity and in the functioning of rural society. The CAP reform for the 2014–2020 EU budget cycle is now under way and will probably put increased emphasis on the compatibility of agriculture and nature conserva-tion objectives, including an increased integration with WFD goals (Henke *et al.*, 2011).

The tight link between water and land use puts the focus on the need to plan and make decisions from the perspective of integrated water and land use management. This requires finding new ways of constructive cooperation between water authorities and the administrative levels that are in charge of defining, funding and implementing most of the sectoral policies that determine land use. These are the autonomous regional governments, who in the current institutional setting of the interregional RBDs are not responsible – and thus not accountable – for WFD implementation. Possibly the key to a successful implementation of the WFD lays in strengthening the link between land and water management and in creating the institutional structures that facilitate co-responsibility and full cooperation between the central state and the autonomous regions.

REFERENCES

Arrojo, P. (2008). Luces y sombras de una legislatura histórica en materia de gestión de aguas [Lights and Shadows over an Historic Legislature for Water Management]. *Panel Científico Técnico de Seguimiento de la Política de Aguas en España*. Fundación Nueva Cultura del Agua – Ministerio de Medio Ambiente [Scientific Panel for the Evaluation of water policy in Spain – New Water Culture Foundation – Ministry for Environment]. Available from: http://www.unizar.es/fnca/varios/panel/22.pdf [Accessed May 15th 2012].

Carceller, T. (2011). Redes de seguimiento del estado de las masas de agua subterráneas. Situación actual, camino recorrido y principales retos. [Groundwater Status Monitoring Networks. Current Situation, Advances and Main Challenges]. Iberian Groundwater Conference: Challenges in the 21st Century. September 2011, Zaragoza, Spain.

EC (European Commission) (2010). An EU policy framework to assist developing countries in addressing food security challenges. COM(2010)127 final. SEC (2010) 379. Brussels, 31.3.2010.

Garrido, A. & Varela-Ortega, C. (2008). Economía del agua en la agricultura e integración de políticas sectoriales [Water Economics in Agriculture and Integration of Sectoral Policies]. *Panel Científico Técnico de Seguimiento de la Política de Aguas en España*. Fundación Nueva Cultura del Agua – Ministerio de Medio Ambiente [Scientific Panel for the Evaluation of Water Policy in Spain – New Water Culture Foundation – Ministry for Environment]. Available from: http://www.unizar.es/fnca/varios/panel/23.pdf [Accessed May 15th 2012].

Hedelin, B. & Lindh, M. (2008). Implementing the EU Water Framework Directive – Prospects for sustainable water planning in Sweden. *European Environment*, 18(6): 327–344.

Henke, R.; Severini, S. & Sorrentino, A. (2011). From the Fischler Reform to the future of CAP. In: *The Common Agricultural Policy after the Fischler Reform*. Ashgate Publishing Ltd., London, UK.

Hernández-Mora, N.; Ferrer, G.; La Calle, A.; La Roca, F.; del Moral, L. & Prat, N. (2011). *La Planificación Hidrológica y la Directiva Marco del Agua en España: Estado de la cuestión* [Water Resources Planning and the Water Framework Directive in Spain: Current situation]. SHAN Series – Seguridad Hídrica, Agricultura y Naturaleza [Water Security, Agriculture and Nature]. SHAN n. 2. Botín Foundation, 2011. Available from: http://www.fundacionbotin. org/monografias_observatorio-del-agua_publicaciones.htm [Accessed May 15th 2012].

Iglesias, M. (2011). Programa de seguimiento y control de la cantidad y calidad del agua subterránea en Cataluña [Programme for Monitoring and Control of Groundwater Quantity and Quality in Catalonia]. Iberian Groundwater Conference: Challenges in the 21st Century. September 2011, Zaragoza, Spain.

INE (Instituto Nacional de Estadística) (2011). *Construcción y Vivienda – Estadística de la Construcción: construcción de viviendas según calificación* [Construction and Housing – Construction Statistics: Housing construction by typology]. Instituto Nacional de Estadística [National Statistics Institute, Spain]. Available from: http://www.ine.es [Accessed May 15th 2012].

La Calle, A. (2008). La adaptación española de la Directiva Marco del Agua [Spanish Transposition of the Water Framework Directive]. *Panel Científico Técnico de Seguimiento de la Política de Aguas en España*. Fundación Nueva Cultura del Agua – Ministerio de Medio Ambiente [Scientific Panel for the Evaluation of Water Policy in Spain – New Water Culture Foundation – Ministry for Environment]. Available from: http://www.unizar.es/fnca/varios/ panel/51.pdf [Accessed May 15th 2012].

López-Gunn, E. (2009). *Agua para todos*: A new regionalist hydraulic paradigm in Spain. *Water Alternatives*, 2(3): 370–394.

Martínez Gil, J. (1997). *La nueva cultura del agua en España* [The New Water Culture in Spain]. Ed. Bakeaz and New Water Culture Foundation.

MIMAM (Ministerio de Medio Ambiente) (2007a). *Precios y costes de los servicios del agua en España: Informe integrado de recuperación de costes de los servicios del agua en España* [Prices and Costs of Water Services in Spain: Integrated Report of Water Services Cost Recovery in Spain]. Equipo de Análisis Económico, Ministerio de Medio Ambiente [Economic Analysis Group, Ministry for Environment].

MIMAM (Ministerio de Medio Ambiente) (2007b). *El uso del agua en la economía española: Situación y perspectivas* [Water Use in the Spanish Economy: Current Situation and Future Perspectives]. Equipo de Análisis Económico, Ministerio de Medio Ambiente [Economic Analysis Group, Ministry for Environment].

Chapter 4

Institutional reform in Spain to address water challenges

Elena López-Gunn¹, Enrique Cabrera²,
Emilio Custodio³, Rosa Huertas⁴ &
Fermín Villarroya⁵
¹ Water Observatory of the Botín Foundation;
  Complutense University of Madrid, Madrid, Spain
² Departamento de Ingeniería Hidráulica y Medio Ambiente,
  Universidad Politécnica de Valencia, Valencia, Spain
³ Department of Geotechnical Engineering and Geo-Sciences,
  Technical University of Catalonia, Barcelona, Spain
⁴ Duero River Basin Organization, Valladolid, Spain
⁵ Department of Geodynamics, Complutense University of Madrid,
  Madrid, Spain

ABSTRACT: This chapter provides on both a review and a discussion of the main elements for institutional reform to address the water challenges of the 21st century. The building blocks analysed here are: the legal and institutional framework, existing economic incentives, the structure of the current water administration, and procedures for water policy and planning. In relation to legislative frameworks the main conclusion is the need to strengthen the implementation of the current norms by increasing regulatory capacity and oversight, while in some cases a reform of the law might be necessary. In relation to economic incentives the need to increase transparency on cost recovery and a broader discussion on who pays for what, and what elements should be subsidized. Also the importance of budgetary control, and a wider debate with actors involved including civil society on financing mechanism and allocation. Finally the chapter discusses the current tensions brought about by an unfolding decentralization process and how this has played out in the water scene. Some suggestions are made to strengthen territorial coordination and supervision, while allowing enough flexibility and space for an effective and accountable decentralization process which involves regional governments as legitimate actors, but also users and their potential rights and responsibilities as important partners in water management. However, a deeper *water democratization* is ultimately underpinned by a healthy and active civil society that has access to relevant information and acts as a final check on the system to provide a stronger overall accountability from all water institutions and the process of water policy and planning.

Keywords: institutional reform, legislative framework, water savings, incentive structure

I INTRODUCTION

Spain is well recognized worldwide for a long tradition and history in water management, nestled in the Mediterranean basin, a cradle for civilization and an area marked by its climate. In this geographic location water scarcity is a defining feature that has triggered innovation throughout history, both in institutional terms and in water infrastructure. This chapter however is not focused on history to address past challenges, but on institutional reforms to address the challenges of the 21st century. The chapter is based on a review of the building blocks in institutional reform, namely: water law, incentives (namely economic instruments), water administration, policy and planning (OECD, 2011). When discussing potential institutional reform it is important to distinguish *rebels without a cause* (Llamas & Cabrera, 2012), i.e. steps that can be taken without the need for structural reform, from areas where deeper (structural) institutional reforms are needed. The chapter aims to contribute to the pending *Pacto del Agua* (Water Pact) in Spain, in order to secure long term interests as matters of state policy in relation to water, considering it both as a productive and an intangible asset.

2   SETTING THE SCENE

2.1   Institutional issues

Spain is a quasi-federal country, with 500,000 km² for a population of 46 million inhabitants, a mean rainfall of about 670 mm/year, which disguises a wide difference between the so called *wet* North, more akin to countries like France, UK or Central Europe, and the *dry* Spain in the interior, with a harsh continental weather, and the Mediterranean coast and the archipelagos, where much of the population is concentrated. Water resources are evaluated at 114,000[1] hm³ of which 47,000 hm³ are used (level of abstractions) [hm³ = cubic hectometre = million m³ = 10^6 m³]. In a context of institutional reform Spain is well positioned to deal with its inherent climate uncertainty and variability, and where the greatest challenge and opportunity is how to play with the advantages and disadvantages of different types of water resources (surface, ground, soil, artificially recharged, reclaimed, and desalinated) and where their complementarities can be bolstered through flexible management, which permits a portfolio of actions. In terms of water management Spain (together with the USA) pioneered the catchment management approach in the last century. The creation of the Ebro river basin authority, in 1926, was followed by 10 river basin organisations (RBOs), as well as two island water administrations for the Balearic and the Canary archipelagos (Custodio, 2011a), covering the whole of the country. In addition, in 1958, Water Commissariats were established. The final stage has been the adaptation of these existing institutions to the *Estado de las Autonomías* (State with Autonomous Governments) (Cabrera & García Serra, 1998) hereinafter designed as Regions. The Spanish Constitution established that water had to be managed at State level for the inter-community basins (those shared by two or more regions), whereas for intra-regional basins (i.e. those located within a single region) water is managed through the creation of regional water agencies. River basin organizations are the executive arm of the central administration, through the Directorate General for Water (DG Water), located in the Ministry of Agriculture, Food and Environment (MAGRAMA),

---

1 hm³ = cubic hectometre = million m³ = 10^6 m³.

where RBOs are responsible for inter-sectorial allocation, water quantity and quality monitoring and enforcement, the authorization of water and discharge permits and water pricing for e.g. agriculture. Similar organizations with the same responsibilities exist in the intra-regional basins.

Few countries have the legal proviso for an overall national water strategy. Spain, together with e.g. Australia, is fortunate to have this option. This provides a platform to deliver a coherent, goal oriented vision through a National Water Plan. In the past, policies for a National Water Plan (the first was approved in 1933) and then in contemporary times in 1993 and 2001, has generated massive mobilizations in Spain both in favour and against the strategic lines set (López-Gunn, 2009; Villarroya et al., 2010). This can be seen as both positive and negative highlighting that water is a special resource that mobilizes people, and negative if this paralyses political decisions.

In Spain, the decentralization process is still in flux and this has also been reflected in the water administration, with a tug of war between specific regions and the state.[2]

2.2 Legislative framework

One of the most interesting and unusual aspects of the Spanish legislative framework is its diversity. This diversity refers on the one hand to a multilevel legal framework, from the supranational level (European Union, EU, Water Directives) through to national laws, regional laws and local byelaws, and on the other hand to water rights, covering the span from fully private to state concessions, and all types of water encompassing not just surface and groundwater but also new regimes for desalinated, reclaimed or artificially recharged waters.

The Spanish *multilevel legal and institutional framework* is underpinned by a series of fundamental guiding principles, like rational use, responsibility, efficiency, sustainability, solidarity, and representation. At the level of international conventions, Spain has signed the bilateral Albufeira Convention with Portugal for shared rivers. Meanwhile, at the supranational level it is bound by EU legislation, in particular by the EU Water Framework Directive (WFD). Like all directives it sets overall goals in relation to water whilst leaving freedom on implementation to the member states. The WFD does not however position itself on the nature of water rights but it does state clearly the equal duty – independent of the juridical nature – to protect water resources (García Vizcaíno, 2011; Huertas, 2011a; Poveda, 2011). At the national level Spain has a national water law, dating back to 1985 (with main reforms in 1999 and 2003), locally modified by the Regions' Statutes of autonomy, and also with some regional water laws[3]. The major changes in the 1999 reform referred to the possibility of water trading, the establishment of public corporations to act as investment agencies and, in line with the WFD, the recognition of the protection of the environment not as a user but as an operational condition.

---

2 For example the case in the Guadalquivir basin, and also between regions like e.g. Valencia and Castille-La Mancha over the Júcar basin, or even between regions hydrologically connected through water transfers like Castille-La Mancha and Murcia.

3 E.g. Canarias (Law 44/2010, 30th December). Also Law 11/2006 from the Basque Country.

In relation to access to water and typology of water rights, Spain has a great *diversity in water rights*, with the whole spectrum from public to private to collective water rights co-existing within the current legislative framework. A key change in the 1985 Water Act was the inclusion of groundwater into the public domain. Water generated through new technology, like desalinated[4] and reclaimed water (which requires a public water concession) is therefore directly or indirectly part of public water resources. Under the Spanish legal system, water is bundled to land (bundled rights), for those rights under 7,000 m³/year and also in the case of irrigation, the main consumptive user. There is an important proviso for *use or lose it*, with an expiry on water rights if these are not used for more than three consecutive years[5] (see Box 1), equally applicable to private and public waters, although this has rarely being enforced. In terms of *water trading*, the 1999 reform opened the possibility to trade water rights for all rights in the Register (including temporary water rights), with some exceptions for private rights in the Catalogue[6] (see Chapter 16 on water trading). For a fully-fledged legislative framework, clear monitoring, enforcement, and sanctioning are a must. *Metering* is now a *legal requirement* for all water right holders and users, since an order was issued in 2009,[7] which applies to all water rights (public or private).[8] For the WFD – and independent of the nature of the rights as stated above – all right holders have an equal duty to contribute to the achievement of the protection of water resources (public trust doctrine).[9]

---

4 Under Art. 2 of the *Texto Refundido de la Ley de Aguas* on desalinization water from the sea, are part of the public domain, independent of who undertakes the desalinization. Therefore in order to be able to use desalinized water there are two steps to follow: a public concession for the public domain to capture sea water and a concession for the desalinized water which can then be used from different ends (public water supply, irrigation, etc.). This procedure was set out as a results of the Water Law reform due to the changes introduced in the Hydrological Plan in 2005, which before this change required as a previous step (no longer necessary) that desalinized water should be incorporated beforehand into a riverbed, a reservoir, etc., before these were considered as part of the public domain.

5 Art. 66.2 TRLA.

6 Art. 343.4 *Reglamento del Dominio Publico Hidraulico* although this is not the case for the Canary Islands.

7 Order MARM 1312/2009, 20th May, known as *Orden de contadores*. This has facilitated that a large number of uses are now metered. In the Duero Basin for example this is the case for 90% of hydroelectricity uses which are now metered, which before had environmental impacts from water diverted or turbinated for both mini-hydropower and large hydropower stations. In irrigation the number metered is still small, however the existence of this rule to meter is a step forward to grant legal security since it clearly sets out metering characteristics, the duties of users and facilitates the control by the administration, and also for other issues liked sanctioning, etc.

8 Under Art. 2 of the *Texto Refundido de la Ley de Aguas* a number of cases in the Duero basin have been processed successfully on this basis. It seems very distant the day in which all the abstraction points (wells) have their metering system.

9 The role of monitoring is twofold: one is to collect information necessary for good management and planning, and the second is to be able to act in case of infringements through sanctioning. The ALBERCA programme is a powerful tool which is used to gather information on how much water is used, where, by whom and for what purpose, and also to help with the requests for water rights or their modifications. This information can then be contrasted with GIS to allow the identification of possible infringements, checked through field visits thus optimizing scarce human resources (Huertas, 2011b). ALBERCA however is not equally effective in all basins, and particularly in the context of inter-community basins.

2.3   A modern water economy?
Economic and financial instruments

In a modern water economy there are three key issues: the first is to decouple economic growth from increased resource use, becoming more efficient and productive in the use of water resources, as shown by Gleick (2003). The second is to devise a clear system on how money is collected or raised, and that (as far as feasible) all costs are internalized e.g. from state water budgets or in revenues from users and polluters, to be re-invested in water management and renewal of ageing infrastructure. The third issue refers to how money is spent in terms of budgetary allocation and responsive budgeting.

Water is an important economic resource because it has no substitute, fulfilling vital ecological functions. In terms of investment and financing, infrastructure has large upfront costs. In Spain, water history has been marked by infrastructures to address natural water scarcity (Bru & Cabrera, 2010), where these large investments were often undertaken by the State. In this context of *cost recovery*, there are at least two major types of instruments to create economic incentives: water pricing and tradable water rights. As Merrey *et al.* (2007: 206) state "with water pricing policies the payment goes to the state or the water agency, whereas with tradable water right payments go to the holder of the rights" (see Chapter 16 on water trading). Prices are crucial because these are powerful signals to trigger behavior change, and the close relationship between water prices and efficient water use (Cabezas *et al.*, 2008).[10] Yet there is increased pressure to start to be more efficient (and productive) in all water uses, providing market signals to both cities and agricultural water users, that consume the bulk of water in Spain. The main stumbling block however is the political economy of water, since water pricing is unpopular and therefore often considered as political suicide. The questions centre on whether water infrastructure and services are paid for indirectly (through taxes) or directly by users themselves (Cabrera *et al.*, 2012). European subsidies over the last 15 years have represented a large investment effort into water infrastructure; it has also allowed keeping the current *subsidized prices* on water (see Box 1).[11]

---

10 For example, in the case of groundwater, full costs are paid for energy (pumping) and for the infrastructure investment (wells, pumps, etc.), which in many ways explains why groundwater use in some regions in average terms is four times more efficient that surface water use (Corominas & Del Campo, 2000; Buchberger & Cabrera, 2010). Since farmers bear all the direct costs, farmers seek to be productive as compared to surface water where costs are born by society, through e.g. state infrastructure. Meanwhile in urban water supply, as shown by Galbiati (2011), current prices e.g. in Catalonia do not cover costs and therefore are subsidized through regional budgets, which has compounded the problem of regional governments' deficits.

11 However these subsidies will not extend over the whole amortization period of long term investments. Equally political prices do not often reflect operation and maintenance costs. In the case of water, compared to other economic goods, scarcity is not necessarily reflected in the price. This can be seen in the higher water tariffs in Northern Europe as compared to Southern Europe. The actual average of urban water supply price in Spain is 1 to 1.5 €/m³. These tariffs are 20% of the tariffs in Northern Europe, because in countries like Germany and Denmark all the investments made are reflected in water tariffs (Cabrera, 2008).

Box 1   The *Water Bubble* (by Enrique Cabrera, translated from *La Burbuja Hídrica*, in *Levante*, 5 February 2011)

Demagogic, provincial and focused on the immediate, politicians have designed debates over water to win votes. The last three legislatures provided all the evidence: the first legislature promoted the Ebro transfer, the second legislature (like a exchange of stickers), –desalination plants for water transfers–, which was stopped, and given the territorial tensions generated, the final legislature was a deep slumber from which we have not yet woken up, facilitated by a series of wet years. However with the current problems on the table, when a new drought triggers a feeling of water scarcity, a point-less debate will start again: that it is better to transfer water, that desalination plants are better, that the river is mine ... rarely has so little given so much. And while this happens, services vital to citizens are close to bankruptcy, because fees barely cover the operation and maintenance costs. Thus the 1,200 million € debt of the Catalan Water Agency has grown steadily. A general malaise, even if other communities do not utter a word. Untimely also because with the current crisis mayors will not raise water tariffs and even less so in an electoral year. No matter what the EU Framework Directive says (since 2010 requires recovery of all costs). But there is no alternative, up until now the system has survived because the necessary investments (sewage, desalination) were paid by the European Commission and to a lesser extent by the State and regional governments. However, Europe's money runs out (2013) and the administration swamped by debt, will do well just to cover its deficits [...]. Even then we are talking more of a cultural than a real problem. Because citizens pay, on average, 75 €/year for water, a miserable figure (0.4% of income per capita) when so much is at stake. It should be noted however, that the increase in water rates is only meaningful if all the money raised is destined to improve services. We must create a regulatory agency to monitor the health of budgets, check contracts, and oversee private operators. As an added bonus, with the upgrade in the price of water, and the consequent improvements in efficiency, regional water wars would come to an end, because there would be water left over, more so if goals are imposed to those uses that continue to be subsidized (agriculture and rural villages). Water is a scarce public good and cannot be squandered.

---

3   MAIN CHALLENGES: DIAGNOSIS AND SUGGESTIONS FOR REFORM

3.1   Changing incentives for sustainable services

The current economic crisis provides a window of opportunity for deep reform on economic and financing instruments (Saleth & Dinar, 2000; Merrey *et al.*, 2007). A potential venue would be a review of *best practice water pricing* (Buchberger & Cabrera, 2010) like sliding price strategies or block pricing, which can differentiate between different uses and their capacity and willingness to pay. In the case of public water supply, water prices do not cover costs, and these hidden subsidies are giving the wrong signal on water service costs, not contributing towards valuing the service provided or educating the consumer (Buchberger & Cabrera, 2010). In the future however, it is more likely that costs will be devolved to users and consumers. In this context, Art. 9 of the European Water Framework Directive specified that by

2010 account or advances made of cost recovery should be implemented. A recent law (2012)[12] states that the relevant mechanisms should be established to assign the costs on services related to water management. This provides an opportunity because of the confluence of the requirement under the WFD for cost recovery and the current debt in state, regional and local public budgets. Achieving cost recovery will need the implementation of economic and financial instruments.[13] Yet how to raise finance is on the table because of the estimated 150,000 million € at current value (or about 3,300 € per inhabitant) needed to renew urban infrastructure (see Chapter 13 on Urban water supply). Equally, at basin level, finance is needed to implement the river basin plans (*Programme of Measures*) to fulfil requirements on good ecological status. On top of these investments, one has to add normal day to day running costs for the whole organizational set up for water management, which has often been accused of lack of capacity, rooted in a lack of financial and human resources. An important area where more research and knowledge are needed to generate a more accurate understanding on responsive and responsible budgeting; in other words, how governments (including regional governments, water agencies and more recently state companies) allocate budgets and monitor the outcomes of budget expenditure (Merrey *et al.*, 2007). This builds accountability on how public funds are spent and creates incentives for sound service delivery. It represents a shift in paradigm away from infrastructural development, towards infrastructural management.[14] An unavoidable element of institutional reform is how to address cost recovery to ensure the long term viability of the water sector, and the institutions needed to support it, while sheltering it from financial crisis and gradual disrepair. In terms of institutional strengthening, there is a need to look for innovative and effective cost recovery mechanisms. A non-exhaustive list of some examples of these mechanisms could be:

i   Annuity-based capital cost recovery like that introduced in Chile and Australia, or like the case in South Africa, which have included some costs for water conservation, management, research, and cost recovery (Saleth & Dinar, 2000).
ii   Experiment with payment for ecosystem services within a green economy or re-discover ecological agriculture, which used to be the traditional agricultural model and where waste becomes a valuable product, whilst reducing and preventing water treatment costs.
iii   Urban water sector decentralization by creating an autonomous and self-dependent utility type organizations for the provision of urban water services, which encourages urban water supply agencies to be autonomous and financially self-dependent.
iv   Introduce responsive budgeting, i.e. a means to examine the priorities reflected in budgets and also the need to use and apply cost-benefit and cost effectiveness

12 Royal Decree-Law 17/2012 of 4 May, on urgent environmental matters.
13 The economic and financial regime is somehow incomplete since it only allows for cost recovery for infrastructure, without including e.g. costs for water quantity and water quality monitoring networks, when in theory cost recovery is a binding principle for all public administrations.
14 However temptation remains because supply based policies are more flamboyant (water transfers, desalination plants, etc.) and thus politicians often opt for this easier path (Cabrera, 2011).

analysis, while ensuring the basic financial viability of key agencies by e.g. being able to raise their own financing.

v   Look at examples from other countries which have better established self-financing mechanisms like the Dutch Water Boards, and have clearer information on who pays and for what.

3.2   An effective legislative framework

Together with securing the economic means to be able to carry out and deliver modern water management, an effective legal framework is also key, setting the institutional *rules of the game* for all actors involved. This section summarizes the main pending challenges in relation to an effective regulatory system.

The first challenge is related to the WFD which to this day, although formally completed into Spanish law, is lacking synchronization with the Regulation that accompanies the water law. Day to day management and activities are still ruled by a Regulation dating back to 1986 which has different terminology and ethos to the WFD.[15]

The second challenge dates back to the mid-1980s, when due to the potential compensation to existing water rights under the Spanish constitution (which forbids expropriation of rights without compensation),[16] a hybrid system developed for groundwater, where private right holders could opt to keep their private rights.[17] In relation to water rights this has remained a management challenge because of the co-existence of private and public rights, with different (sometimes conflictive) approaches on how to address this co-existence. One option formulated argues for devising a system to absorb and migrate the existing private rights into state concessions with proper compensatory schemes (see Poveda, 2011), whereas a different approach focuses on the implementation side, and its strengthening, focusing on guaranteeing compliance with the duties imposed by the WFD to all rights holders are applied (independent of their juridical nature).

The third challenge refers to bringing water books up to date. This refers on the one hand to registering existing rights, where for example, it is estimated that in the case of groundwater only 30% of wells are registered in Spain, and in surface water, where e.g. in many cases traditional irrigation communities have not been issued formal water concessions. On the other hand it refers to the much trickier issue of informal water use (see Box 2), and how to find solutions to incorporate these into existing registries through negotiation, or if necessary have the institutional capacity (and support), to close all relevant abstractions to ensure that state capacity and legitimation is not undermined.

A fourth challenge refers to the solid regulation of water markets. A pre-requisite before a rush to market solutions lies precisely in perfecting the regulatory framework as witnessed in the latest financial crisis, and de-regulatory trends. Clarifying

---

15 For example, the reformed water law refers to groundwater bodies while the Regulation on the Water public domain still refers to hydrogeological units, which are no longer the management unit for planning under the WFD.

16 Art. 33.

17 Once owners accredited well ownership and use before 1985, including them in the Catalogue of Private Waters.

regulatory mechanisms (who sets the rules and what are the rules) and the capture of benefits (who wins or loses in imperfect markets) are key issues (Merrey *et al.*, 2007; Bruns *et al.*, 2005). Water markets require as pre-requisite clear water rights, possibly unbundled from land (e.g. like in Australia: McKay, 2010), and the adequate administrative or regulatory capacity to monitor and enforce rules. Therefore an essential ingredient for a functioning water trading system lies in both clear water property rights, and a strong regulator.

Box 2 Review of existing water rights, Duero catchment (based on Huertas, 2011b)

From 2008 there has been an evaluation to update registered rights compared to current use for two case study areas (Páramo de Cuéllar and Los Arenales) currently identified as potentially over-used (after the 1985 Water Law) or as groundwater bodies in poor status (WFD). The study reviewed 6,354 uses, of which 17% are not in use, and thus procedures to extinguish these rights are now in motion which have revoked some rights or are in the process, 30% are used under the same conditions as those originally granted and registered in the Catalogue of Private Rights, and 53% will be now moved to State water concessions due to substantial changes in the right. Of this 53%, 17% is due to changes in ownership, 56% is due to changes in the use and irrigated area (usually to increase it) and the remaining 27% is due to changes in both. This highlights the important gap between registered rights and actual use of the right. In other areas and catchments in Spain it has been very difficult or impossible to extinguish water rights like in the case of the Pirineo Oriental in the 1980s, in the area around Doñana in the 1990s, or in the Canary Islands.

However, effective regulation includes not just the quality of the law, but also the capacity to regulate and the political will and skills to ensure compliance. In this context there are two aspects that could be strengthened.

The first key aspect is the monitoring capacity to oversee implementation,[18] which takes into account the duty by all users to monitor and report water use, and the parallel duty for authorities to oversee and monitor this legal duty. An example on how to increase regulatory capacity and monitoring is by exploring options and opportunities to reach covenants with users via agreements.[19] In Spain, in the case of groundwater there are examples of co-management experiences, with agreements written jointly between users and the administration, like the Eastern Mancha aquifer, Catalonia and the Duero basin (Huertas, 2011a). It represents an important shift in mentality for both the administration and users, developing co-responsibility mechanisms where the administration delegates some its duties, whilst users act as mature water managers with both rights *and* responsibilities.

The second key aspect is the sanctioning regime for water resources marked by the Water Law,[20] which typify the type of infringements and sanctions. The sanctioning

---

18 E.g. through a plan on how monitoring will be achieved, resources invested, timeframes involved and for what purposes.

19 Granting water users the relevant financial and institutional support a pre-requisite or operating condition for co-management.

20 Title VII of the TRLA.

regime could be contrasted with other environmental laws like e.g. waste, coast, and biodiversity, where penalties are higher[21] and therefore have a higher dissuasive power. It seems that some steps are being taken in this direction. An area that is already being strengthened refers to the valuation of damages to establish a clearer and more precise system for calculating penalties to avoid insecurity and arbitrariness. This is signalling the basic ethical principle that it has to be more beneficial to comply with the law than to break the law, and rewards compliance, while offering a potential income stream from a solid sanctioning regime which could be hypothecated into strengthened regulatory oversight.

Third, issues related to institutional capacity and strengthening (discussed in more detail in the next section), like ensuring available data are processed in a manner suited for water planning purposes.[22] Technical progress like satellite technology, information technology, and computer based water control, like smart devices, help to reduce the transactions costs of institutional reform (Saleth & Dinar, 2000).

Box 3   Legal hotspots for reform: Informal water use (modified from Bravo, 2011)

Independent of the type of ownership right (public, private) lies the question whether the use is legal or a-legal. These refer to differentiating and bringing up to date on the one hand, users that –25 years after the passing of the 1985 water law – have not requested to register their right, either as water concessions or private water rights, from those that continue to abstract water after having had their water right refused by the water authority. Which of these uses are legal or a-legal? Which of these water abstractions should be sanctioned? If they are legal, how much water can they abstract since these are not officially registered? How much area can be irrigated? Who determines these figures? For example, in relation to the duty to install a water meter, it would be difficult to monitor or sanction, if use is not registered, since it is unclear what the authorised volume or area that can be irrigated is, and therefore it is difficult to decide if the area has been increased, when no area is registered in the first place. It is even more problematic in equity terms, when this *use* co-exists with other registered users which have installed a water meter, with a maximum volume to be abstracted and where the administration can sanction abstractions above authorised limits. The question of who has to prove the right to be registered, judgements state that: a) the administration has a duty to sanction and fine those that have not yet registered their rights; and b) the duty falls on the owner of the right to accredit the right with a series of documentary evidence. In cases of dispute it goes to provincial courts, and then all the way up to the Supreme Court. This has created a slow limbo for both water users and the administration, which has left the system paralysed. This is further complicated by an apparent de-synchronised system between different laws, rulings, legal systems (civil and criminal) and sanctioning regimes.

Finally, some important aspects of the 1999 reform, namely to establish private and transferable water use rights, grant full financial autonomy to water authorities,

---

21 E.g. in the case of water the maximum fine is 600,000 € as compared to 2 Meuros.

22 Particularly when decisions have to be taken on the basis of enough financial and personal resources, with the exercise of legitimate authority to control and sanction, with specific objectives, well defined and supported by knowledge backed up by solid data from fully functioning and comprehensive monitoring networks (Custodio, 2011b), complemented by *ad hoc* or periodic studies.

make the construction of new projects dependent on users' prior agreements to pay full costs, and how to encourage the participation of the private sector in construction, distribution, sewage treatment and pollution control, did not fully materialize and could now be re-visited (Saleth & Dinar, 2000).

3.3   Reforming the water administration

This section outlines potential reform lines without the intention of being prescriptive, but rather to put the spotlight on key possible intervention points in the organizational architecture and existing institutional inertias that have to be taken into account. In terms of organizational set up and reform there are at least three issues to consider, the first related to a territorial vision, the second on the effectiveness of the river basin organizations and finally, the overarching vision for water as state policy in a quasi-federal and highly diverse country.

The first aspect refers to ensuring a general integrated vision, present in the origins of the water administration. This has been weakened to make this integrated vision fit with a legitimate desire by regional communities' to play an active role in the management of water resources that are either generated or rise within their own territories. The decentralization process has generated a large element of innovation and even healthy competition between water authorities and regional agencies, but it has also come possibly at the cost of a loss of general vision. Transboundary issues at domestic level have in some cases fallen prey to an insular vision. For a country formed by autonomous regions, each with their own Statute, close to a federal state, a key area is to establish the full potential of the principle of subsidiarity, well established under EU law, and how this blends in terms of cooperation and co-responsibility when applied to water management (Custodio, 2011a). An idea put forward has been the potential creation of a *National Water Agency* as described in Box 4, similar to other agencies like ONEMA in France or the Environment Agency for England and Wales. As Custodio (2011a) argues, the advance in water management on a catchment basis, one of the fundamental tenets of Integrated Water Resources Management, has been threatened by its accommodation and adaptation to a decentralized autonomous system where regions play a key role. The current situation can be partly attributed to the use of water for political (short term) gain (López-Gunn, 2009).

Box 4   The creation of a National Water Agency (based on Cabrera, 2005, and Custodio, 2011a)

No one questions the need to reform the water administration, and that this passes necessarily through the creation and establishment of a National Water Agency (NWA). This NWA would play a coordination and regulatory role, in charge of overseeing water use. This agency would impose the principle of cost recovery and design and collect information on a battery of indicators that can highlight efficiency in water use. This NWA could also ensure coordination and negotiation at national level, searching for consensual decisions, where all parties agree to abide by the same rules, applicable to inter- and intra-communitarian catchments, and which takes into account potential impacts. This agency would provide an overall general vision, combined with the capacity to coordinate and undertake new general agreements, with a legitimacy recognized by all parties. In view of the current economic crisis and concerns over the multi-layered and overtly complex administrative framework in Spain, this agency should be small, with basic infrastructure, and able to commission strategic studies needed to take decisions based on best available knowledge. It could be born from existing institutions, or created from scratch. This NWA would be supported by a Council of Experts, from all sectors and across parties, based on prestige and well versed in the current problems. This agency should be based on the consensus of all bodies with responsibility for the management of water resources, i.e. both river basin organization and regional water agencies, with a solid base amongst civil society and users, with enough stability and independence to provide a long-term vision. Its creation and existence should be regulated by the Water Act, as part of the expected and necessary reform expected and necessary. This Council of Experts would be made up of people from the country's most prestigious, independent, experienced, impartial and knowledgeable experts, from the many facets of water resources in Spain, with regional, national as well as European and international experience, capable of balancing possible political connections, with impartiality, with the selection of its members through a competitive, open, and public process in response to specific and testable merits and criteria. The NWA should have the support of specialized bodies to carry out their work, either on their own or preferably with other existing bodies with proven excellence in their field. It would take advantage of what there is on offer, providing a balanced symbiosis between political society, civil society and adequately represented users, focusing on achieving outcomes and on sound governance. Thus, it would be autonomous, but recognized and respected for its own authority and with access to all information, and whose formal agreements have a degree of binding value to the executive. The current National Water Council, as designed, does not meet these conditions. Further work of this Council and Agency is to facilitate inter-sectorial coordination and transparency in information, while improving access to existing information, which is abundant and rich, commissioning specific studies for pertinent policy questions, while promoting efficient use of accumulated knowledge thus avoiding loss of time and effort. Spain could look at the experience of countries like France, Australia or the UK and draw lessons on national water agencies (see for example BOM, 2008; Cabrera *et al.*, 2012; Guerin-Schneider & Nakhla, 2010; McKay, 2010).

The second aspect refers to the internal set up and functioning of RBOs. The *Water Commissariats*, when they were created in 1959 were kept separate from both infrastructure and planning, although this was understood only as planning water works. This is summarized in the motive for their creation and interestingly it echoes similar institutional reforms undertaken in the UK (late 1990s), Australia (beginning of the 21st century), and in France. This reflects the raison *d'être* of the Water Commissariats and the importance to have a clear separation of roles from e.g. Technical Directorate of the State RBOs, which have tended to absorb larger budgets and thus resources.[23]

Meanwhile the other changes necessary are in terms of a shift in mentality and a much needed – and largely absent – state vision and strong political will and skill. The vision is focused that the real issues in Spain in relation to water no longer turn around resources themselves, but rather in management and allocation decisions, where a step forward is now pending completion. This means a shift away from an

---

23 In the process potentially *starving* water commissariats of much needed independence and resources to carry out its regulatory role along the philosophy of separation of powers.

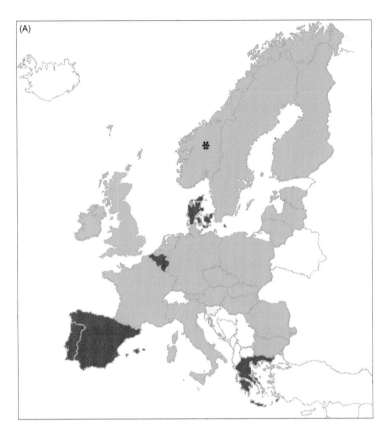

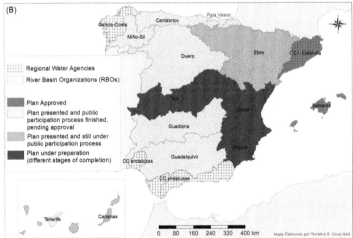

Figure 1a and Figure 1b Situation of the application of the European Water Framework Directive in the EU (Source: EC February 2011) and Spain (April 2012) (Source: Kindly prepared by Terrativa S.A.). http://ec.europa.eu/environment/water/participation/map_mc/map.htm (Updated 22/12/2011).

Note: The map refers to the state of adoption of river basin plans, however it does not provide information on whether these plans are in accordance with the EU Water Framework Directive.

administration that was traditionally and historically geared towards infrastructure development, towards a modern administration whose main role is focused on management rather than infrastructural development. This however represents a substantial change in the organizational culture.[24]

The third aspect refers to a strong political will, that shies away from the instrumental use of water for political gain, and seeks the *general interest* enshrined under Spanish law, but which now has a much more complex and nuanced meaning beyond building or managing infrastructure. The last few years have witnessed a paralysis in terms of decision making, like e.g. the approval of catchment plans, because of the potential political costs to the ruling party.[25] Finally, it should be accepted that in a country like Spain with a rich tapestry of regions, climates and history, there has to be some well-established coordination mechanisms between regional authorities and river basin organizations (see Chapter 3).[26]

The EU has already ruled against the Kingdom of Spain for failure to submit water plans, with the exception of Catalonia (Infraction 2010/2083) (Custodio, 2011a) Therefore the image provided is of lack of overall coordination. Most catchment plans were ready in 2009 for public consultation, and it is thought that many of them were shelved due to perceived political risks. Only a few of them have been opened to public debate, leaving the Spanish water administration in limbo since 2009 when the plans were due to be released. The end result is that Spain, after considerable administrative effort, is one of the few EU member states that has not met the WFD deadlines. An added layer of complexity is that in Spain – as compared to other EU member states – these catchment plans have a normative status. This is even more worrying considering that Spain had a considerable first mover advantage as the member state with the longest tradition in catchment planning in the EU, well before the WFD. The administration has been caught in the quagmire of politics and decentralization, a case of politics getting in the way of decision making, like the case of the boundaries of the Jucar basin which are now pending a decision by the Constitutional Court. This is partly because of the difficult – and often unpopular – decisions that had to be taken in relation to water allocation and cost recovery, in the context of a modern democratic country which has to honour reasonable rights and

24 The incoming government is taking some steps and decisions that seem to be strengthening this state vision inside the inter-community basin organisations, though organizational changes that mark clearly the hierarchical relationships between the different public organizations involved in water management, for example making the Water authorities and the State water companies depend on the general Water Directorate (before this were hierarchically equal). Also some legal modifications are proposed in order to clarify that the function of policing and control of the public domain is a state competence.
25 The unwillingness to broker a difficult – but necessary – negotiated vision in effect has been a delay tactic and passing the buck to the next elected government, thus inhibiting and emptying public institutions of their essence, making institutions themselves a pawn or instrumental to prioritizing short term electoral gain (or lack of electoral losses).
26 A key element is to coordinate water planning with land use planning (Custodio, 2011a) (see Chapter 11), showcased in agriculture, very often the largest consumer of land and water, and also associated with a remarkable power to influence policy or decisions regarding irrigation as the dominant water user, forest cover, how to establish protected areas, nitrate vulnerable zones or how to accommodate urban and residential purposes (Custodio 2011b), but also to cancel out potential blockages like those referring to nitrate vulnerable zones, or the coordination of agricultural and irrigation policies.

claims, while striking a balance between legitimate private interests and social ends (Custodio, 2011a). However as the Head of Water Planning in one of the Spanish river basin authorities claims, it just translates in less time for the actual implementation since a delay in submission of the plans does not change the overall planning framework applicable to all EU member States, where dates are and remain fixed. Thus the incoming new government has a litmus test in the final approval of pending plans to bring Spain up to speed with the rest of Europe.

Finally, Spain has a strong untapped potential though both a deeper user participation commitment and a system more open for civic society and citizen science. Promoting deeper user participation would require the establishment of clearer responsibility and accountability mechanisms.[27] Together with the National Water Agency described in Box 4, water users represent the balancing opposite. This is based on the functional distinction and balanced action between decentralized arrangements needed for user participation and some centralized mechanisms needed for coordination, enforcement, and conflict resolution. Spain has a millenary tradition in user participation and this historical legacy now offers an opportunity to help address 21st century challenges by developing a structured programme of *water democratization*. As stated in chapter 17, user participation is a longstanding principle in Spanish water policy. As public law corporations, water user groups have a public-private nature, a public role in administering a public good (water) being able to administer, distribute and impose fines and sanctions. It also has a wealth of entrepreneurial initiative and data from private users and thus becomes an important piece for institutional strengthening.

At a more fundamental level, the concept of user itself has to be widened so that it is not only consumptive users but also new users, like e.g. the tourism sector, have a platform and these can be formally incorporated into the main representative structures. A clearer delimitation is needed on the rights and duties of each actor, e.g. between the water administration, the regional government, users and ultimately consumers (Saleth & Dinar, 2000). Meanwhile, equally important, the current civil society in Spain is often damaged by an excess in politics, which affects its organizational capacity, marginalizing it and making it less effective and more disfranchised. Yet it is civil society that can act as a loud voice for the underpinning ethical principles that guide water policy based on general principles like equality, solidarity, precaution, and other practical procedural principles like subsidiarity (Villarroya *et al.*, 2010). Legal rules are not immutable, nor are the principles underlying administrative decisions except ethical ones. In the de-synchronization of the legal system and its implementation, which can lead to its review by constitutional means, the missing lynchpin is centred on the broad involvement of civil society. What now seems the right thing might not be so in the future and therefore must regularly connect or link up with the underlying ethical and moral principles of a strong civil society (Custodio, 2011b). Both decentralization and centralization forces have to be supported by a healthy and active civil society demanding and expecting accountability from users and the administration, while assuming its dual role both as citizens and consumers. As Saleth & Dinar (2000) state: "the key to this centralisation-decentralisation dilemma lies in carefully crafting institutional

---

27 In Spain an opportunity for institutional reform would be to make subsidies conditional on taking up this role, with subsidies based on targets and/or delivery and completion of allocated pre-determined tasks, and for a given duration.

arrangements at different geographical levels so as to achieve local flexibility and regional coordination with an overall global strategic and goal setting".

4   CONCLUSION: REFORMING INSTITUTIONS

Spain is in a moment of deep change, due to a confluence of external and internal factors, from internal changes in government to a change, review or reform of some key policies at EU level like agriculture, water or energy that will have a direct and indirect knock on effect on Spanish water policy and planning. The necessary ingredients for successful reform are measures that are economically attractive, technically feasible, and politically acceptable. Rolling back the State however does not mean an elimination of the role of the State. On the contrary, in the current economic crisis, its role is both essential and crucial in its regulatory function and its enabling function, guaranteed by strong and capable institutions. Spain as can be seen in many of the chapters in this book faces current challenges in water management, and has been relatively innovative in searching for solutions to these challenges. Institutional reform is a fundamental aspect to make sure that many of the potential solutions to address achieve their full potential. There is window of political opportunity (Kingdom, 1995; Kartin, 2000) because of a series of endogenous and exogenous factors (Saleth & Dinar, 2000) to gradually strengthen institutions, by making the implementation of the legal framework more effective, by bolstering the agencies in charge of water management with clear allocation of roles and responsibilities, and also – most important – the economic and financial instruments to undertake these tasks as outlined here, which currently are insufficient. In the context of deep reforms, the water sector is only a small part of the overall reform process. However, the water sector can benefit from the overall impetus for structural reform in Spain, where the shift in focus away from infrastructure and towards medium and long term planning would slot in. Water management can now be focused on identifying the overarching goals, according to specific clear priorities, regulated and supported though a water planning process. This would provide a road map and overall direction for adaptive management (Custodio, 2011b).

Reform often happens in times of crisis, where e.g. droughts, a recurrent feature of the nature of our climate, provide stress tests to the resilience and adaptability of the institutional system. Pro-active institutional reform prepares deep changes before the opportune moment comes. The current situation offers an opportunity for reform brought by an economic crisis and deep political change which is having a knock on effect in the water sector. A different economic era means that the politically easy and economically costly supply solutions of the past are unlikely to be able to materialise due to their large economic price tag, combined with less water availability (decrease in flows, potential impacts of climate change, etc.) thus unable to save the political day. If reforms are not introduced, the Spanish mature water economy will not be fully functional and ready to weather inherent climate variability, thus knocking the productive model and key economic sectors like tourism, industry, etc. out of kilter. It is easier to introduce solutions and reforms gradually, so that these are in place when the time comes. The *water bubble* (Cabrera, 2011) that went hand in hand with the *construction bubble*, has burst and the re-invention of the construction sector can also come from a re-invention of the water entrepreneurial sector. The main challenge is how to create a competent and competitive sector, not bailed out or inflated by public finances but rather competitive due to its technical competence and know-how, supported by reformed and strengthened institutional frameworks. Coming of age in water management requires clear leadership, assuming the structural changes needed, where planning needs substantial investments, affecting the economy of the state, and which requires both bold action and careful accommodation and agreements between the parties. The ultimate challenge thus lies with society at large – which ultimately sways politicians to act or not to act – and its awareness and acceptance that changes, particularly those that have been delayed over a long time, will be hard at first, and yet, like a sour testing medicine just what the doctor ordered for a healthy and thriving political water economy. In a similar line to other deep reforms currently on the table – like the labour market or the banking sector–, crisis brings along opportunities for (democratic) change and bold leadership.

REFERENCES

BOM (Bureau of Meteorology) (2008). *The Water Act 2007 and Water Regulations 2008.* Water Information; Information Sheet 2.

Bravo, G. (2011). La naturaleza jurídica de las aguas subterráneas con la directiva marco de aguas: tendencia hacia la demanialización [The legal nature of groundwater in the Water Framework Directive: a tendency toward demanialization] *Jornada jurídica de aguas subterráneas. November 2011.* AEUAS, Ministerio de Medio Ambiente, Rural y Marino.

Bru, C. & Cabrera, E. (2010). Water, history and sustainability, a complex trinomial hard to harmonise in Mediterranean countries. In: E. Cabrera & F. Arregui (eds.), *Water engineering and management through time. Learning from history*. CRC Press Balkema.

Bruns, B.R.; Ringler, C. & Meinzen-Dick, R. (2005). *Water rights reform: lessons for institutional design*. Washington IFPRI.

Buchberger, S. & Cabrera, E. (2010). Water and city in the 21st century. A panoramic vision. In: E. Cabrera & F. Arregui (eds.), *Water engineering and management through time. Learning from history*. CRC Press Balkema.

Cabezas, F.; Cabrera, E. & Morell, I. (2008). *El agua: una cuestión de Estado. Perspectiva desde la Comunidad Valenciana* [Water: as a State Matter. View from Valencia] Asociación Valenciana de Empresarios, Valencia, April 2008].

Cabrera, E. (2005). *El agua, ¿un problema de escasez?* [Water, a problem of scarcity?] Cuadernos de sostenibilidad y patrimonio natural. Fundación del Banco Santander Central Hispano. N° 8: El protagonismo de las materias primas.

Cabrera E. (2008). *El suministro de agua urbano en España* [Urban water supply in Spain]. Panel Científico Técnico de Seguimiento de la Política de Aguas en España. Fundación Nueva Cultura del Agua – Ministerio de Medio Ambiente [Scientific Panel for the Evaluation of Water Policy in Spain – New Water Culture Foundation – Ministry for Environment]. Available from: http://www.unizar.es/fnca/varios/panel/31.pdf [Accessed August 8th 2012].

Cabrera, E. (2011). La burbuja hídrica [The water bubble]. *Levante – El Mercantil Valenciano* 5 February 2011, p. 12.

Cabrera, E. & García Serra, J. (1998). Las estructuras de gestión del agua en España e Israel. Dos modelos contrapuestos [The structures of water management in Spain and Israel Two opposing models]. *Iberian Congress on water management and planning*. Zaragoza (Spain). 14 to 18 September 1998.

Cabrera, E.; García Serra, J.; Cabrera, E. Jr.; Cobacho, R. & Arregui, F. (2002). Water Management Paradox in Southern Europe. *The case of Spain*. *5th International Conference on Water Resources Management in the era of Transition*. Athens (Greece). 4–8 September 2002.

Cabrera, E.; Pardo, M.A.; Cabrera, E. Jr. & Arregui, F.J. (2012). Tap water costs and service sustainability. A close relationship. Submitted to the *Water Resources Management Journal*. Under review.

Corominas, J. & Del Campo, A. (2000). *El papel económico de las aguas subterráneas en Andalucía* [The economic role of groundwater in Andalusia]. Papeles de la Fundación Botín. No. 8. Available from: http://www.fundacionbotin.org/file/10455/[Accessed 15th March 2012].

Custodio, E. (2011a). *Desafíos para la planificación y gestión del agua subterránea en el siglo XXI* [Challenges for planning and groundwater management in the XXI century]. AIH–GE, Zaragoza, September 2011: 1–20.

Custodio, E. (2011b). *¿Tiene España una política estatal de recursos hídricos y de aguas subterráneas?* [Does Spain have a state policy of water resources and groundwater?] AIH–GE. Zaragoza, September 2011: 190–195.

European Commission (2011). River basin management plans 2009–2015 – Information on availability by country. Available from: ec.europa.eu/environment/water/participation/map_mc/map.htm. [Accessed 15th January 2012].

Galbiati, L. (2011). El plan de gestión del agua de Catalunya los nuevos planes de cuenca según la DMA. [The water management plan in Catalonia the new basin plans under the WFD] *1st National Seminar 'The new river basin plans under the EU Water Framework Directive* Nov. 2010. Available from: http://www.fundacionbotin.org/1er-seminario-nacional-los-planes-hidrologicos_observatorio-del-agua_actividades.htm [Accessed 15th March 2012].

García-Vizcaíno, M.J. (2011). Régimen jurídico de utilización de las masas de agua subterránea: problemática y propuestas de modificación [Legal status of use of groundwater bodies: problems and proposals for amendments]. *Jornada jurídica de aguas subterráneas. November 2011*. AEUAS, Ministerio de Medio Ambiente, Rural y Marino.

Gleick, P.H. (2003). Water Use. *Annual Review of Environmental Resources*. Annual Reviews Publisher, Palo Alto, California, USA: 275–314.

Guerin-Schneider, L. & Nakhla, M. (2010). Emergence of an innovative regulation mode in water utilities in France: between commission regulation and franchise bidding. *Journal of Law, Economics and Organization*, Vol. 29(3), pp. 1–23.

Huertas, R. (2011a). Retos y oportunidades en la gestión colectiva de las aguas subterráneas. [Challenges and opportunities in the collective management of groundwater] *4th National Seminar, Botin Foundation Water Observatory*, March 2011. Available from: www.fundacionbotin.org/agua.htm [Accessed 15th April 2012].

Huertas, R. (2011b). La naturaleza jurídica de las aguas subterráneas con la Directiva Marco del Agua: tendencia hacia la demanialización [The legal nature of groundwater in the Water Framework Directive: a tendency toward demanialization]. *Jornada jurídica de aguas subterráneas*. November 2011. AEUAS, Ministerio de Medio Ambiente, Rural y Marino.

Kartin, A. (2000). Factors inhibiting structural changes in Israel's water policy. *Political Geography*, 19: 97–115.

Kingdom, J.W. (1995). *Agendas, alternatives and public policies* (2nd edition). New York: Harper Collins College Publishers.

Llamas, M.R. & Cabrera, E. (2012). *Los conflictos del agua en España: rebeldes sin causa. El pacto del agua es ahora posible*. [Water conflicts in Spain: Rebels without a cause. The water agreement is now possible] Available from: http://www.efeverde.com/contenidos/blogueros/la-blogosfera-de-efeverde/cdo-creadores-de-opinion/los-conflictos-del-agua-en-espana-rebeldes-sin-causa.-el-pacto-del-agua-es-ahora-posible-por-m.-ramon-llamas-y-enrique-cabrera [Accessed 9th January 2012].

López-Gunn, E. (2009). Agua para todos: A new regionalist hydraulic paradigm in Spain. *Water Alternatives*, 2(3): 370–394.

McKay, J. (2010). Some Australian examples of the integration of environmental, economic and social considerations into decision making; the jurisprudence of facts and context. In: *Global Justice and Sustainable Development*. D. French (ed.). ICD Publishers, the Netherlands.

Merrey, D.; Meinzen-Dick, R.; Mollinga, P. & Karar, M. (2007). Policy and institutional reform: The art of the possible. In: D. Molden (ed.), *Water for Food, Water for Life: Comprehensive Assessment of Water Management in Agriculture*. Earthscan, London, UK.

OECD (2011). Water Governance in OECD countries: a multilevel approach. OECD Studies on Water. OECD Publishing, Paris, France.

Poveda, J.A. (2011). La naturaleza jurídica de las aguas subterráneas con la Directiva Marco de Aguas: tendencia hacia la demanialización. Jornada jurídica de aguas subterráneas [The legal nature of groundwater in the Water Framework Directive: a tendency toward demanialization]. *Jornada jurídica de aguas subterráneas*. November 2011. AEUAS, Ministerio de Medio Ambiente, Rural y Marino.

Saleth, R.M. & Dinar, A. (2000). Institutional Changes in Global Water Sector: Trends, Patterns and Implications. *Water Policy*, 2(3): 175–199.

Villarroya, F.; López-Gunn, E. & De Stefano, L. (2010). Los paradigmas de la gestión del agua en España: de la misión hidráulica regeneracionista a la Nueva Cultura del Agua [The paradigms of water management in Spain: the regeneracionista hydraulic mission to the New Water Culture]. *XVI Simposio Enseñanza de la Geología*, 20 años de AEPECT. Alcalá, L. & Mampel, L.(eds). Fundamental n° 16: 291–296.

Part 2

Metrification of water uses

Chapter 5

Towards an Integrated Water Resource Management (IWRM)

Maite Aldaya & M. Ramón Llamas
Water Observatory of the Botín Foundation; Department of
Geodynamics, Complutense University of Madrid, Madrid, Spain

ABSTRACT: This chapter analyses the pros and cons of key tools and methods for achieving a more Integrated Water Resource Management (IWRM), with special reference to the application of the concepts of virtual water and water footprint. IWRM is widely recognized as a good idea but its practical guidance and implementation has hardly begun. IWRM requires consideration of the tangible (measurable or quantitative) and intangible values of water resources (difficult to quantify, such as cultural, spiritual values or intergenerational equity) and related socio-economic and environmental aspects both from the production and consumption perspective. The water footprint tool, coupled with other socioeconomic and environmental data, can be a good tool providing a transparent and multidisciplinary framework for informing and optimizing water policy decisions and to facilitate the IWRM for the analyses of nations, regions, basins or products. It generally provides an easily communicable framework for sensibilisation and is usually a good tool to deal with the stakeholders. Nevertheless, it is necessary to keep in mind its limitations, such as data constraints or comparability limitations. By extension it seems that perhaps the most important issue to solve the global water problems is to achieve a more fair and equitable regulation of the food (virtual water) trade.

Keywords: integrated water resource management, water accounting, tangible and intangible values, water footprint, virtual water

I INTEGRATED WATER RESOURCE MANAGEMENT

During the last two decades the concept of Integrated Water Resource Management (IWRM) has become very popular. The main reason for this has been the pervasive awareness that water resources need to be managed across subsectors at the basin level. Since the 1990s water management has expanded to cover efficient water use, equitable sharing of benefits and environmental sustainability aspects. IWRM is a process which promotes the coordinated development and management of water, land and related resources, in order to maximize the resultant economic and social welfare in an equitable manner without compromising the sustainability of vital ecosystems (GWP, 2000). IWRM involves collection and management of natural resources information, the understanding of the interactions that occur in the use of these resources, together with the implementation of policies, practices and administration structures, which enable the resources to be used. Water problems cannot be solved if they are considered from only one scientific or institutional perspective. It is necessary to focus them from a multidisciplinary perspective and it is not feasible that only one institution or governmental authority deals with all the uses of water. Participation and coordination among different institutions and stakeholders is necessary. The good governance of freshwater resources requires equilibrium between the utilitarian (measurable or quantitative) and the intangible values (difficult to quantify, such as cultural, spiritual values or intergenerational equity). It was at the UNESCO Working Group meeting on the Ethics of Fresh Water Uses in the year 2000 when the need of this equilibrium was stated (Llamas & Delli Priscoli, 2000).

2　WATER ACCOUNTING AND ASSESSMENT: CHALLENGES AND OPPORTUNITIES

In most parts of the world, the development of consistent and systematic water accounting systems both from the production and consumption perspective are in their infancy, but rapidly developing. There is a need to quantify and account for water flows within the economy (including for environmental needs) and related impacts in the appropriate time and spatial scales, to enable transparent information systems which could be used to develop robust allocation and management systems that underpin a green economy.

The better informed the decision-makers are, the more likely they are to make the right decisions. For water managers this means being able to provide reliable information about where and when water is available, of what quality, where and how it is used, what happens to wastewater, how much water leaves the country in exports of goods that use water in their production (virtual water) and how much enters the country in imports, impacts on the social, economic and environmental sectors, and the intangible values.

From the production perspective, water balances (i.e. inventories or registers) represent the fundamental approach to accounting for the flow of water into and out of a system.

Consumption-focused instruments present difficulties in linking consumption back to impacts in specific river basins at specific times. The impacts of consumers on water resources are generally indirect and linked to long supply chains, not only related to human activities in the same watershed but also via inter-basin exchanges and international trade. In addition these supply chains and interdependencies are not restricted to single sectors (e.g. agriculture, industry or urban water supply) but evolve into interdependencies between sectors.

This will be a challenge for water managers in most countries, which lack the necessary measurements and do not systematically collect the necessary data. When the information is available, it will be possible to calculate the country's water balance and the water footprints (volume of water used) of various users. Using this information, water managers can advise decision-makers in other sectors of the feasibility of their plans and the implications for water.

To provide a robust basis for analysis and decision-making, such assessments must meet certain criteria.

3   THE VIRTUAL WATER AND WATER FOOTPRINT CONCEPTS

The virtual water concept was coined in the 1990s by Professor Tony Allan (Allan, 2011). The virtual water of a product or service is the volume of freshwater used to produce the product or service. The water footprint (WF) is an indicator that looks at both direct and indirect water use of a consumer or producer (Hoekstra, 2003). A water footprint can be calculated for a process, a product, a consumer, group of consumers (e.g. municipality, province, state or nation) or a producer (e.g. a public organization, private enterprise). All components of the water footprint are specified both geographically and temporally (Hoekstra *et al.*, 2011). The blue water footprint refers to consumption of blue water resources (surface and ground water) along the supply chain of a product. The green water footprint refers to consumption of green water resources (rainwater stored in the soil). The grey water footprint refers to pollution and is defined as the volume of freshwater that is required to assimilate the load of pollutants based on existing ambient water quality standards. This is still an interesting, complex, and controversial concept. Since this book is intended mainly for the water policy makers and not for the academic community, the grey water concept is not usually included.

There is a growing need to integrate nature conservation, social equity and economic growth into the process of decision making. For the time being and almost in the entire world, water footprint analysis has focused on hydrological aspects. A significant innovation of this work is to consider the economic, social, ecological and intangible aspects. The water footprint combined with other socio-economic and environmental methods, as suggested in this book, seems to be a useful tool that provides a transparent and multidisciplinary framework for informing and optimizing water policy decisions and the needs of economic sectors and healthy ecosystems. This is being developed in a progressive way. For instance, already in 2008 in the Guadiana Basin analysis, the economic values related to water uses were already included (see Chapter 9). This was probably the first time that this approach was used in the world. In the study of the Guadalquivir basin (see Chapter 8), also probably for the first time, the uses of green and blue water by the natural ecosystems were included.

Traditionally, countries formulate national water plans by evaluating how to satisfy water users. Although countries consider both options to reduce water demand and to increase supply, they generally do not include the global dimension of water management. Many countries import agricultural or industrial goods without determining whether imported products cause water depletion or pollution in producing countries. Governments could engage with consumers and businesses to work towards sustainable consumer products. National water footprint accounting could be one of the components in national water statistics, supporting the formulation of national water plans and river basin plans.

In this context, Spain has been the first country in the world to adopt the water footprint evaluation in governmental policy making. In September 2008 the Spanish Water Directorate General approved a regulation that includes the analysis of the water footprint of the different socio-economic sectors as a technical criterion for the development of the River Basin Management Plans, that all EU Member States will have to accomplish every six years as part of the requirements of the Water Framework Directive (Official State Gazette, 2008).

3.1   Strengths of the water footprint

- When coupled with other tools, the water footprint is a comprehensive and transparent tool providing the *big picture* for strategic planning purposes.
- **Easily understood by non-technical audiences.** The water footprint can be an effective public awareness-building tool. It has also been a useful tool to deal with the different stakeholders, mainly the farmers' associations or lobbies.
- **The influence of technology in the yield of crops.** The type of agricultural technology available has a significant role in the determination of the virtual water of crops (m^3/t of product) [t = tonne = 10^3 kg]. It may be more significant than the conventional calculations of the water use per crop (m^3/ha), and this technology usually improves with time (see Garrido *et al.*, 2010 and Chapter 6).
- Water footprint and virtual water trade assessments are a relevant input into various governmental policy areas, such as national, state, river basin or local water policy; environment; agriculture; industry/economic policy; energy; trade; foreign policy and development cooperation.
- Hitherto, decision-makers have focused on water issues related to production within the related territories, without considering the **virtual water flows** linked to trade in agricultural and industrial products. Water challenges and opportunities are often tied to the structure of the global economy. By looking only at water use within their territories, decision-makers have a blind spot to the issue of sustainability of consumption.
- **Improving water efficiency.** The water footprint provides new dimensions of water use efficiency: user, basin, and trade. At the user level, technology, education and pricing play a key role. There are opportunities to significantly improve agricultural, domestic and industrial sector water use efficiencies. Overall, food supply chains should be made more water efficient. At the basin level, water allocation efficiency is required to ensure water use is sustainable, equitable, and that appropriate *value* is derived from the resource. All sector requirements must be considered holistically, including environmental water requirements. At national, regional, and global scales, virtual water trade can be used as a tool to improve overall water use efficiency by considering the comparative advantages of certain water uses in particular regions. A good example is seen in the analysis of the Guadalquivir basin (see Chapter 8).
- In industrialized and emergent economies the water footprint, associated with other instruments, has become a good tool to inform water rights re-allocations in order to move from the motto of *more crops and jobs per drop* to a motto of *more cash and care of nature per drop*. In other words, to achieve a win-win type solution (see Guadalquivir analysis, Chapter 8).

3.2   Weaknesses of the water footprint

The water footprint tool presents the following difficulties:

- **Terminology confusion:** Different water footprint studies use different terminology to refer to the same concepts. The recent analyses of the WF done by the Spanish Government for some Spanish basins are an example of this. Therefore, it seems important to use the same terms in such studies. This is why in this book a glossary has been included.

- **Problems with data availability**: The available data do not always fit to the requirements of the water footprint studies. A clear example is the territorial scope of the data. Some data are available at provincial, regional or national level but do not agree with the watershed boundaries (e.g. trade data). Adjustments and simplifications are necessary, which introduce uncertainties and errors.
- **Inadequate data**: A lack of sufficient data about climate, soils and growing periods of crops is in many cases the greatest factor limiting the ability to provide meaningful information on the water consumption of crops or other products. This is most often due to inadequate databases, or lack of access to existing data, and this causes a cascade of errors in the final estimation of the consumptive water uses per crop and surface.
- **The concept of grey water.** This concept used as an assessment of pollution can be misleading (e.g. sometimes wrongly understood as an approach to dilute pollutants). This relatively new concept, created in 2006, is currently being developed and refined. For the moment, natural improvement is not taken into account, which may be important for some pollutants.
- **The water footprint, when not connected to the socio-economic and environmental values related to the different uses, provides partial information and may be misleading.** The emphasis is placed on the concept of water self-sufficiency. This causes a *hydrocentric* unrealistic approach.
- **Dry matter.** When estimating the water footprint of a product (m^3/t), based on available crop yields, we are considering fresh weights. However, from an ecological viewpoint, primary production is measured as dry matter yield. In this line, results would change when taking the dry matter into account; the fresh weight is mostly water in vegetables but cereals have a much higher dry matter.
- The usefulness of the water footprint in water planning; **the water footprint linked to other data.** The water footprint *per se* generates little direct practical conclusions for planning purposes, except the knowledge of the environmental impact in time and space of water use by the inhabitants of a region. However, from the planning point of view, it can be very useful if water footprint data are linked with other data consistent in space and time, including:

    a   Socio-economic indices (e.g. gross value added, profit, employment) in each economic sector. Socio-economic data can be used along with the water footprint (€/ha; €/m³; employment/m³), providing information on the apparent productivity and efficiency of water use, so that the effects of the water use in human activities are assessed.

    b   Hydrological and environmental indices. The water footprint can be compared to the potential resource available and environmental water requirements in terms of both green and blue water.

    c   A balance between utilitarian (quantifiable or measureable) and intangible values (cultural, emotional or religious aspects, which are harder to quantify) is also necessary.

In sum, the water footprint linked to other data (hydrological, socio-economic, environmental, intangibles) could contribute to awareness-raising and transparency development, which could contribute to the proper use of water resources and support the water planning process, particularly in water resource allocation, which is related to socio-economic development, reducing consumption and transfers of resources, environmental improvement and the achievement of social objectives, and in summary, to a suitable and efficient use of scarce resources. This fits perfectly with the main objectives of the European Water Framework Directive.

4   CONCLUSIONS

  i Integrated Water Resources Management (IWRM) requires consideration of the tangible (measurable) and intangible values of water resources and related socio-economic and environmental aspects both from the production and consumption perspective.
 ii The water footprint tool, coupled with other socio-economic and environmental data, is a good tool providing a transparent and multidisciplinary framework for informing and optimizing water policy decisions and to facilitate the IWRM for the analyses of nations, regions, basins or products.
iii Nevertheless, it is necessary to keep in mind the limitations of the tool (e.g. cascade of errors in large territories, data constraints).
 iv The water footprint is usually a good tool to deal with the stakeholders, mainly with the farmers' associations or lobbies. This is relevant because the farmers are usually the main human water users.
  v The analysis of Spain, the most arid country in the EU, seems to show that, due to the recent advances in science and technology, and mainly to globalization, many of the current Spanish water conflicts can be solved in the short or middle term. A crucial issue for that will be achieving a water pact between the main political parties so that water issues are not used as a political means to win votes for the next election.
 vi By extension it seems that perhaps the most important issue to solve the global water problems is to achieve a more fair and equitable regulation of the food (virtual water) trade.
vii This better regulation would allow the reallocation of water rights or uses in the arid and semiarid regions and the use of the scarce water resources for the more profitable uses.

REFERENCES

Allan, T. (2011). *Virtual Water: tackling the threat to our Planet's most precious resource*. I.B. Tauris Publishers. 384 p.
Garrido, A.; Llamas, M.R.; Varela-Ortega, C.; Novo, P.; Rodríguez-Casado, R. & Aldaya, M.M. (2010). *Water Footprint and Virtual Water Trade in Spain. Policy Implications*. Series: Natural Resource Management and Policy, Vol. 35. Botín Foundation and Springer, New York, USA. 153 p.
GWP (Global Water Partnership) (2000). *Integrated Water Resources Management*. Technical Advisory Committee (TAC) Background Paper No. 4. Global Water Partnership, Stockholm, Sweden.

Hoekstra, A.Y. (ed.) (2003). *Virtual water trade*. Proceedings of the International Experts. Meeting on Virtual Water Trade, 12–13 December 2002. Value of Water Research. Report Series No 12, UNESCO-IHE, Delft, the Netherlands. Available from: www.waterfootprint. org/Reports/Report12.pdf [Accessed May 12th 2012].

Hoekstra, A.Y.; Chapagain, A.K.; Aldaya, M.M. & Mekonnen, M.M. (2011). *The Water Foot-print Assessment Manual: Setting the Global Standard*. Earthscan, London, UK.

Llamas, M.R. & Delli Priscoli, J. (2000). *Report of the UNESCO Group on the Ethics of Freshwater Uses*. Papeles del Proyecto Aguas Subterráneas [Groundwater Project Papers], Botín Foundation, Santander, Spain. A Series, 5: 58–99.

Official State Gazette (2008). Approval of the Water Planning Instruction. Ministry of the Environment and Rural and Marine Affairs. Official State Gazette, 229, 22nd September 2008. Available from: http://www.boe.es/boe/dias/2008/09/22/pdfs/A38472-38582.pdf [Accessed May 15th 2012].

Chapter 6

Overview of the extended water footprint in Spain: The importance of agricultural water consumption in the Spanish economy

Daniel Chico & Alberto Garrido
Water Observatory of the Botín Foundation;
CEIGRAM, Technical University of Madrid, Madrid, Spain

ABSTRACT: The Spanish economy is increasingly decoupling from water use. It has grown and progressed during 1996–2008 using less national water resources. Agriculture, the economic sector that consumes most water, provides only a small part of the GDP and employs a decreasing part of the labour force. Water has therefore a relative importance for the economy at a national level. To understand the drivers of water use, we do an analysis of the trends of agricultural production. In Spain, crop production is changing, linked to increases in irrigated land, redistributions and concentrations of production and increased international trade. As a result, its water consumption is increasing primarily in association with the import of agricultural products. Alongside, the exports of national products are increasing, and as a country, Spain is exporting ever more valuable products with less virtual water (VW) content, while increasing its imports and thus reallocating its water footprint (WF).

Keywords: national water footprint, virtual water trade, agriculture, livestock

1   INTRODUCTION

This chapter summarizes recent evaluations of the virtual water (VW) trade and water footprint (WF) in Spain, with special attention to the agricultural sector. It offers revised computations with respect to the pioneering work of Garrido *et al.* (2010), and uses longer and newer data sets and revised methods. The WF indicator makes it possible to extend the analysis of water use by the different economic sectors. By including direct and indirect uses, and taking into account both irrigation (so-called blue) water and soil (green) water, it gives a much broader picture of water consumption than the traditional water abstractions. These advantages are shown in this chapter, as well as in the following ones. These evaluations indicate the economic activities in Spain where water is consumed and virtually traded.

The chapter is organized as follows. First, we provide an overview of the agricultural sector's main economic figures and its contribution to the Spanish economy. The second section describes the methodological innovations and databases that have been used in this study with reference to Garrido *et al.* (2010). The third part analyzes the WF of the agricultural sector and its evolution over the period 1995–2009. An evaluation of the VW trade of the agricultural sector is offered in the fourth part. Finally we draw conclusions about the evolution of the WF of the sector and its VW trade.

2   MAIN FIGURES OF THE SPANISH ECONOMY AND RELATED WATER USE

Table 1 shows the relative importance for the economy and the labour force of the general sectors of the Spanish economy.

It provides information from the National Statistics Institute (INE) with the share of Gross Domestic Product (GDP) and labour force. Data for GDP and employment in the tourism sector was obtained from the Ministry for Industry, Tourism and Commerce (MITIC 2011), and subtracted from the service sector. The estimation of the water consumed in the industrial and construction sectors and the tourism sector is based on the study carried out by the Spanish Environment Ministry on the WF of Spain, applying the coefficients used for the transmission of the WF from the primary sectors to the rest of the economy (MARM, 2011a). Water consumed by the livestock sector includes the water consumed in the production of feed and forage crops.

The Spanish agricultural and fisheries sectors contributed to 2.3% of the Gross National Product (GNP) in 2009, down from 3.8% in 2000 (see Table 1). The related agri-food industry contributed with 1.77% down from 2.11% in the same period. The share of employment in 2007 in the agri-food industry is approximately 2.6% of the national total, mainly linked to the bread and pastry industries (46% of employment), the meat (14%) and the beverage industries. These industries also generate most of the added value in the agricultural sector, with the meat industry accounting for 18%, beverages 17%, and bread and pastries 12%.

Despite its importance for the rural economy, agriculture has a small importance in economic terms for the country as a whole. The importance of agriculture is linked to its local effects, its shaping of the rural society and landscape, and its provision of food security. As was already pointed out in Garrido et al. (2010), Spain has achieved a decoupling of its economic growth from water consumption and VW flows: increasingly less water is used to produce one euro of output. Nevertheless, agriculture remains the main water consumer with over 75% of the total WF.

Within the agricultural sector, vegetable production represents 60% of the GNP and animal products 35%, the remaining 5% includes forestry and fisheries production. Between 1995 and 2008 crop production grew by 4% in real terms whereas animal production grew by only 1% (Garrido et al., 2011). Final Agricultural Production in Spain rose steadily from 1996 to 2003, but has fallen since then, with slight upturns in 2007 and 2008 as a result of price rises in agricultural products. However, the evolution of agricultural income, in constant euros, increased from 1980 to 2003, but since 2003 it has gone down steadily, reaching by 2008 similar levels as in the mid-1990s (Garrido et al., 2011). The reasons for this downturn are the downward trends of agricultural prices in real terms and the increase of input prices since 2003.

Between 1995 and 2008 there was a general decrease of the cultivated rainfed surface, as well as that of irrigated arable crops and vegetables. By contrast, the irrigated surface of trees and vineyards had a tendency to increase until the 2003 reform

Table 1 Percentage of GDP, labour force and water consumption of the Spanish economic sectors.

	Sector	2000	2001	2002	2003	2004	2005	2006	2007	2008	2009
% of GDP	Agriculture, livestock and fisheries	3.8	3.7	3.5	3.4	3.1	2.7	2.4	2.5	2.3	2.3
	Industrial	18.8	18.3	17.7	17.2	16.7	16.3	15.8	15.6	15.6	14.6
	Construction	9.3	9.9	10.5	10.9	11.4	12.1	12.6	12.4	12.5	12.1
	Tourism	11.6	11.5	11.1	11.0	10.9	10.8	10.9	10.8	10.5	10.0
	Services (except tourism)	46.9	47.3	48.0	47.8	47.8	47.4	47.2	48.6	50.8	53.2
% Labour	Agriculture, livestock and fisheries	6.1	5.8	5.8	5.6	5.4	5.1	4.7	4.5	4.4	4.3
	Industrial	19.4	19.0	18.5	17.9	17.3	16.8	16.3	15.8	15.6	13.6
	Construction	11.2	11.8	12.1	12.5	12.8	13.3	13.6	13.8	12.6	11.1
	Tourism	11.6	11.5	12.0	12.12	12.37	12.78	13.3	13.6	13.8	10.8
	Services (except tourism)	51.8	51.8	51.6	51,9	52.1	52,0	52,1	52.3	53,6	53,8
Water consumption (hm³)	Crop production	27,206	28,855	28,795	29,126	30,899	21,037	25,819	30,681	33,077	25,145
	Livestock	40,839	42,301	42,952	43,733	44,343	44,008	42,969	49,331	42,995	42,563
	Industrial and construction	2,081	1,874	1,870	1,892	2,007	1,366	1,677	1,993	2,148	1,633
	Tourism	518	467	466	471	500	340	418	496	535	407
	Services (Urban except tourism)	1,735	1,874	2,012	2,078	2,047	2,178	2,077	1,983	1,921	1,965

Sources: INE (2011a, 2011b), MITYC (2011), MARM (2011a).

[hm³ = cubic hectometre = million m³ = 10⁶ m³].

of the Common Agricultural Policy (CAP) (Ruiz *et al.*, 2011). The subsequent reforms of the CAP in 2003 and 2006 have consolidated a trend initiated with the McSharry reform of 1992 to shift farm income support from price support to decoupled direct payments. With the exception of some crops, like cotton, tobacco, olives and sugar beet, which kept a variable percentage of support via prices, the majority of cultivated land is now indirectly supported with direct payments to the farmer. Ruiz *et al.*, suggest that, as a consequence of the reform, there has been some land redistribution among crops, with cereals occupying lands previously dedicated to industrial crops like tobacco or cotton, which have seen their profitability going down significantly. There has been an increase in the average yields of cereals, also linked to a price effect and to the abandonment of the least productive areas (see Chapter 11).

Barley and wheat cover 23% and 15% of the country's agricultural surface, respectively. Overall, cereals occupy 46% of the surface, but only account for 21% of the production and 14.2% of the economic value. Land devoted to rice and maize diminished after 2003, as CAP payments were totally decoupled from production. Olive trees and vineyards occupy 16% and 7.4% of the surface, respectively. Both crops more than doubled their irrigated surface during 1995 and 2008 (146% in the case of vineyards), and either maintained (olives) or decreased (vineyards) their rainfed surface. Vegetables dropped their rainfed area, maintaining its area in national terms. If we look at the value of the agricultural production, vegetables, fruits and citrus, and pork meat, account for roughly half of the total value (18.0, 14.5 and 12.5% respectively). Significant increases in production value are identified in vegetables, pork and olive oil. Cereals, grapes and citrus fruits grew from 1995 to 2000, but their value remained about the same between 2000 and 2008.

As for the animal sector, its main production, pork meat, is followed by bovine meat, milk and dairy products. Each year the sector produces more than 5.5 Mt [Mt = million tonnes = 10^9 kg] of meat, 60% of which is pig meat and 25% chicken. The largest growth in numbers of animals has been in these two species: 58% in the case of pork and 12% in poultry. The sector has grown in efficiency as well, since production of meat per animal improved, as well as the value produced per animal.

3   METHODOLOGY AND DATA

The method used to estimate the WF and VW trade of the Spanish agricultural sector builds on and improves the methodology used in Garrido *et al.* (2010). We estimated crop water consumption as a function of crop evapotranspiration, based on FAO ETO methodology (Allen *et al.*, 1998). However, in relation to the work by Garrido *et al.* (2010), our estimations have been refined to expand the studied period, as well as for better measuring water consumption by permanent crops. Significantly, an estimate of the deficit irrigation was applied, approximating the irrigation water consumed to the crop net water requirements defined by each river basin authority (RBA) as the upper limit.

The databases used included detailed annual crop production, area and economic value at provincial level, obtained from the Ministry of the Environment and Rural and Marine Affairs (MARM, 2011b). Information on the number of slaughtered animals, production and economic value was also obtained from MARM (2011b). Climatic data were obtained from the National Meteorological Agency (AEMET, 2010) and the farmers' irrigation service (*Servicio de Asesoramiento al Regante*, MARM, 2011c). Data on volume, economic value and origin of international trade was taken from DATA-COMEX, the database of international trade by the Ministry of Industry, Tourism and Commerce (MITYC, 2011). Crop net irrigation requirements in each river basin were obtained from the RBAs as given in the Water Framework Directive official documents.

To calculate the VW content of animal production, the methodology of the Water Footprint Manual (Hoekstra *et al.*, 2011) was followed. The WF of livestock products calculation followed the same methodology. This WF accounts for: 1) the water consumed in animal servicing (water used in the farm); 2) the water drank by the animals; and 3) the VW included in the animal feed. This diet refers to a mixed diet of extensive and intensive production systems. These diets were designed as explained in Rodríguez-Casado *et al.* (2009). Drinking and service water used for pigs and poultry was refined from the sources used in Garrido *et al.* (2011), as well as data on age and weight at slaughter.

Trade statistics distinguishing between green and blue water, were obtained from Mekonnen & Hoekstra (2010a, 2010b) for the imported products. The VW content of the imported products (e.g. coffee) was calculated by weighing the average of the product's VW content depending on the country of origin. The VW content of exported products was considered equal to the water content consumed in the production of those products within Spain. This is mainly because Spain does not generally re-export products.

In comparison with Chapter 7, which focuses on groundwater use, the estimations of water consumption are slightly higher for this chapter (15% on average). In Chapter 7 an effort was made to work at a smaller scale (municipal and district level) in order to be capable of differentiating the origin (surface or groundwater) of the water. In this chapter the data were homogenized to the provincial level, thus introducing some level of error. Still, the estimates are in relative agreement with those showed in Chapter 7.

The evaluation of the WF of olives, olive oil and tomato production, reported in Chapter 10, was much more precise, and included in the case of olive production a limitation on the available water. Therefore, the WF calculated for olive production shows higher green water content per unit of product (m^3/t) [t = tonne = 10^3 kg] and a smaller WF of the production (hm^3) [hm^3 = cubic hectometre = million $m^3 = 10^6\ m^3$] than in Chapter 10. Discrepancies in the results are also related to the modelling of irrigation applications and water balance. The need for modelling the whole range of production for the country in a consistent way led to some simplifications. This may have altered the calculations by as much as 35% and affected the proportion between green and blue water. Nevertheless, the relationship of water consumption and crop types, and the trends, are maintained.

4 AGRICULTURE'S WATER FOOTPRINT

As we saw in Table 1, agriculture is the main water consumer and merits a more in-depth analysis. In this section we present the results obtained on the WF and VW water trade of the sector.

Figure 1 plots the national WF of the agricultural sector along with the VW trade flows, the water needed to produce the traded goods. Crop production WF

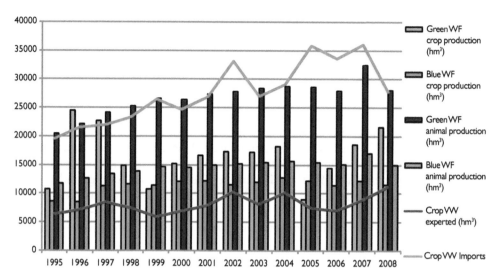

Figure 1 The internal green and blue WF (blue and green) of crop and animal production and VW flows (hm³/year). (Source: Own elaboration).

(green and blue) has grown by 5% on the period to around 30,000 hm³/year. Animal production WF grew by 30% to around 45,000 hm³/year, including the WF of forage and feed. A relevant fact is the proportion of green and blue water used in agricultural production. In crop production, blue water is gaining relevance, in crops like cereals and vineyards, which increased the relevance of blue water by 20%. Olive production decreased its proportion of green water from 92 to 84%.

VW trade flows have also increased, though at a different pace. VW imports amounted to around 32,000 hm³ at the end of the period, and nearly tripled the VW exports (9,000 hm³); they have increased by 55% during the period of investigation and maintained a more or less stable share of green water. On the other hand, crop VW exports have also increased (23%), but the share of green water in these VW flows has decreased. The share of traded blue water is slightly gaining importance in the crop production (with the exception of citrus crops), whereas for the animals, the green water is completely prevalent, related to increasing imports of feeds. Spain imports a much higher volume of water than it exports. Virtual imported volumes are about the same than those needed to grow its own crops. Net imports have followed a clearly upward trend. The imports are mainly green water (90%). The exports have a much higher share of blue water (40%) but it is also diminishing.

Table 2 shows an overview of the WF per crop type at the beginning and end of the period of study. Cereals (38%) and olive trees (20%) account for the largest share of the national agricultural WF, followed by citrus, fruits and industrial crops. This is related to their large share in the cultivated area, and to their higher WF per unit (m³/t). Both cereal and olive production rely mainly on green water and have a predominantly rainfed production. Nevertheless, blue water is becoming increasingly more important as the irrigated surface increases (see Chapter 8 for the case of olive sector in the Guadalquivir basin). Both olive and vineyard production doubled their

Table 2 Share of the national WF (%), WF per unit (m³/t), WAP (€/m³ green and blue water, at nominal prices), and blue water WAP (€/m³) per type of crop, at the beginning (av. 1995–1996) and end of the period (av. 2007–2009).

Type of crop	Aver. (1995–1997)				Aver. (2007–2009)			
	Percentage of national WF	Volumetric WF (m³/t)	WAP (€/m³)	WAP blue water (€/m³)	Percentage of national WF	Volumetric WF (m³/t)	WAP (€/m³)	WAP blue water (€/m³)
Cereals	33.9	555	0.27	0.96	39.7	534	0.33	0.84
Pulses and tuber	2.7	196	0.84	1.75	2.1	283	0.82	1.63
Industrial crops	10.3	297	0.30	0.47	4.9	273	0.33	0.59
Forage crops	4.8	46	2.22	4.79	7.2	69	1.44	2.68
Vegetables	4.8	123	2.92	3.58	5.1	123	3.71	4.30
Citrus	5.4	301	0.92	1.28	6.6	330	0.70	1.11
Other Fruit trees	8.2	674	0.60	1.33	7.4	685	0.78	1.59
Vineyard	7.2	454	1.00	10.23	6.9	357	1.03	3.98
Olive trees	22.6	1801	0.36	7.27	20.1	955	0.45	3.65

Source: Own elaboration.

WAP: Water apparent productivity.

irrigated surface over the period. Citrus and cereals increased theirs by 15% and 20%. The largest reductions in WF are related to the decrease in the surface of industrial crops, non-citrus fruit trees and pulses and tubers, whereas forages, vegetables and cereals increased their importance in the national crop WF by 51%, 145% and 16% respectively.

The products holding the largest share of the national WF (cereals and olives) have a lower water apparent productivity (WAP) whereas vegetables, vineyard and forage crops exhibit the highest WAP. As for their trend, WAPs have had a different behaviour than the WF. Fruit trees, cereals, vegetables and olives have increased their WAP (32% in the case of fruit trees), whereas that of pulses, citrus and forages decreased. Vineyards maintained their WAP.

Crop exports account for a small part of the total production but are significant in economic terms. Spain exports a small part of the total production, but at a relatively high price and at an increasing value. Agricultural exports show a positive trend. On the other hand, Spain is importing more in volume, at increasingly less value per unit, a small amount of goods compared to what is produced. This leads to positive terms of trade making agriculture a source of income for the country.

Analysing the water consumption of this trade, the VW exports grew slower than the value of the products and so the WAP of the exports has increased. VW imports are much less relevant economically than in volume (hm³) terms. This is mainly because imported products on average have a high WF per unit (m³/t) and low value. We may also see that the growth of the WAP of the exports is higher than the growth of the WAP of national production and the imports. Spain is increasing the competitiveness of its water exports, obtaining increasingly more value for its water consumption.

The main products exported are citrus, vegetables, wheat, wine, olive products and pork meat.

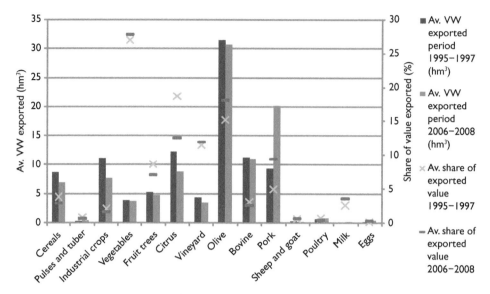

Figure 2 Average VW exported (hm³/year, left axis), and average share of the exported value at the beginning (1995–1997) and end of the period (2006–2008) (right axis). (Source: Own elaboration).

Figure 2 disaggregates per product type the share of economic value and absolute WF (hm³/year) of the exports. We see how pork and olives and olive oil have greatly increased their importance in exports over the period. In contrast, citrus fruits have greatly decreased their share. Vegetables and wine have maintained their share of the value alongside a small and stable part of the VW exported. Other productions decreased the VW exported (cereals, industrial crops and citrus mainly). The country is increasingly concentrating its exports in a range of products like wine, olives, vegetables and pork, which in turn are (with differences in the degree) improving their WF per unit (m³/t) and their WAP (€/m³).

As for the VW imports, most of them are associated with cereals (maize, wheat and barley) and oil seeds and their products (soya bean, soya cake, sunflower seeds and cassava). Annual agricultural VW imports increased from 23,500 hm³ in 1995 to 36,300 hm³ in 2008. Their economic value increased from 6,000 M€ to 11,234 M€, an 87% increase. They are composed of 90% green water, and originate principally from South America (Argentina and Brazil), France and the USA.

5 DISCUSSION

Imports represent less economic value per imported VW and are almost exclusively composed of green water (more than 90%). Agricultural imports are overwhelmingly linked to the imports of cereals and oilseeds (soya). These two productions are even increasing their volume (t) imported, but show different trends. Cereals seem to be increasing their share in imported economic value, because of their growth in imports, although at lower economic relative value (€/m³). Consequently they are still gaining a larger share of the country's external WF. Oil seeds, which account for a third of the value and 40% of the VW imported, increased their volume of imports (t) less than their value, and more than their WF. Thus, there is a tendency to import more expensive oilseeds with less WF per unit (m³/t). Should the prices of the oilseeds remain high in international markets, the downward trend followed by the internal WF of oil crops may slow down too. This growth in VW imports is related to the growth in meat production and exports, as many of the imported cereals and oilseeds are used for animal feed. Pork production has increased in the period, while its WF tripled. Pork meat exports tripled their value and their VW exports in the period. This is related to the high VW content per unit of product. A similar effect is seen in the rest of the animal production.

The increases in the WAP of the exports have not come that much from the meat sector despite its significant growth. It is the olive production that increased the most in value in the exports, but also increased significantly its WAP. The olive sector had this performance at the same time as it increased the (still low) proportion of blue water in its production (see Chapter 10). Wine on the other hand decreased its importance in the VW exports, but maintained its value, and managed to triple its WAP. Both productions also managed to lower by 30% their WF per unit of product exported. They have become more effective in the use of water for generating value. Vegetables maintained their importance, and remain the most important category of exports. They maintained their high WAP, by decreasing slightly their WF per unit of product.

While the change in CAP in 2003 reduced existing incentives to intensify production, there has been a concentration (in space and products grown) of Spanish agriculture which has brought sustainability issues in certain regions. International crop trade in Spain has increased in the period. The country is trading increasingly more, and improving its terms of trade. A certain degree of specialization has occurred in the traded goods, narrowing the range of exported products and improving the productivity of the water consumed. Agriculture, and subsequently water, is increasingly linked to international markets, and thus to its trends and growth. In consequence, the need to balance the market forces with the available local resources is becoming increasingly important.

6 CONCLUDING REMARKS

This chapter analyzed the importance of water in the Spanish economy, and in more depth the agricultural sector. The Spanish economy has gradually decoupled from water consumption, using less water per euro of output. Agriculture and agro-industry, the main water consumers, have a relatively minor importance for the economy and employment, but form the basis of food security. Spanish agriculture has experienced significant changes in the last 15 years. Main changes in crop production are giving blue water consumption an increasing importance, especially in the fastest growing sectors like vineyards and olive trees. On the contrary, in the animal sector the WF is turning greener, as the imports of feedstuff have increased significantly, to attend the increase in production and exports.

The relation between the economy and water resources has potentially damaging effects at local and basin levels. An important complement to the study presented here is the impact of agriculture on water quality. Agricultural production has evolved differently between the crop types, and accordingly between regions. Not every region is capable of competitively producing the most valuable crops, and thus, as international trade increases and competition arises, the biggest producing regions will increase their production. A view of this may be found in the analysis of the olive and tomato sectors in Chapter 10. Together with this concentration and specialization, there are issues of over-exploitation and water quality that have arisen in many of these regions. Other chapters analyze the growth in groundwater use (Chapter 7), the problem of agricultural diffuse pollution (Chapter 12), and land use changes (Chapter 11).

REFERENCES

AEMET (2010). Data supplied by the National Meteorological Agency. [http://www.aemet.es/es/portada].

Allen, R.G.; Pereira, L.S.; Raes, D. & Smith, M (1998). *Crop evapotranspiration – Guidelines for computing crop water requirements*. FAO Irrigation and drainage, paper 56. Food and Agriculture Organization, Rome, Italy.

Garrido, A.; Llamas, M.R.; Varela-Ortega, C. *et al.* (2010). *Water footprint and virtual water trade in Spain. Policy implications*. Springer, New York, USA.

Garrido, A.; Bardají, I.; de Blas, C. *et al.* (2011). *Primeros indicadores de sostenibilidad en la agricultura y ganadería en España* [Preliminar sustainability indicators of Spanish agriculture and animal sectors]. Plataforma de la Agricultura Sostenible, Madrid, Spain. Available from: http://www.agriculturasostenible.org/v_portal/informacion/informacionver.asp?cod=1788&te=414&idage=2198 [Accessed on 6th May 2011].

Hoekstra, A.Y.; Chapagain, A.K.; Aldaya, M.M. & Mekonnen, M. (2011). *The water footprint assessment manual: Setting the global standard*. Earthscan, London, UK.

INE (2011a). *Contabilidad Nacional de España* [National Accounting of Spain]. National Statistics Institute. Available from: http://www.ine.es/jaxiBD/menu.do?L=0&divi=CNTR&his=0&type=db [Accessed on 6th May 2011].

INE (2011b) *Encuesta sobre el suministro y saneamiento del agua en España* [Survey of Spanish Water Suuply and Water Treatment]. National Statistics Institute. Available from: http://www.ine.es/jaxi/menu.do?L=0&type=pcaxis&path=%2Ft26%2Fp067%2Fp01&file=inebase [Accessed 14th December 2011].

MARM (2011a). *Huella hídrica de España: Sostenibilidad y territorio*. [Spanish Water Footprint: Sustainability and Territory]. Spanish Ministry of the Environment, and Rural and Marine Affairs, Spain. Available from: http://www.marm.es [Accessed 12th 2011].

MARM (2011b). *Agricultural and Statistics Yearbook*. Spanish Ministry of the Environment, and Rural and Marine Affairs. Available from: http://www.magrama.gob.es/es/estadistica/temas/estad-publicaciones/anuario-de-estadistica/default.aspx [Accessed 9th November 2011].

MARM (2011c). *Sistema de Información Agroclimática para el Regadío (SIAR)*. [Agroclimatic Information System for Irrigation]. Spanish Ministry of the Environment, and Rural and Marine Affairs, Spain. Available from: http://www.mapa.es/siar/Informacion.asp [Accessed 30th June 2011].

Mekonnen, M.M. & Hoekstra, A.Y. (2010a). *The green, blue and grey water footprint of crops and derived crop products*. Value of Water Research Report Series, UNESCO-IHE, 1, no. 47. Delft, The Netherlands.

Mekonnen, M.M. & Hoekstra, A.Y. (2010b). *The green, blue and grey water footprint of farm animals and animal products*. Value of Water Research Report Series, UNESCO-IHE, 1, no. 48. Delft, The Netherlands.

MITYC (2009). *Tourism employment Yearbook*. Spanish Ministry of Industry, Tourism and Commerce. Available online at: [http://www.iet.tourspain.es/es-ES/estadisticas/frontur/informesdinamicos/paginas/anual.aspx]. Accessed on January 2012.

MITYC (2011). *Datacomex, Estadísticas de comercio exterior* [Datacomex. Statistics of Foreign Trade]. Spanish Ministry of Industry, Tourism and Commerce. Available from: http://datacomex.comercio.es [Accessed 4th November 2011].

Rodriguez Casado, R.; Novo, P. & Garrido, A. (2009). *La huella hídrica de la ganadería española* [Water footprint of the Spanish livestock sector]. Papeles de Agua Virtual, 4, Botín Foundation, Madrid, Spain.

Ruiz, J.; Bardají, I.; Garrido, A. & Iglesias, E. (2011). *Impacto de la reforma de la PAC 2003 sobre la agricultura española* [Impact of the 2003 CAP reform over Spanish Agricultural production]. In: Asociación Española de Economía Agraria (AEEA). VIII Congreso de Economía agraria: El sistema agroalimentario y el mundo rural en una economía innovadora y sostenible. 14–16 September 2011, Madrid, Spain.

Chapter 7

An overview of groundwater resources in Spain

Lucia De Stefano[1], Luis Martínez-Cortina[2] &
Daniel Chico[3]
[1] *Water Observatory of the Botín Foundation; Department of*
Geodynamics, Complutense University of Madrid, Madrid, Spain
[2] *Spanish Geological Survey (IGME), Madrid, Spain*
[3] *Water Observatory of the Botín Foundation;*
CEIGRAM, Technical University of Madrid, Madrid, Spain

ABSTRACT: In Spain, as in most arid and semiarid countries, during the last half century the *silent revolution* of intensive groundwater use has provided important socio-economic benefits. Nonetheless, traditionally water management has focused on surface water and has paid little attention to groundwater. The European Water Framework Directive (WFD) planning process has resulted in significant advancements in the knowledge of groundwater resources and their use in Spain. However, data on groundwater resources are still partially incomplete and an official country-wide overview of groundwater resources (and their uses) is still not available. At present the estimated groundwater demand is about 7,000 million m³/year, mainly for irrigation purposes. Intensive groundwater use has contributed to the degradation of this strategic resource, which is expected to be partially remediated by the WFD implementation. Previous studies in Andalusia found that in irrigated agriculture groundwater use was economically more productive than surface water. This was attributed to a series of factors, chiefly groundwater resilience to long dry spells, and it was suggested that this could apply also to other regions in Spain. The data presented in this chapter seem to question this former idea, since no clear correlation could be found between the source of water and its apparent water productivity in irrigated agriculture. This is an issue that merits further study, including combining local and country-wide data to refine the calculations.

Keywords: groundwater, economic uses, water quantity, water quality, Water Framework Directive, groundwater body

I INTRODUCTION

Groundwater (GW) has often been called the *hidden* resource: by nature groundwater is *out of sight* and therefore often also *out of mind* for policy-makers, lay people, and, to a lesser extent, water practitioners. Nonetheless, this resource has allowed for a significant socio-economic development in many regions of the world, and plays a strategic role in many countries, especially in arid and semiarid regions. Spain, with a semi-arid climate in most of its territory, is no exception to these trends and the key role played by groundwater in several spheres of Spanish society (urban supply, economic uses or groundwater-dependent ecosystems) makes it crucial to have a good knowledge of this resource.

Studies of specific aquifers are of great value for collecting high-resolution data and understanding local dynamics. At the same time, country-wide studies of groundwater resources and their uses have the value of helping decision-makers, practitioners and scientists to grasp the magnitude of groundwater challenges at a country level and to frame local water problems into a *bigger picture*. Thus, country-wide analyses allow for a better understanding of inter-linkages, similarities and differences between challenges at different scales; and the identification of possible links between the evolution of groundwater use and other water-related phenomena such as changes in groundwater quality, the appearance of new water uses, or the creation of different forms of groundwater user associations. Moreover, the assessment of the monetary value of groundwater uses provides an overview of their economic relevance, informing decisions related to water rights reallocation or investments in new water infrastructure.

At present, no recent general overview of Spanish groundwater resources and uses is available. The Groundwater White Book (MIE & MOPTM, 1995), the Water White Book (MMA, 2000) and the book by Llamas *et al.* (2001) are the only comprehensive studies on this subject that have been undertaken in Spain during the past two decades. Dumont *et al.* (2011) and Molinero *et al.* (2011) made a first attempt to fill this gap with an overview of groundwater uses and status based on the RBDs planning documents available in year 2010. In terms of economic value, regional and local studies on the value of groundwater uses provide interesting insights into the subject (e.g. Hernández-Mora & Llamas, 2001; Aldaya & Llamas, 2008; Salmoral *et al.*, 2011), while the most recent works on the economic value of water uses at national level do not distinguish between ground and surface water (MMA, 2007; Garrido *et al.*, 2010).

This chapter aims at providing this missing overview at national level in Spain. A key source of information for the elaboration of the present analysis has been the official documentation produced by the River Basin Organisations (RBOs) as part of the new planning process required by the European Water Framework Directive (WFD, see Chapter 3). Due to delays in the WFD planning process, part of the data for some River Basin Districts (RBDs) (e.g. Tagus, Ebro) were still unavailable at the termination of this chapter.

This chapter starts with an overview of current estimates of groundwater resources, their quantitative and qualitative status, and continues with a snapshot of groundwater uses by the different sectors, with special emphasis on the economic value of irrigated agriculture. It concludes with considerations of the challenges ahead in improving the knowledge of this strategic resource.

2 GROUNDWATER RESOURCES

During the last decade, the regulatory system of the Spanish groundwater sector has experienced several changes, mainly due to the approval and transposition of the WFD (Directive 2000/60/EC, transposed into Spanish law in 2003) and the associated Directive for the protection of groundwater (Directive 2006/118/EC, transposed in 2008).

In terms of water planning, the WFD involves changing the basic groundwater management unit, from hydrogeological units (HUs) to groundwater bodies (GWBs). A groundwater body includes one or several aquifers (or portions of them) whose waters have common characteristics and are confronted with similar challenges – either qualitative or quantitative. During the WFD planning process, 730 GWBs have been identified and characterized across the country (Figure 1). The shift from hydrogeological units to GWBs has meant almost doubling the number of groundwater management units relative to the former 411 HUs. The HUs covered around a third of the area of Spain (over 175,000 km²) while the GWBs now include almost two thirds of the whole territory (about 350,000 km²). This change of management units has implied including in the planning process aquifers that locally play a key role in water supply (e.g. for some small urban areas) and that previously were not considered as HUs because of their limited water yield. Moreover, it has entailed an important effort in the definition and characterization of the new GWBs and increased management efforts in terms of monitoring and implementation of measures to ensure good status for a larger number of aquifers.

The WFD River Basin Management Plans (RBMPs) contain two main sections assessing groundwater resources: the inventory of natural water resources, which includes hydrological series of aquifer recharge and groundwater flow (and other variables); and the assessment of the available resources, which in turn is estimated using the concept of renewable resource.

The WFD and the IPH (*Instrucción de Planificación Hidrológica*, or technical instructions to guide the hydrological planning process; MARM, 2008) issued by the former Spanish Ministry for Environment, Rural and Marine Affairs (MARM)

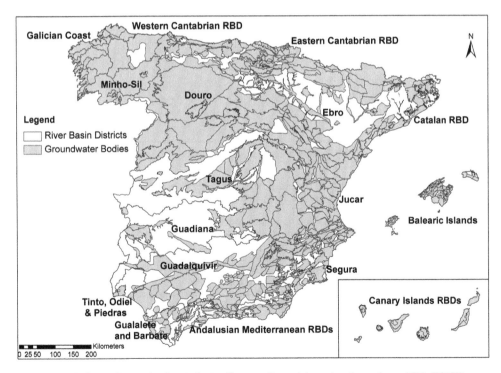

Figure 1 Groundwater bodies in Spain. (Source: Own elaboration from data of SIA (2012)).

provide guidelines to homogenize the approach of River Basin Organisations to the assessment of groundwater. Nonetheless, the application of IPH is compulsory only in inter-regional RBDs[1], and even in these ones, there has been a certain heterogeneous interpretation of some of the established guidelines (see also Dumont *et al.*, 2011). This should be taken into account when interpreting the overall and RBD-specific figures presented in the next pages.

Table 1 summarizes the estimates of recharge, renewable resources and available resources, as provided in the RBDs planning documents made available at the end of

Table 1 Number of groundwater bodies (GWB), renewable resource and available resource by River Basin District (RBD).

River Basin District	# GWB	GWB area (km²)	Recharge[a] (mm/year)	GW renewable resource[b] (hm³/year)	Available GW resource (hm³/year)
Galician Coast	18	12,988	242	3,869	3,471
Minho-Sil	6	17,602	ND	3,774	3,193
Western Cantabrian RBD	20	13,875	301	4,217	3,328
Eastern Cantabrian RBD	14	3,472	386	1,273	1,090
Douro	64	75,885[c]	71	3,737	2,990
Ebro	105	54,125	ND	ND	ND
Catalan RBD	39	11,254	70	1,930	1,141
Tagus	24	21,866	ND	ND	ND
Guadiana	20	22,484	34	550	623
Jucar	90	40,135	61	3,355	2,327
Segura	63	18,500	36	700	535
Guadalquivir	60	35,609	ND	2,700	1,962
Tinto, Odiel & Piedras	4	1,018	56	66	46
Guadalete & Barbate	14	1,927	ND	166	52
Andalusian RBDs	67	10,395	80	833	676
Balearic Islands	90	4,737	ND	410	181
Canary Islands[d]	32	7,425	2–370	ND	360
Total	730	353,297	–	–	21,975[e]

Source: Own elaboration from RBDs planning documents publicly available in December 2011.

Notes
The autonomous cities of Ceuta and Melilla are not included. ND = No Data. In the case of Tagus, Ebro and some of the Canary RBDs, the lack of data is due to a delay in the planning process.
[hm³ = cubic hectometre = million m³ = 10⁶ m³].
a In the RBMPs there is some terminological confusion about groundwater recharge, rainfall infiltration and renewable resource: the same terms not always refer to the same concepts, and different terms sometimes refer to the same concept.
b In some cases these figures represent the *potential resource* or *natural GW resource*. The meaning of these values can be slightly different even when the term used in the RBMP is *renewable resource*.
c Surface area corresponding to the lower groundwater units.
d Figures for the Canary Islands summarize the results of their 7 RBDs (one for each island).
e Tagus and Ebro are not included in this figure.

1 Inter-regional RBDs are those shared by several regions (Autonomous Communities) and are managed by the central government through River Basin Organisations called *Confederaciones Hidrográficas*. RBDs located entirely in one region are managed by RBOs called *Agencias Autonómicas del Agua*.

2011 and shows that the overall available groundwater resources are calculated to be approximately 22,000 hm³/year [hm³ = cubic hectometre = million m³ = 10⁶ m³]. This represents a significant share of the regulated Spanish water resources, as the storage capacity of surface water reservoirs for consumptive use is about 55,400 hm³ and the average surface water reserve for consumptive use during the last ten years was approximately 33,400 hm³ (MAGRAMA, 2012).

The estimation of the available groundwater resources is particularly sensitive and challenging because it defines the amount of water that is actually available for economic uses and determines the quantitative status of a groundwater body. According to the WFD definition, which was translated literally into the IPH, the available groundwater resource "means the long-term annual average rate of overall recharge of the body of groundwater less the long-term annual rate of flow required to achieve the ecological quality objectives for associated surface waters … to avoid any significant diminution in the ecological status of such waters and to avoid any significant damage to associated terrestrial ecosystems" (WFD, art. 2).

According to this definition, all the inputs to a GWB are accounted as renewable resource (including irrigation return flows), which can lead to some resources being taken into account more than once if, for example, GW resources estimates for each GWB are added together to obtain a global figure for the whole RBD (see also Martínez Cortina et al., 2011). Moreover, the quantification of water needs for the associated surface water bodies and terrestrial ecosystems is very challenging, since there are no widely accepted criteria to define them (see also Chapter 11).

Undoubtedly the different characteristics of each RBD and GWB make it difficult to strictly apply the same criteria in all of them. Nonetheless, it seems that in the WFD planning process there are also some terminological and conceptual uncertainties that have led to a heterogeneous interpretation of some concepts, mainly in relation to renewable and available resources (Box 1).

Box 1   Some examples of criteria for the calculation of the available GW resources

- In Minho-Sil the renewable resource is considered to be equal to the rainfall infiltration estimated using a hydrological model. The available resources are at least 10% of the renewable resource, with additional requirements in protected river stretches.
- In Douro the RBMP estimates the *total natural resource*, which takes into account both lateral groundwater transfers and the so-called rejected recharge (when infiltration is larger than what the GWB can store). The available resource is 80% of the total natural resource, plus the return flows and the artificially recharged volumes.
- The Balearic Islands RBMP uses the term *potential GW resource* referring to all the system inputs (infiltrated rainfall). The available resources are those abstracted in 2006 (estimate based on water demands).
- In Guadalquivir the renewable resource corresponds to the aquifer recharge. The available resource is 80% of that figure, except in specific GWB where it is 50%.

(Source: based on RBDs planning documents publicly available in December 2011)

3 GROUNDWATER STATUS

The WFD has shifted the focus of groundwater management from only satisfying water demands to achieving good chemical and quantitative status[2] of groundwater bodies, as well as protecting the associated aquatic and terrestrial ecosystems. In Spain more emphasis has traditionally been set on quantitative problems than on the deterioration of groundwater quality, although in many cases this does not reflect the real challenges in the medium or long term (see Chapter 12). The WFD gives equal weight to these two aspects of groundwater protection and underscores that they are closely inter-twined. Indeed the Directive requires European Union (EU) Member States to assess both the chemical and the quantitative status of groundwater bodies and establishes that the global status of a particular GWB is determined by the poorer of the two.

The WFD requires that all the GWBs be in (at least) good status by 2015, although it is possible to request time extensions to 2021 and 2027 or to set less stringent objectives (LSO) for those GWBs where good status cannot be met. According to the RBDs planning documents, 392 of the 730 groundwater bodies are currently in good status (54%); 583 groundwater bodies will meet the objective of good status by 2027 (about 80% of the GWBs), while less stringent objectives have been set in 26 GWBs (4%). In the remaining 121 GWBs (17%) there are insufficient data to predict the achievement of good status by 2027 (Figure 2).

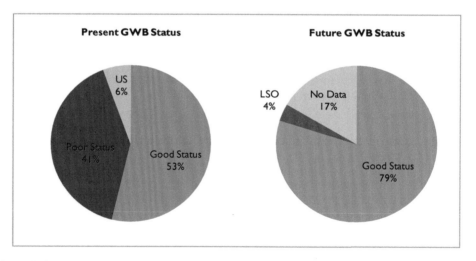

Figure 2 Overview of groundwater bodies status. US: Under Study; LSO: Less Stringent Objectives. Data by RBD can be found in Table A-1 in the Appendix to this chapter. (Source: Own elaboration, based on the RBDs planning documents publicly available in December 2011).

2 Good groundwater chemical status is achieved when the chemical composition of the water body: is not affected by salt intrusion; meets the established quality standards; does not prevent the associated surface waters from achieving the established environmental objectives; and does not cause significant damage to the associated terrestrial ecosystems.

Good groundwater quantitative status is achieved when: the available groundwater resource is not exceeded by the long-term annual average rate of abstraction, and the GWB is not affected by anthro-pogenic alterations that can cause water salinization or other intrusion or prevent the associated surface waters from achieving the established environmental objectives.

The RBDs with a higher number of GWBs in poor status are Guadiana (75% of the GBWs) and Segura (68%), while the Northern RBDs barely have any GWBs in poor status (see Table A-1 in the Appendix). It should be noted that the magnitude of the challenge ahead may also be seen by using parameters other than the number of GWBs in each status category. For example in Guadalquivir, 19 GWBs (32%) have a poor quantitative status and those GWBs provide about 75% of the groundwater abstracted in the RBD (Dumont *et al.*, 2011). However, overall figures have to be interpreted with caution, as the actual situation in each RBD depends on the specific socioeconomic context and the resulting anthropic pressure on groundwater resources.

Pollution, mainly by nitrates, is the main cause of non-compliance with the objectives of good status: out of 297 GWBs currently in poor status, 75% do not comply with the required quality standards. Similarly, qualitative problems are the main reason for establishing *less stringent objectives* for 2027 in 26 GWBs.

The WFD process has required investments in the improvement of the knowledge of groundwater resources, especially by increasing the density of the groundwater monitoring network. The diagnosis of groundwater status presented in Table A-1 of the Appendix is based on data from a monitoring network that are still under development and consolidation, thus in some GWBs the data may not fully reflect the real status of the water body.

4 GROUNDWATER USES

Compliance with the WFD requires adequate knowledge about the pressures on groundwater bodies in terms of quantity and quality. As in the case of groundwater resources assessment, a first difficulty in obtaining a global estimate of groundwater uses in Spain is the terminology applied in the RBMPs to refer to water use (use, demand, consumption, gross- or net-withdrawal). In some cases, some terms are utilised as synonyms although they actually refer to very different concepts. A second important obstacle is the generalized absence of direct water use measurements. As a consequence, methods to estimate groundwater abstraction vary across RBDs.

In most of the RBDs, the planning documents estimate water demands, which are expressed in *Units of Demand* (a spatial polygon with a specific water demand) for the main water uses (urban supply, agriculture, industry). The demands of each *Unit* are obtained using a combination of direct (e.g. water meters, supply surveys), mixed (e.g. remote sensing) and indirect methods (e.g. statistical data on population or crop areas multiplied by an average water demand rate). Sometimes, official water rights registries are also used as a source of information to estimate water demands, although the RBMPs acknowledge that those registries often do not reflect the situation on the ground. For example, in some of the northern RBDs the sum of the granted water rights doubles or quadruples the estimates made through the *Units of Demand*[3], possibly due to the fact that some water demands have decreased or that water rights were granted overestimating the actual demands. In the Guadiana RBD, groundwater rights sum 923 hm³/year (Rodríguez-Cabellos, pers. comm., January 2012), but the estimated abstraction is only 500 hm³/year, due to the current legal restrictions on groundwater use.

---

3 In W. Cantabrian RBD: 769 *vs.* 474 hm³/year for all uses and sources of water. In E. Cantabrian RBD: 407 *vs.* 112 hm³/year.

A decade ago, groundwater demand was estimated to be about 5,500 hm³/year (MMA, 2000). According to the most updated planning documents, the estimated overall groundwater demand is now about 7,000 hm³/year. This represents about 22% of Spain's total water demand (31,500 hm³/year).

Most groundwater demand (over 80%) occurs in 7 of the 25 existing RBDs (Figure 3). In absolute terms, the RBD with the highest groundwater withdrawals is Jucar (1,600 hm³/year), followed by Douro and Guadalquivir with close to 1,000 hm³/year, while estimated extractions are around 500 hm³/year in Segura, Guadiana, the Catalan RBD and the

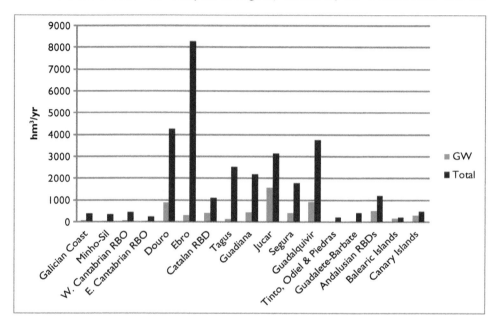

Figure 3 Share of groundwater demand over the total water demand by River Basin District (RBD). (Source: Own elaboration based on RBDs planning documents publicly available in December 2011).

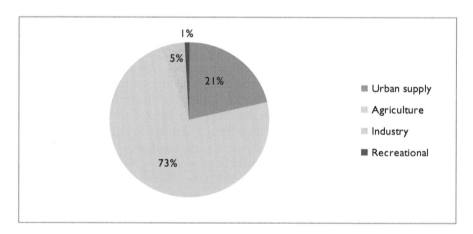

Figure 4 Estimated overall groundwater demand by sector. (Source: Own elaboration based on RBDs planning documents publicly available in December 2011).

Andalusian RBDs. The highest shares (over 40%) of groundwater withdrawals (relative to the RBD total) can be found in the Canary and Balearic Islands, Jucar and the Catalan and Andalusian RBDs (see Table A-2 in the Appendix).

Overall, the main groundwater use is agriculture (73%). At a country level, groundwater supplies only 21% of the domestic water demand (and industrial uses connected to the urban water supply network) (Figure 4). Nonetheless, groundwater plays a key role in the urban supply in some RBDs such as the Canary Islands, Balearic Islands, Jucar, Catalonia (where industrial use is especially important), and several RBDs in northern Spain.

Box 2   Surface water, groundwater and conjunctive use

In this chapter water resources estimates are presented distinguishing between surface water and groundwater. However, the reality on the ground is far more complex, and water users often strategically combine all the resources that they have to find the best formula (in terms of availability, price, quality and timing) for their activity. Hence, conjunctive use of surface and groundwater is the rule in many regions of Spain. The Llobregat area, in Catalonia, is one of the best examples of successful conjunctive water use in Spain. As another example, in Campo de Cartagena aquifer (Segura basin), irrigated agriculture is supplied by a mix of surface water transferred from the Tagus basin (through the Tagus-Segura Transfer, TTS in Spanish) and groundwater. During droughts, the transferred volumes are restricted and farmers complement surface water by pumping groundwater. Thus, during dry periods, groundwater represents 70% of the supply for irrigation, while during wet periods it is only 30% (see Figure 5).

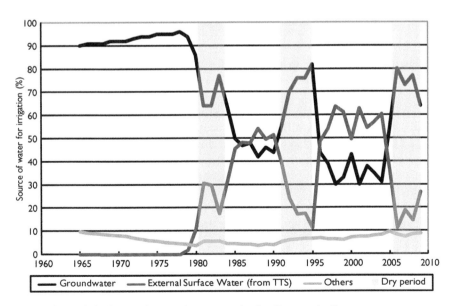

Figure 5 Surface and groundwater use in the Campo de Cartagena area. (Source: Cabezas (2011))

5   THE ECONOMIC VALUE OF GROUNDWATER
USE IN AGRICULTURE

The planning documents elaborated by the RBOs provide figures about the overall value of water use by different economic sectors, including irrigated agriculture. However, these data are rarely broken down by source of water (surface, groundwater, desalinated, reused, conjunctive use), thus making it difficult to understand the contribution of different water sources to the economy of the RBD and therefore identify some of the trends driving their use. This section makes a first attempt to fill this gap for irrigated agriculture.

5.1   Methodological approach

As the data available in the RBDs planning documents consider agriculture as a whole and do not associate GW extraction or consumption to a specific crop, it has been necessary to build an ad hoc 5-step methodology based on the extended water footprint approach of Garrido *et al.* (2010).

The origin of the irrigation water (groundwater, surface, conjunctive use[4]) was obtained for different water management units (Agricultural Demand Units; Water Exploitation Systems; Regional Irrigation Inventory) provided by the RBOs, while the yearly irrigated crop surface was available at the municipal level from the Ministry for Environment, Agriculture and Rural Affairs. The attribution of crop areas to a specific water origin was obtained by crossing the municipal boundaries and the water management units, using geographical information system tools (Figure 6, Step 1).

To estimate water consumption by crop and year, the blue and green water consumption[5] was calculated applying the methodology of Garrido *et al.* (2010) and extracted only the blue component, which corresponds to the annual irrigation water consumption per crop. In a second stage, the crop net irrigation requirements defined by each RBO were used as upper limits for the blue water consumption (Step 2). This approach helps adjusting the estimated consumption to practices on the field, which include deficit irrigation.

The area of crops irrigated by each water source was multiplied by the crop water consumption rates to obtain crop consumption by water source (crop blue water footprint, WF, Step 3). The apparent land productivity (€/ha) was estimated from the crop yields (t/ha) [t = tonne = 10^3 kg] and crop market prices (€/kg) obtained from MARM (2010). The ratio between the apparent land productivity and the crop water consumption yield the apparent water productivity (€/m^3) by water source.

This study shed light on several shortcomings of the data available to calculate the economic value of irrigated agriculture by water source. First, at the conclusion of this study there were no up-to-date country-wide data about the sources of water for irrigation by crop type and at high-resolution geographical level. The last

4 Desalinated and reused waters were not considered due to their small significance in absolute terms of irrigation volumes and also due to the lack of systematic data on these two sources. Conjunctive use of surface and groundwater was calculated only in the Jucar RBD, where official data were available.
5 Blue water is the consumed irrigation water, while green water is the water that the plant obtains from the soil water content due to precipitation.

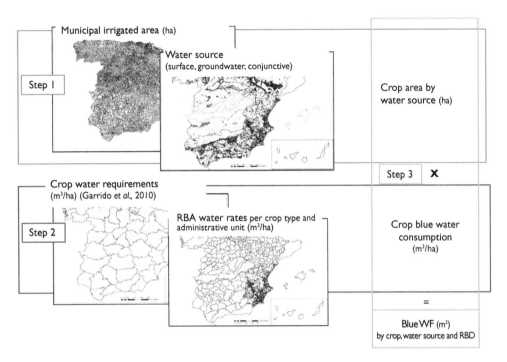

Figure 6 Steps 1 to 3 of the methodology followed for obtaining the crop blue water consumption by crop, origin of water and RBD. WF: Water Footprint. (Source: Own elaboration).

agricultural census recording this information dates back to 1999, and the 2009 census released in July 2011 did not include data on origin of water. Second, there are no country-wide fully reliable data on the area of irrigated crops on a yearly basis and at certain levels of detail – municipal or similar. There are several data sources but often figures either differ significantly depending on the provider or are not available for a long period of time or at high spatial resolution.

5.2 Main findings

Using the available official data, it was estimated that the consumption of groundwater for irrigation is approximately 3,200 hm³/year over an area of 1 Mha [Mha = million hectares = 10^6 ha], which represents about one third of the total irrigated surface (3.3 Mha) and one fourth of the total water consumption (approx. 12,000 hm³/year) (Table 2).

In the RBD planning documents groundwater demand for irrigation is estimated to be about 5,000 hm³/year. This figure refers to gross demand, i.e. including losses and return flows, which range from 10% to 40% of the applied water depending on the crop and irrigation technique. Taking into account this important difference in the considered variable (demand *vs.* consumption), the overall figures obtained in the analysis are fairly consistent with those used in the RBD documents. Nonetheless, discrepancies between the RBMPs data and figures in Table 2 can be detected at the scale of some RBDs. This is due to the fact

Table 2 Irrigation consumption and its economic value at RBD level (average 2005–2008).

RBD	GW irrigated area (ha)	Total irrigated area (ha)	GW consumption (hm³/year)	Total consumption (hm³/year)	Value of the GW production (M€)	Total value of irrigated production (M€)
Douro	136,073	402,035	544	1,582	302	941
Ebro	28,022	652,338	107	2,385	111	2,681
Northern RBDs	3,770	20,759	14	80	41	224
Tagus	13,772	190,590	60	834	42	599
Catalan RBD	29,102	60,061	105	211	124	257
Guadiana	143,636	326,784	377	1,191	512	1,604
Jucar	160,546	490,849	535	1,655	408	2,257
Segura	70,123	202,024	271	803	584	1,450
Guadalquivir	239,481	753,776	665	2,285	921	2,407
Andalusian RBDs	84,834	211,099	308	755	1,283	2,461
Balearic Islands	12,536	12,536	69	69	61	61
Canary Islands	22,889	23,408	166	169	338	354
Total	944,784	3,346,259	3,221	12,018	4,728	15,300

Source: Own elaboration.

Note: Due to data and resources constraints, estimates for the Northern RBDs (Western Cantabrian RBD, Eastern Cantabrian RBD, Minho-Sil, Galician Coast) and Tagus, were calculated with a simplified version of the methodology. In Jucar, the values obtained for conjunctive use were attributed to surface and groundwater on an equal share (Estrela, pers. comm., January 2012). The 2008 regional irrigation inventory for Guadalquivir and the Andalusian RBDs is considered to be more reliable than the MARM agrarian statistical data, especially in relation to olive production (see Chapter 8). Therefore figures for olive production were adjusted taking into account data from that inventory.

that RBAs often combine data from different sources for estimating irrigated crops consumption while the present study used only the official agrarian statistical data[6]. Table 2, however, provides a comparison between RBDs using the same set of data and gives an order of magnitude of groundwater irrigation consumption and its economic value.

A 1999 study undertaken for Andalusia showed that the apparent productivity of groundwater (GWAP) in irrigated agriculture was significantly higher than that of surface water (SWAP) (Hernández-Mora & Llamas, 2001). This trend was confirmed by Corominas (pers. comm.), who found that in 2008 the apparent productivity of the Andalusian groundwater irrigation on average was more than twice that of surface water (1 €/m³ *vs.* 2.60 €/m³). According to Hernández-Mora & Llamas (2001) reasons for this could be found in the greater control and supply guarantee that groundwater provides mainly during droughts, and the greater dynamism that characterize farmers who seek their own sources of water and bear the full direct costs of drilling, pumping and distribution. Llamas (2003) suggested that this could apply also to other regions

6 Figures for Guadalquivir and the Andalusian RBDs in the 2008 regional irrigation inventory are higher than those obtained with the nation-wide data: Guadalquivir: GW irrigated land: 321,233 ha; GW production value: 1,009 M€.

Andalusian RBDs: GW irrigated land: 101,902 ha; GW production value: 1,770 M€ (Corominas, pers. comm., January 2012).

of Spain and remarked that similar trends can be observed also in other countries like India. The data produced in this study allow analysing further the relationship between crops water apparent productivity (WAP) and the source of water used.

Table 2 shows the economic value of agricultural production using groundwater, which is about 4,700 M€/year or 30% of the total value of Spain's irrigated crop production. When considering all crops and all RBDs it can be observed that the overall GWAP is on average between 30% and 50% (depending on the year) higher than the SWAP. Nonetheless, this trend is not evident when looking at specific RBDs and crops (Figure 7).

Figure 8 shows the linear fit between average apparent water productivity of crops (average of SWAP and GWAP) and the percentage of groundwater used to

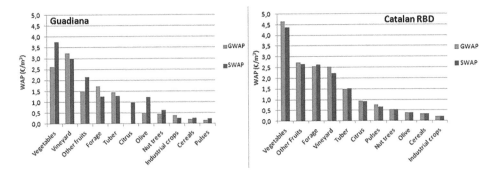

Figure 7 Surface and groundwater apparent productivity (SWAP, GWAP) in Guadiana and the Catalan RBD. (Source: Own elaboration).

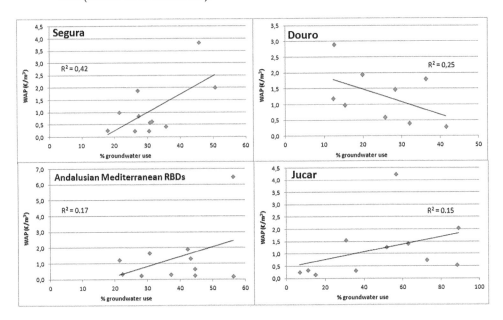

Figure 8 Linear fit between average apparent water productivity (WAP) and percentage of groundwater use by crop in selected RBDs. Each diamond represents a crop type. (Source: Own elaboration).

irrigate crops in several RBDs. It can be observed that the correlation of those two variables is rather poor, and in some cases (e.g. Douro) the RBD average water apparent productivity is higher for crops irrigated with surface water.

These preliminary results are in line with the study of the Guadiana basin by Aldaya & Llamas (2008), where no major differences were found in the WAP with surface and groundwater, and suggest that the trends detected in Andalusia should be extrapolated to other regions with caution. Indeed, these findings seem to indicate that the reliability of groundwater and entrepreneur attitude attributed to groundwater users are only two among several factors that determine WAP of crops. Other factors like the availability and reliability of surface water supply, climatic and soil conditions, advanced irrigation systems or technical know-how could be equally important, thus reducing the comparative advantage of groundwater relative to surface water. Some of these factors are outlined hereafter.

First, at present farmers willing to shift to irrigation or increase their access to water often can only tap into groundwater because no additional surface resources are available in most of the RBDs (most of the economically and environmentally viable reservoirs have already been built in the 20th century). This means that farmers may opt to access groundwater even if their irrigated crops provide tight benefit margins and have a limited WAP. Second, in some basins multiannual reservoirs ensure a high reliability of surface water also in case of droughts, thus providing similar supply guarantee as groundwater. Third, the modernization of irrigation systems has contributed to the optimization of use (see Chapter 19), thus increasing the control and guarantee of surface water to a degree similar to groundwater self-supply. Fourth, improvements in irrigated agriculture as a whole (irrigation advisory services, better access to irrigation and production technology, etc.) contribute to a progressive shift of surface irrigation toward more productive crops. This shift has also been favoured by changes in the EU Common Agriculture Policy incentives and by the increase of international trade in crops. Finally, farmers increasingly use a combination of surface and groundwater (and to a lesser degree other sources like reused or desalinized water, if available) to ensure water supply. This diversification of source on one side makes the boundaries between surface and groundwater uses more difficult to draw, and on the other side, increases the water guarantee of irrigation, independently of its predominant water source.

6 CONCLUSIONS

The WFD planning process has produced important advancements in the knowledge of groundwater resources and their use in Spain. Nonetheless, pitfalls in the available data and the methodological heterogeneity across the country necessitate interpreting the resulting snapshot with caution. Moreover, while it is now possible to access data by RBD, an official up-to-date overview of groundwater resources and their uses at national level is still not publicly available.

As part of the WFD planning process, 730 GWBs covering an area of over 353,000 km² have been defined. The order of magnitude of the available groundwater resources in Spain is about 22,000 hm³/year (Tagus and Ebro are not included). Currently, 54% of the GWBs are in good status and the WFD implementation process is expected to increase this percentage to 80% by 2027, while for 17% of the GWBs there are no sufficient data to model their status by 2027. Reasons for poor status are both quantitative water problems and pollution, especially due to excess nitrates and salinization.

According to the RBDs planning documents, annual groundwater demand is about 7,000 hm^3 (or 22% of the total), which suggests an increase relative to previous 2000 official figures (5,500 hm^3/year). Unsurprisingly, the main water user is the agricultural sector (73% of groundwater demand), although groundwater plays a strategic role in urban water supply in several RBDs. Groundwater consumption is estimated to be about 3,200 hm^3/year for irrigated agriculture, 300 hm^3/year for urban supply and industry connected to the urban water network[7], and 60 hm^3/year for self-supplied industry.

In terms of groundwater productivity for irrigated agriculture, no clear difference between the apparent productivity of surface and groundwater in irrigated agriculture could be found in most of the RBDs. This differs from the results of previous regional studies and possible reasons have been suggested in this chapter. Surely this is an issue that merits further study, also combining local and country-wide data to refine the calculations, both on water consumption and economic productivity.

As a final consideration, it is important to highlight the uncertainty of all the estimates presented in this chapter, which is mainly due to the limited quality data on groundwater demand and consumption. In particular, there is an urgent need for more and better data on actual water consumption and the economic value of irrigated agriculture, differentiated by water source.

ACKNOWLEDGEMENTS

The authors thank all the Spanish River Basin Organisations, the Ministry for Environment, Rural and Marine Affairs, and the Agricultural Department of the Canary Islands for their prompt reply to their data requests for the elaboration of the study summarizes in section 5. The authors are thankful to Joan Corominas, María del Carmen Cabrera, Victor Arqued, José Ángel Rodríguez Cabellos, Loreto Fernández and Juan Antonio López Geta. Credits for data processing are due to David Mateos and Pedro Zorrilla (Terrativa) as well as to Mario Ballester and Irene Paredes.

REFERENCES

Aldaya, M.M. & Llamas, M.R. (2008). *Water Footprint analysis for the Guadiana Basin*. Papeles de Agua Verde (PAV Series), Vol 3. Botín Foundation, Santander, Spain. 112 pp. Available from: http://www.waterfootprint.org/?page=files/Publications [Accessed 25th April 2012].

Cabezas, F. (2011). Explotación de las aguas subterráneas en la cuenca del Segura [Explotation of groundwater in the Segura basin]. In: Villarroya, F.; De Stefano, L. & Martínez-Santos, P. (coords.), *El papel de las aguas subterráneas en la política del agua en España* [The role of groundwater in Spain's water policy] SHAN Series Vol. 3. Botín Foundation. 103 pp. Available from: http://www.fundacionbotin.org/monografias_observatorio-del-agua_publicaciones.htm [Accessed 25th April 2012].

---

7 Assuming a return flow of 80% of the supplied water.

Dumont, A.; De Stefano, L. & López-Gunn, E. (2011). El Agua Subterránea en España según la Directiva Marco del Agua: una visión de conjunto. VII Iberian Conference on Water Management and Planning. Iberian Rivers +10. Looking at the future 10 years after the WFD. 16–19 February 2011, Talavera de la Reina, Toledo, Spain: 1–8.

Garrido, A.; Llamas, M.R.; Varela-Ortega, C.; Novo, P.; Rodríguez-Casado, R. & Aldaya, M.M. (2010). *Water footprint and virtual water trade in Spain. Policy implications.* Springer, New York, USA.

Hernández-Mora, N. & Llamas, M.R. (eds.) (2001). *La economía del agua subterránea y su gestión colectiva.* Mundi-Prensa Ed. and Marcelino Botín Foundation, Madrid, Spain.

Llamas, M.R. (2003). Lessons Learnt from the Impact of the Neglected Role of Groundwater in Spain's Water Policy. In: Al Sharhan & Wood (eds.), *Water Resources Perspectives: Evaluation, Management and Policy.* Elsevier Science, Amsterdam, the Netherlands: 63–81.

Llamas, M.R.; Fornés, J.; Hernández-Mora, N. & Martínez Cortina, L. (2001). *Aguas Subterráneas. Retos y Oportunidades* [Groundwater. Challenges and opportunities]. Mundi Prensa Ed. and Marcelino Botín Foundation, Madrid, Spain.

MAGRAMA (2012). Sistema de Indicadores del Agua: Almacenamiento en embalses [Water Indicators System: Storage in reservoirs]. Available from: http://servicios3.magrama.es/ BoleHWeb/accion/cargador_pantalla.htm?screen_code=60000&screen_language=&bh_ number=29&bh_year=2012 [Accessed 12th May 2012].

MARM (2008). Orden ARM/2656/2008, de 10 de septiembre, por la que se aprueba la instrucción de planificación hidrológica. BOE número 229 de 22/9/2008, páginas 38472 a 38582.

MARM (2010). *Agricultural and Statistics Yearbooks.* Spanish Ministry of the Environment, and Rural and Marine Affairs. Available from: http://www.mapa.es/es/estadistica/pags/ anuario/introduccion.htm [Accessed 12th November 2011].

Martínez Cortina, L.; Mejías, M.; Díaz Muñoz, J.A.; Morales, R. & Ruiz Hernández, J.M. (2011). Cuantificación de recursos hídricos subterráneos en la cuenca alta del Guadiana. Consideraciones respecto a las definiciones de recursos renovables y disponibles [Quantification of groundwater resources in the upper Guadiana basin. Considerations about concepts of renewable and available resources]. *Boletín Geológico y Minero*, 122 (1): 17–36.

MIE & MOPTM (1995). *Libro Blanco de las Aguas Subterráneas* [Groundwater White Book]. Ministerio de Industria y Energía and Ministerio de Obras Públicas, Transportes y Medio Ambiente. Spain.

MMA (2000). *Libro Blanco del Agua en España* [White Book of Water in Spain]. Secretaría de Estado de Aguas y Costas, Dirección General de Obras Hidráulicas y Calidad de las Aguas, Ministerio de Medio Ambiente, Spain.

MMA (2007). *Precios y costes de los servicios del agua en España: Informe integrado de recuperación de costes de los servicios del agua en España* [Prices and costs of water services in Spain: integrated report of water services cost recovery in Spain]. Equipo de Análisis Económico, Ministerio de Medio Ambiente, Spain.

Molinero, J.; Custodio E.; Sahuquillo, A. & Llamas M.R.(2011). Groundwater in Spain: Legal framework and management issues. In: Findikakis & Sato (eds.) *Groundwater Management Practices.* CRC Press.

Salmoral, G.; Dumont, A.; Aldaya, M.M.; Rodríguez-Casado, R.; Garrido, A. & Llamas, M.R. (2011). *Análisis de la huella hídrica extendida de la cuenca del Guadalquivir* [Analysis of the Water Footprint of the Guadalquivier Basin]. SHAN Series. Vol. 1 Botín Foundation. Available from http://www.fundacionbotin.org/monografias_observatorio-del-agua_publicaciones. htm [Accessed 25th April 2012].

SIA (2012). Sistema de Información del Agua [Water Information System]. Available from: http://servicios2.marm.es/sia/visualizacion/descargas/dma.jsp [Accessed 25th April 2012].

APPENDIX

Table A-1 Groundwater status in Spain.

River Basin Districts (RBD)	# of GWB	Present good status (# of GWB)	Present poor status (# of GWB)			Environmental objectives	
			Quantitative	Chemical	Global	Projected good status	LSO (# of GWB)
Galician Coast	18	18	0	0	0	—	0
Minho-Sil	6	5	0	1	1	GS in 2021: 6	0
Western Cantabrian RBD	20	20	0	0	0	—	0
Eastern Cantabrian RBD	14	13	0	1	1	GS in 2015: 14	0
Douro	64	48	5	14	16	GS in 2021: 47; GS in 2027: 50	14
Ebro	105	82	1	23	23	GS in 2027: 103	2
Catalan RBD	39	14	6	23	25	GS in 2015: 18; GS in 2027: 39	0
Tagus	24	9	2	14	15	ND	ND
Guadiana	20	5	11	13	15	GS in 2027: 20	0
Jucar	90	48	34	22	42	GS in 2015: 53; GS in 2021: 67; GS in 2027: 67–90	0 (ND:23)
Segura[a]	63	13	43	16	43 (US:7)	ND	ND
Guadalquivir	60	32	19	16	28	GS in 2015: 35; GS in 2021: 48; GS in 2027: 60	0
Tinto, Odiel & Piedras	4	2	0 (US: 1)	2	2	GS in 2015: 4	0
Guadalete & Barbate	14	5	3 (US: 8)	7 (US: 2)	7 (US:2)	GS in 2021: 7; GS in 2027: 10 (US: 2)	2 (US:2)
Andalusian RBDs	67	27	32	35	40	GS in 2015: 41; GS in 2021: 52; GS in 2027: 62	5
Balearic Islands	90	46	18	36	44	GS in 2015: 64; GS in 2021: 75; GS in 2027: 87	3
Canary Islands[b]	32	US	US	US	US	ND	ND
TOTAL	**730**	**392**	**174**	**223**	**297**	**GS in 2027: 583 ND: 121**	**26**

Source: Own elaboration, based on data from RBDs planning documents publicly available in December 2011. Data for the autonomous cities of Ceuta and Melilla are not included.

Notes
GWB: groundwater bodies; LSO: less stringent objectives; GS: good status; US: under study; ND: no data.
a Data available in the report about Important Water Management Issues have been interpreted as follows: no risk = good status; proven risk = poor status.
b Figures for the Canary Islands summarize the data available for their 7 RBDs (one for each island).

Table A-2 Estimated groundwater extraction (mostly based on water demand estimates) and estimated total water demand by use. Figures in hm³/year.

River Basin Districts (RBD)	Urban supply[a]		Agriculture[b]		Industry[c]		Recreational uses[d]		TOTAL[e]	
	GW	Total	GW	Total	GW	Total	GW	Total	GW	Total
Galician Coast	49	274	38	84	1	45	ND	1	88	404
Minho-Sil	34	81–114	30	206–306	6	15	ND	4	73	369–439
Western Cantabrian RBD	91	239	6	70	15	162	ND	3	112	474
Eastern Cantabrian RBD	24–79	48–157	1	2–4	5	61–245	ND	1	77	112–407
Douro	68	332	800	3,900	50	60	6	8	746–924	4,300
Ebro	48	358–494	237–252	7,681	47	208	8	9	338	8,190
Catalan Basins	198	632	197	388	59	110	ND	8	454	1,138
Tagus	30	787	100–145	1,713	11–30	24–62	ND	ND	141–150	2,636
Guadiana	46	151–200	438	1,995	4	19–44	SD	ND	488–509	2,212–2,239
Jucar	323	548	1,178	2,474	100	124	9	10	1,610	3,156
Segura	ND	217–237	412–478	1,432–1,662	ND	23–30	ND	14	485	1,700–1,960
Guadalquivir	104	436	833	3,329	11	36	ND	ND	948	3,802
Tinto, Odiel & Piedras	4	56	31	151	0	46	2	2	31–37	253
Guadalete & Barbate	20	122	39	320	ND	ND	4	6	63	448
Andalusian RBDs	140	336	377	838	3	23	19	28	539	1,225
Balearic Islands	142	174	49	68	3	3	0	8	194	253
Canary Islands[f]	123	219	209	247	8	17	12	25	352	508
TOTAL (approx.)	1,500	5,200	5,000	25,000	300	1,100	65	130	7,000	31,500

Source: Own elaboration based on data from RBDs planning documents publicly available in December 2011. It does not include the autonomous cities of Ceuta and Melilla.

Notes
a These figures also include water use by the industry supplied through the urban distribution network.
b This includes irrigation and animal breeding.
c Industry not connected to the urban distribution network.
d Mainly golf courses.
e Often summing up figures for sectorial uses does not yield exactly the overall figures provided by the RBO (last column).
f Figures for the Canary Islands summarize the results of their 7 RBDs (one for each island).

Chapter 8

The extended water footprint of the Guadalquivir basin

Aurélien Dumont[1], Gloria Salmoral[2] &
M. Ramón Llamas[1]
[1] *Water Observatory of the Botín Foundation; Department of*
  Geodynamics, Complutense University of Madrid, Madrid, Spain
[2] *Water Observatory of the Botín Foundation;*
  CEIGRAM, Technical University of Madrid, Madrid, Spain

ABSTRACT: This chapter analyzes the Extended Water Footprint (EWF) of the Guadalquivir basin in south of Spain. An innovative aspect is that not only the use of blue water for direct human use (irrigation, urban and industrial supply) has been taken into account but also the use of green water for the mentioned uses and the natural ecosystems; the latter amounts to 291 mm/year. The results show that agriculture is the main consumer (192 mm/year), 34% being blue water and 66% green water. Economic productivity fluctuates between less than 0.40 €/m^3 for the most traditional crops (cereals, maize, cotton and rice) and values reaching 2 €/m^3 for olives and more than 4 €/m^3 for vegetables. But the highest economic productivity is tourism (more than 200 €/m^3) and industries such as thermo-solar energy (50 €/m^3). A better water management could be achieved thanks to a reallocation of water resources between the different uses. This reallocation may occur without social conflict with the farmers since the quantities of blue water required constitute 1–2% of the current total blue water use. However, this process is much more complex since a large number of economic, social, political and environmental factors need to be considered.

Keywords: water footprint, water productivity, hydrological cycle, groundwater, irrigation

1 INTRODUCTION

This chapter synthesizes the key issues of a monograph published by the Water Observatory of the Botín Foundation (Salmoral *et al.*, 2011), where details on methodology, data sources and references can be found. Guadalquivir basin, located in south of Spain (Figure 1), is a semiarid region (535 mm/year of rainfall), in which water repartition among economic sectors and the environment implies a relevant and controversial issue for water resources management. It has an area of 57,527 km^2, with a population of approximately 4.1 million, and the irrigated area reached 8,460 km^2 (846,000 ha) in the year 2008. The present study analyzes the Extended Water Footprint (EWF) of the Guadalquivir basin, considering both the traditional Water Footprint (WF) accounting in terms of water consumption, and the associated economic value. The study focuses on the quantitative components of the green (rainwater stored in

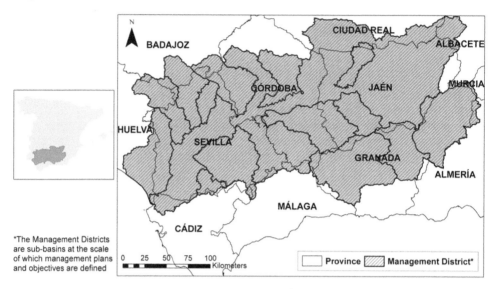

Figure 1 Localization of Guadalquivir basin and its different management districts. (Source: Own elaboration).

the soil) and blue (surface and groundwater) WF but does not estimate the grey colour component (freshwater required to assimilate load of pollutants).

The study presents some innovations in the methodology and results in comparison to previous WF studies:

–   The EWF on groundwater is calculated at the basin scale.
–   The hypothesis that crop irrigation requirements are fully met is not considered.
–   The WF of dams (evaporation) is taken into account.
–   The WF accounting of the different economic sectors and main land uses (including green water used by forests and pasture) has been balanced at basin scale and included within the hydrological cycle.

2   METHODOLOGY AND DATA

2.1   Water footprint of agriculture and livestock

The agricultural WF was estimated for the time period 1997–2008 and the blue and green components were distinguished. Green water was calculated as the minimum between effective rainfall and crop water requirements. In semi-arid regions like the Guadalquivir, farmers often have to cope with restrictions on water availability, thus for blue water accounting we considered that irrigation water requirements were not fully met. Blue water was calculated on the basis of the water allocated for each crop group within each *management district* and considering an additional irrigation restriction depending on the level of drought according to the Drought Plan of the Guadalquivir River Basin Authority (GRBA). Rainfed and irrigated crop areas were obtained from regional statistics at municipal level. In the case of groundwater, the total amount of abstractions for agriculture was obtained from the GRBA.

For livestock, the direct WF is the volume of water required for animal drinking and farm management. The indirect WF refers to virtual water embedded into animal feed coming partially from the agrarian production (already accounted in the agricultural WF), pastures within the basin and feed imports. The assessment of the livestock WF on the resources of the basin only considers the direct consumption.

2.2 Water footprint of the industry, domestic supply, energy, tourism and dams

Urban, tourism and industrial WFs were estimated based on data provided by the GRBA. To obtain the water consumption for domestic and industrial sectors, return flows of 72% and 44% (CHG, 2010) respectively were subtracted from each sector's total water abstractions. Specific data on groundwater abstractions are obtained from the same source. The volume of water evaporated from dams was calculated using the same method as Hardy & Garrido (2010), considering that all reservoirs are artificial lakes. Thus the evaporation is considered as a WF. This WF was not attributed to a specific use as the objectives of a dam are numerous.

2.3 Balance of green and blue water at basin scale

In order to make the WF assessment meaningful for water planning purposes, we have integrated the different WFs calculated within the mean annual water balance of the basin (Table 1). Data on mean annual precipitation, total run-off (surface plus groundwater run-off) and actual evapotranspiration was obtained from the GRBA and we used the values of the WFs calculated for an average climatic year (2003) on the following basis:

– Total run-off is the sum of the blue WF and the remaining water flow running along the streams and through groundwater bodies.
– Basin actual evapotranspiration represents the sum of the green WF consumed by crops and pastures and the water consumed by forest ecosystems. The water

Table 1 Components of the hydrologic cycle and the balance of green and blue water at the scale of the basin.

Rainfall*	Run-off**	Evapotranspiration**
563 mm/year	107 mm/year	456 mm/year
100%	19%	81%
32,042 hm^3	6,087 hm^3	25,955 hm^3
	Blue WF + Blue water flows	Green water (agriculture, pasture, forests ecosystems)

Source: Salmoral et al. (2011).

[hm^3 = cubic hectometre = million m^3 = 10^6 m^3].
* For the reference year 2003.
** The repartition between run-off and evapotranspiration is given by CHG (2010).

Table 2 Economic indicators used to assess the economic water productivity.

Sector	Valuation factor
Agriculture	Water and land productivities: production value in real € (year 2000) per unit of water consumed or land cultivated.
Domestic supply	Tariff of tap water for urban users.
Industry	Average tariff of water supplied to the industry.
Tourism	Evaluation of the economic returns of tourism for the local economy.
Energy	Energy tariff (€/kWh) multiplied by the amount of production obtained depending on the type of plant (expressed in kWh/m³).

demanded by forest ecosystems is a difficult variable to calculate. We estimated this volume by subtracting total run-off and green WF of crops and pastures from annual precipitation.

2.4   Economic water productivity

The economic assessment of the WF is based on a series of indicators adapted to the economic sector considered (see Table 2). For the purpose of this study, only blue water was evaluated in economic terms.

3   RESULTS

3.1   The Extended Water Footprint (EFW) of the Guadalquivir basin

3.1.1   The EWF of agriculture and its evolution over time

Between 1997 and 2008 the total WF (green and blue) of agriculture production ranged between 4,200 hm³ (year 1999) and 7,400 hm³ (year 2001) [hm³ = cubic hectometre = million m³ = 10^6 m³]. These variations are mainly ascribed to the irregular pattern of precipitations within the basin, which have a high influence on the green WF (Figure 2). The slightly greater green water footprint in 1999 in comparison to 2005, despite rainfall having been much lower in the former, is because of olive tree expansion among these years. During the period, 69% of mean annual WF in the Guadalquivir basin was green and the remaining 31% was blue, including both surface and groundwater. Overall, olive orchards consumed the largest proportion of green and blue water (72% and 31% of the total WF, respectively).

The economic water productivity in the basin rose from 0.70 €/m³ to 1.40 €/m³ between 1997 and 2007, however water productivity differs considerably among crops (Figure 3). Between 1997 and 2007, 46% of the blue water consumption belongs to crops that generate less than 0.40 €/m³, mainly cotton, rice and maize. Crops generating more than 1.50 €/m³ only account for 10% of total blue WF (vineyards, open air vegetables, winter fodder and strawberry). In other words, the largest proportion of blue water resources is allocated to produce low water economic productivity crops

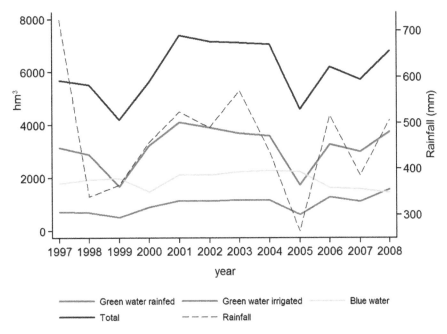

Figure 2 The water footprint of crop production. (Source: Salmoral *et al.* (2011)).

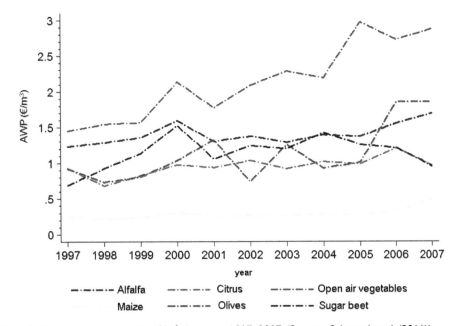

Figure 3 Blue water productivity (€/m³) between 1997–2007. (Source: Salmoral *et al.* (2011)).
Only crops which comprise more than 5% irrigated production over the time period are represented.

(see Chapter 6). For the whole basin, the land productivity associated to irrigated production is twice (1997) to four (2005) times greater than the one generated by rain-fed agriculture.

3.1.2   The EWF of groundwater in agriculture

According to the GRBA, annual groundwater abstractions in the basin reach 900 hm³. Considering an average global irrigation efficiency of 85%, the WF of groundwater in the Guadalquivir basin is around 770 hm³. However, if we calculate the WF using the regional statistic datasets of irrigated areas and we assume there is sufficient groundwater to fully meet irrigation crop requirements, we obtain a WF close to 1,060 hm³ (see Table 3). Differences encountered among both WFs values are due to the fact that farmers may be irrigating below the full crop irrigation requirements.

Olive groves hold the largest share of the WF (65% when considering that irrigation requirements are not fully met), mainly because the irrigated surface of this crop has experienced the largest increase over the last years, with a WF of 120 hm³ in 2002 and 490 hm³ in 2008 (see Chapter 10). Vegetables and fruits account for 21% of the WF and industrial crops and cereals about 14%. As groundwater is mostly used to irrigate crops of higher value, the mean economic productivity is higher for groundwater (1.15 €/m³) than for surface water (1.02 €/m³). However, this is a small difference, probably not significant given the uncertainties in the numbers at the basin scale (see Chapter 7).

3.1.3   Synthesis of the EWF of the Guadalquivir

Table 4 summarizes the EWF of the different socioeconomic sectors within the basin. The reference year for agriculture is 2003 (normal climatic year without irrigation water restrictions), and 2007 for the remaining socioeconomic sectors. Overall, agriculture represents the largest WF proportion (93% of the total, 80% considering only blue water). Evaporation from reservoirs is also important since it comprises 11% of the blue WF. Sectors such as tourism and golf have a much lower share of the blue component (<1%) in spite of their greater water productivity.

Table 3 Extended water footprint and share of groundwater irrigated surfaces, year 2008.

	Industrial crops	Cereals	Olive	Vegetables	Fruits	Total
Irrigated surface with groundwater (ha)	10,754	21,529	245,571	17,839	25,540	321,233
Water footprint* (hm³)	60.8	42.6	794.8	102.4	60.7	1,061.3
Water footprint** (hm³)	60.8	42.6	491.1	102.4	60.7	757.7
Water productivity*** (€/m³)	0.47	0.49	1.17	1.22	2.06	
% of area irrigated with groundwater	11.5	19.8	53.3	31.9	21.3	38.3

Source: Salmoral et al. (2011).

* Crop water requirements are fully met; irrigation water for olive groves is 3,200 m³/ha.
** Crop water requirements are not fully met; irrigation water for olive groves is 2,000 m³/ha.
*** Assuming that requirements are not fully met.

Table 4 Synthesis of the extended water footprint for the Guadalquivir basin. Based on Salmoral et al. (2011).

Water Footprint		hm³	%ᵃ	%ᵇ	Economic indicator	Comments
Agriculture	Green	Rainfed	3,690	48.8		²LP$_{rain}$	Winter cereals: 540 €/ha; Sunflower: 630 €/ha; Olive groves: 1,610 €/ha
		Irrigated	1,190	15.7			
		Total	Min.: 2,200 (in 1999) Max.: 5,345 (in 2008)				
	Blue (surface and groundwater)		2,240	29.6	80	³WP	Cotton: 0.20 €/m³; Rice: 0.40 €/m³; Citrics: 1 €/m³; Olive groves: 1.90 €/m³; Vineyard: 2.50 €/m³; Vegetables: 2.90 €/m³; Strawberry: 11 €/m³
			Min.: 1,470 (in 2008) Max.: 2,290 (in 2004)			⁴LP$_{irr}$	Maize: 2,320 €/ha; Olive groves: 2,840 €/ha; Vegetables: 12,650 €/ha
						⁵Δ€/m³	Sunflower: 0.20 €/m³; Olive groves: 0.80 €/m³; Vegetables: 2.10 €/m³
	Blue (groundwater)¹		750	9.9	26.8		
	Total		7,120	93.2			
Livestock	Blue		19	0.3	0.7		Only direct consumption (drinking and exploitation management)
Urban supply	Blue (surface)		83	1.1	3	1.23 €/m³ (tariff)	Without including industry
	Blue (groundwater)		13	0.2	0.5		
Industry	Blue (surface)		35	0.5	1.2	1.40 €/m³ (tariff)	A portion is own supply, the rest from urban supply
	Blue (groundwater)¹		9	0.1	0.3		
Tourism without golf	Blue		4	0.1	0.1		Included in urban supply
Golf	Blue		6	0.1	0.2	331.3 €/m³	Total economy generated
Dams	Blue		315	4.2	11.2	–	
Electricity generation	Blue		31	0.4	1.1		
	Thermo-solar plants		3.6	0.0	0.1	47 €/m³	Prediction 2015: 11.3 hm³
	Other (thermic)		27	0.4	1	140 €/m³	
Total			7,631	100			
Total without green water			2,751		100		

Figures in italics are included within their respective broader figure.
¹ Year 2008; ² LP$_{rain}$: Land productivity in rainfed systems; ³ WP: Water productivity; ⁴ LP$_{irr}$: Land productivity in irrigated systems; ⁵ Δ€/m³: Increase of water productivity of irrigated systems in relation to the rainfed one; ᵃ %: Including green water in the total WF; ᵇ %: Without including green water in the total WF.

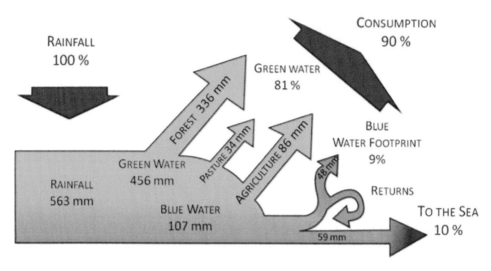

Figure 4 The WF within the hydrologic cycle (inspired by Falkenmark, 2009).
(Source: Salmoral *et al.* (2011)).

The high water productivity of sectors like industry, energy generation and tourism advocates for prioritizing their use in comparison to agriculture. For instance, thermo-solar plant development should be encouraged as their use of water shows a high productivity (47 €/m³) and represents a little share of the total WF. But this prioritization should not result in a rise of the overall basin WF and should be accompanied by an assessment of potential social and environmental impacts. It should also be kept in mind that the presented values are an average and reallocation of water or the prioritization of particular uses should be based on local estimations of the marginal value for water.

3.2   The WF and its integration within the hydrological cycle

Figure 4 summarizes the water balance of the Guadalquivir basin. More than 80% of precipitation turns into green water and only 20% is available downstream in rivers and aquifers as blue water. The majority of the green water is consumed by forests (74%), while the direct human appropriation of green water (WF of agriculture and pastures) amounts to 26%. Regarding blue water, 50% of the total run-off is consumed annually (blue WF) and the other half discharges into the ocean, after contributing to sustaining the ecological functioning of aquatic ecosystems on its way to the river mouth.

4   CONCLUSIONS

An innovative aspect of this study is that not only blue water for direct human use (irrigation, urban and industrial supply) has been taken into account but also green water for human use and nature. In a normal year, forests consume 20,000 hm³ (340 mm), which represents two third of the annual precipitation. Water consumption related to direct human appropriation (WF of the different economic sectors) represents 24% of the precipitation, being the largest part linked to crop production, with 144 mm (15% of precipitation). Thus, green water management through land use planning is a key point to attaining ecosystem conservation and agriculture development (see Chapter 11). The impact of the blue WF on aquatic systems can also be questioned as more than half of the blue water is consumed, with a tendency for increasing groundwater consumption in the head of the basin mainly because of olive groves in Jaén province (see Chapter 10).

Over the period 1997–2008, 46% of the blue water resources were assigned to low value crops (≤0.40 €/m³). The study suggests that a better water management could be reached thanks to water reallocation in activities of greater economic value such as thermo-solar energy. Although additional social and environmental indicators need to be incorporated in the analysis, this reallocation may occur without social conflict with the farmers since the quantities of blue water required for these high-value uses constitute 1–2% of the current total blue water use. However, this process is much more complex since a large number of factors need to be considered (food market, water priorities, water rights, rational water use, social issues and environmental constraints). In the meanwhile, the Government should promote a win-win solution, facilitating for farmers the change towards more productive and less harmful crops. This is the way that the new motto *more cash and care of nature per drop* could be achieved.

REFERENCES

CHG (2010). *Propuesta de Proyecto de Plan Hidrológico de la Demarcación Hidrográfica del Guadalquivir* [Draft Hydrological Plan of the Guadalquivir River Basin District]. Confederación Hidrográfica del Guadalquivir [Guadalquivir River Basin Authority]. Seville, Spain.

Falkenmark, M. (2009). Water for a starving world: time to grasp the 1977 warning. [Presentation]. *4th Water Seminar: Water and food security. International Seminars of the FMB Water Observatory*. Available from: http://www.fundacionbotin.org/file/10123/ [Accessed 25th May 2011].

Hardy, L. & Garrido, A. (2010). *Análisis y evaluación de las relaciones entre el agua y la energía en España* [Analysis and evaluation of the water-energy nexus in Spain]. Papeles de Agua Virtual (PAV), No. 6. Fundación Marcelino Botín, Santander, Spain.

Salmoral, G.; Dumont, A.; Aldaya, M.M.; Rodríguez-Casado, R.; Garrido, A. & Llamas, M.R. (2011). *Análisis de la Huella Hídrica Extendida de la Cuenca del Guadalquivir* [Analysis of the Extended Water Footprint of the Guadalquivir River Basin]. Papeles de Seguridad Hídrica y Alimentaria y Cuidado de la Naturaleza (SHAN), No. 1. Fundación Botín, Santander, Spain.

Chapter 9

The extended water footprint of the Guadiana river basin

Maite Aldaya & M. Ramón Llamas
Water Observatory of the Botín Foundation;
Department of Geodynamics,
Complutense University of Madrid, Madrid, Spain

ABSTRACT: The present analysis is focused on the Spanish side of the semi-arid, transboundary Guadiana river basin. In the basin, the main consuming sector for green water (rainwater stored in the soil as soil moisture) and blue water (surface water and groundwater) is agriculture, accounting for about 95% of total water consumption. Within this sector, high virtual-water crops with low economic values are widespread in the studied Upper and Middle Guadiana regions, particularly cereals with low blue water economic productivity. In particular, the Upper Guadiana basin is among the most significant in Spain in terms of conflicts between agriculture, with almost no food (virtual water) import, and the conservation of rivers and groundwater-dependent wetlands. On the other hand, in the Lower Guadiana basin and Tinto, Odiel and Piedras (TOP) river basins domain, where vegetables and crops are grown under plastic, the blue water economic productivity values are much higher, using both surface water and groundwater resources. The quantity of crops and the amount of employment generated in the whole Guadiana basin are already producing *more crops and jobs per drop*. The aim now is to move towards a policy of *more cash and nature per drop*, especially in the Upper and Middle Guadiana basin.

Keywords: Guadiana river basin, extended water footprint, green water, blue water, virtual water trade, economic water productivity

1 INTRODUCTION

This chapter presents the general and synthetic view of the authors on the use of the extended water footprint for the assessment of the water policy in the Guadiana Basin in Spain. Only three references are provided, where more detailed information can be found.

Today most water resources experts admit that water conflicts are not caused just by water scarcity, but are mainly due to poor water management. Water resources management has inherently a socio-political and economic nature. In this context, virtual water and water footprint analyses link a large range of sectors and issues and provide a transparent framework to find potential solutions, and contribute to a better management of water resources, particularly in water-scarce countries. As the most arid country in the European Union, water use and management in Spain is a hot political and social topic. The semiarid and transboundary Guadiana river basin, located in South-central Spain and Portugal, is among the most significant in Spain in terms of conflicts between agriculture and the environment (Aldaya & Llamas, 2009).

The present study deals with the analysis of the water footprint and virtual water trade of the Spanish part of the Guadiana river basin. The study considers the ways in which both green and blue (ground and surface) water are used by the different economic sectors. This could provide a transparent interdisciplinary framework for policy formulation and ultimately facilitate a more efficient allocation and use of water resources facilitating the achievement of the European Water Framework Directive objectives.

2   STUDY AREA

The Guadiana basin has an area of 66,800 km² (83% in Spain and 17% in Portugal). For practical purposes, the basin has been divided into four areas (Figure 1). These are: a) the groundwater-based Upper Guadiana basin (entirely located in the Castilla-La Mancha Autonomous Region); b) the mainly surface water-based Middle Guadiana basin (comprising part of Extremadura but not the small fraction of Cordoba); c) the Lower Guadiana basin (including the part of the basin in Huelva); and d) TOP domain (comprising the Tinto, Odiel and Piedras river basins).

The Upper Guadiana basin is one of the driest river basins in Spain. In this part, UNESCO recognized the collective ecological importance of 25,000 ha of wetlands in 1980, when it designated the *Mancha Húmeda* Biosphere Reserve. These wetlands provided crucial nesting and feeding grounds for European migrating bird populations and were home to rare animal and plant species. Today, the Tablas de Daimiel National Park (2,000 ha), a Ramsar Site, which used to receive the natural discharge from the

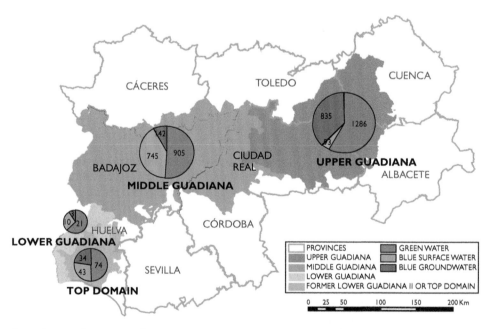

Figure 1 Agricultural use of water resources in the Guadiana (hm³/year) (2001). (Source: Own elaboration).

Western Mancha aquifer, survives artificially, in a kind of *ecological coma*, thanks to the water transfers that have been coming from the Tagus-Segura Aqueduct since 1988 and to the artificial pumpage of groundwater from the aquifer that keeps about 5% of the 2,000 ha of wetlands in the Park flooded. In recent years, the situation has improved slightly since the 2010/2011 high rainfall years and the increase in electricity costs.

3 METHODOLOGY

The water footprint and virtual water trade are calculated using the methodology developed by Hoekstra and co-workers at the University of Twente. The method and data sources are described in more detail in Aldaya & Llamas (2008; 2009). The green and blue water are analyzed for each section of the basin (Upper, Middle, Lower Guadiana and TOP domain) over three different time periods – during an average year (2001), a dry year (2005), and a humid year (1997).

4 HYDROLOGIC RESULTS

4.1 Total water use in the basin

As in most arid and semi-arid regions, in the Guadiana river basin the main sector consuming green water and blue water is agriculture, which accounts for about 95% of total water consumption in the basin as a whole (Table 1). The next-largest blue water user is the urban water supply, which uses less than 5% of the amount of water used in agriculture, most of it returning to the system with lower quality. However, the security of this supply is extremely relevant from a political and economic point of view. Concerning the Andalusian part (Lower Guadiana and TOP domain), agriculture consumes a smaller proportion of water (about 75–80%), which accounts for the increase in the proportion attributed to the urban water supply. The industrial sector, even if it is the smallest water user, represents the highest economic value in terms of Gross Value Added (GVA), which refers to the value of goods and services produced in an economy (see Glossary). Agriculture is also a significant economic activity in the Guadiana river basin, being the most

Table 1 Water footprint of the Guadiana river basin from the production perspective (2001).

Population	Water footprint related to production	Green	Blue	Total	Per capita (m³/cap/year)	GVA (M€)	Water economic productivity (€/m³)
		hm³/year					
1,417,810	Agricultural	2,212	1,827	4,039	2,849	1,096	0.60
	Livestock		22	22	16	286	12.74
	Urban		130	130	91	128	0.99
	Industrial		20	20	14	1,557	77.90
	Total	2,212	1,999	4,211	2,970	3,068	1.53

Source: Aldaya & Llamas (2009).

[hm³ = cubic hectometre = million m³ = 10⁶ m³].

important share of the GVA after the industrial sector (Table 1). Thus, even if urban and industrial uses have an obvious economic and social relevance, agriculture, as the highest water consumer in the basin, is the key to water resources management in the area.

As shown in Figure 1, the blue water consumption in the Upper Guadiana basin is mainly based on its groundwater resources, whereas the Middle Guadiana basin uses its surface water resources, mainly coming from large surface water reservoirs. The Lower Guadiana basin and TOP domain combine both ground and surface water strategies.

4.2   Water footprints of crops (m³/t)

In the basin, 19% of the crop area is devoted to irrigated agriculture; cereals, vineyards and olive trees dominate in the Upper and Middle Guadiana basins, whereas citrus trees and vegetables are more prevalent in the Lower Guadiana and TOP domain. When looking at rainfed agriculture, mainly cereals, olive trees and vineyards are grown in the basin.

When looking at the water footprint of crops, among the studied crops, industrial crops (e.g. sunflowers), grain legumes, grain cereals (1,000–1,300 m³/t) [t = tonne = 10^3 kg] and olive trees (1,000–1,500 m³/t) show the highest values. In humid years, however, olive trees are mainly based on green water resources. Even if olive trees (and vineyards) were traditionally rainfed crops, in recent years the irrigated area seems to have significantly increased for both crops.

It is widely believed that maize and vegetables are water-wasteful in terms of m³/ha. Nevertheless, when looking at the virtual water content (m³/kg), these crops consume less water than is generally believed. Among the studied crops, vegetables (100–200 m³/t) show the smallest virtual water content figures, probably due to the high yields they have.

Vineyards have intermediate virtual water contents, of about 300–600 m³/t. This is largely due to high yields. However, cereals are harvested with high dry matter and low (real) water content, whereas in the case of melons and tomatoes the real water content values are very high.

Despite the semi-arid nature of the Guadiana basin, in the Upper and Middle Guadiana basin, irrigated grain cereal production was widespread in the year 2001. Aside from cereals, vineyards and olive trees were the most widespread crop in the basin that year. Two reasons may explain this trend. First, vineyards are significantly water-efficient (vineyards are traditionally considered dryland crops) and second, irrigated vineyards provide quite high economic revenue per hectare.

In the Lower Guadiana basin and TOP domain on the other hand, irrigated citrus trees and vegetables account for the largest part of the irrigated area and represent the highest total economic values in this region.

5   ECONOMIC RESULTS

5.1   Agricultural economic productivity (€/ha)

As it is widely known, the economic productivity of irrigated agriculture, defined as the market value per hectare, is higher than that of rainfed agriculture. In the case of the Guadiana basin this is true for any type of year (average, humid or dry). From a socio-economic perspective, irrigated agriculture not only provides a higher income, but also a safer income. This is due both to the higher diversification it allows and to the reduction of climate risks derived from rainfall variability. Concerning the agricultural economic productivity per crop of irrigated agriculture, vegetables have the highest revenues per hectare (5,000–50,000 €/ha), followed by vineyards (about 4,000–6,000 €/ha), citrus in the Andalusian section (3,000–5,000 €/ha), potatoes (2,000–6,000 €/ha), and olive trees (about 1,000–3,000 €/ha). Finally grain cereals, grain legumes and industrial crops have productivities of less than 1,000 €/ha.

5.2 Economic blue water productivity (€/m³)

The Andalusian part (i.e. Lower Guadiana and TOP), with a joint surface and groundwater use, has the highest agricultural water economic productivity due to the high economic value of the vegetables which are widespread in this region. The groundwater-based Upper Guadiana basin has intermediate values, whereas the surface water-based Middle Guadiana shows the lowest water economic productivities. Nevertheless, Upper and Middle Guadiana present similar values in dry years. Probably this small difference is due on the one hand to the water irrigation security provided by the existing large surface water reservoirs in the Middle Guadiana, and on the other, because the use of groundwater in the Upper Guadiana basin has serious legal and political restrictions, at least in theory.

When looking at the productivity per crop type in the Guadiana basin (Figure 2), vegetables (including horticultural and greenhouse crops) present the highest economic value per water unit (amounting to 15 €/m³ in the Andalusian part, i.e. the Lower Guadiana and TOP domain). With lower values vineyards (1–3 €/m³), potatoes (0.50–1.50 €/m³), olive trees (0.50–1 €/m³) and citrus trees (0.30–0.90 €/m³) show intermediate values. Finally, with remarkably lower values, grain cereals, grain legumes and industrial crops display an average productivity of less than 0.30 €/m³. These data show that the problem in the Guadiana basin is not water scarcity but the use of large amounts of blue water for low value crops (about 40% in 2001).

The nutritional productivity provides a different picture, wheat having a significantly higher calorific value (2–2.5 Kcal/L) than 1 kg of potatoes (2 Kcal/L) and vegetables (0.3–1 Kcal/L).

5.3 Agricultural trade

Concerning trade in tonnes, euros and virtual water, it is noteworthy that Ciudad Real (Upper Guadiana) is a net exporter, mainly of wine, and barely imports any commodity. During the period studied, this province has relied on its own food production without depending on global markets. This has probably been at the cost of using its scarce water resources.

The province of Badajoz (Middle Guadiana) is a net canned-tomato exporter, while importing other commodities such as cereals (Aldaya & Llamas, 2009). The increase in cereal imports in drier years has to be highlighted.

Huelva (Lower Guadiana and TOP domain) also imports virtual-water-intense commodities, such as cereals, whereas it exports low virtual water content fruits. The drier the year, the higher the cereal imports. In hydrologic terms, importing virtual water through cereals saves 1,015 hm³ in Huelva, whereas growing vegetables for

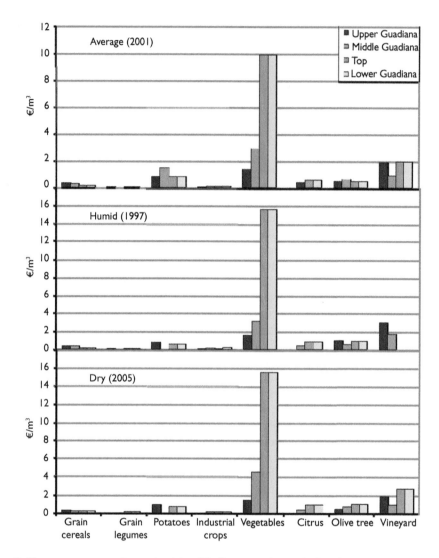

Figure 2 Blue water economic productivity (€/m³) concerning agricultural water consumption by crop and year in the Upper, Middle and Lower Guadiana and TOP domain. (Source: Aldaya & Llamas (2009)).

export uses just 100 hm³. Even if in terms of tonnes and water consumption cereal imports remarkably surpass fruit exports, in economic terms fruit exports are much more important than cereal imports.

Virtual water imports, and in particular cereal imports, can play a role in compensating for the water deficit and providing water and food security in the Middle Guadiana and Andalusian part (Lower Guadiana and TOP domain). Finally, the concept of virtual water trade could be very relevant for this region. Local planning and regional collaboration incorporating the notion of virtual water *trade* could result in

the exchange of goods, diversification of crops, diet awareness raising or crop replacement measures.

6   POLICY IMPLICATIONS

It is important to consider the long-term impact of water management practices in
the Guadiana river basin following the WFD emphasis to attain a *good ecological
status* of surface and groundwater bodies by 2015. The 2008 regulation transpos-
ing the WFD in Spain, including the water footprint assessment of the different
socio-economic sectors in the River Basin Management Plans (Official State Gazette,
2008), could provide a transparent framework for a more integrated water resources
management in the basin. In any case, the ongoing water management problems in
the basin, particularly in the Upper Guadiana, mean that the goals set by the WFD are
unlikely to be achieved within the required timescale.

To find a solution to the problems in the basin, the Ministry of the Environment developed a Special Plan for the Upper Guadiana Basin (PEAG). Chapter 20 on The Tablas de
Daimiel deals in more detail with this topic. The plan is endorsed with 5,500 M€ over
about 20 years (2008–2027), and incorporates actions such as the purchase of land and
water rights from irrigators, or the closure (or legalization) of unlicensed wells. Complementary measures include reforestation and dryland farming. The plan devises a water
consumption scenario that is compatible with a mid-term water table recovery (before
2027) and identifies water management tools to deal with the Upper Guadiana groundwater crisis. Among the mechanisms for the rearrangement of the water use rights aimed
at environmental remediation of aquifers, the PEAG approves the acquisition of private
rights of water use. In theory part of the rights acquired by the Hydraulic Administration
will be for environmental uses (70%) and the rest (30%) transferred to the Autonomous
Community for regulating this situation. Initially the target has been for social purposes.
Unfortunately for the moment the Plan has not been fully implemented due to the economic crisis Spain is undergoing and the rights used for the environment have been less
than 30%, meaning that any significant short-term progress is doubtful. In this context,
policy makers will most probably do what their constituency wants. It is therefore crucial
to focus on raising awareness and capacity-building among the population.

7   CONCLUSIONS

In the Guadiana river basin the main green and blue water consuming sector is agriculture, with 95% of total water consumption. Urban water supply and industry values
have higher blue water economic productivity values than agriculture. The multifunctional value of agriculture, however, has to be taken into account. Rainfed agriculture
has a high relevance in the Guadiana basin in terms of total hectares. Agricultural
economic productivity (t/ha) and total production (t/year) of rainfed agriculture are
notably lower than that of irrigated agriculture. Thus, even if it is less extensive, irrigated agriculture produces more tonnes and euros than rainfed agriculture.

As a whole, high-virtual-water low-economic value crops are widespread in the
Upper and Middle Guadiana regions. For instance, cereals exhibit virtual water values of

1,000–1,300 m³/t, and maize and vegetables (mainly tomatoes and melons) present the smallest values with around 600 and 100–200 m³/t respectively, due to their high yields.

The economic productivity of blue water use in the Upper and Middle Guadiana basin ranges between 0.10–0.20 €/m³ for low-cost cereals and 1.50–4.50 €/m³ for vegetables. These values are relatively small in comparison with the ones obtained in the Andalusian region (Lower Guadiana and TOP domain). In this region, for vegetables (including horticultural and crops under plastic) using both surface water and groundwater resources, this value can amount to 15 €/m³. Even with lower figures, vineyards (1–3 €/m³) and olive trees (0.50–1 €/m³) seem to be profitable crops. As a matter of fact, it is widely known that farmers are currently changing their production to vineyards and olive trees. It could be interesting to examine these trends in the near future.

Nevertheless, we should not over-simplify the issue by assuming that all the water that is not used for vegetables or trees is wasted water. Factors such as risk diversification, labour, market access or other environmental, social, economic and agronomic reasons have to be taken into account in order to find a balance. The major environmental challenge of agriculture is the preservation of the environment without damaging the agricultural sector economy. The quantity of crops and the employment generated in the whole Guadiana basin is producing *more crops and jobs per drop*. The aim now is to achieve the paradigm *more cash and nature per drop*. The present results, indicating the low water consumption and high economic value of vegetables, followed by vineyards, is one of the factors that has to be taken into account in order to achieve an efficient allocation of water and economic resources.

Finally, the report provides a first estimation of trade in agricultural products by considering international imports and exports at a provincial level. The Upper Guadiana basin is a net exporter, mainly of wine, barely importing any food commodity; while the Lower Guadiana and TOP domain import low-value, high water-consuming cereals, and export high-value, low virtual-water content crops such as fruits. This reduces the demand on local water resources that can be used to provide ecological services and other more profitable uses.

These results and concepts are in line with the actions and measures envisaged in the Special Plan for the Upper Guadiana Basin. Their implementation such as rearrangements of the water use rights aimed at environmental remediation of aquifers, would make possible a mid-term water table recovery.

REFERENCES

Aldaya, M.M. & Llamas, M.R. (2008). *Water Footprint analysis for the Guadiana Basin*. Papeles de Agua Verde, 3. Fundación Botín, Santander, Spain. Available online from: http://www.waterfootprint.org/?page=files/Publications [Accessed May 15th 2012].

Aldaya, M.M. & Llamas, M.R. (2009). *Water Footprint Analysis (Hydrologic and Economic) of the Guadiana River Basin*. World Water Assessment Programme, Side Publication Series.

López-Gunn, E.; Zorrilla, P. & Llamas, M.R. (2011). The impossible dream? The Upper Guadiana System: aligning changes in ecological systems with changes in social systems. *Water Front*, 2: 117–128.

Official State Gazette (2008). Approval of the water planning instruction. Ministry of the Environment and Rural and Marine Affairs. *Official State Gazette*, 229. 22nd September 2008. Available from: http://www.boe.es/boe/dias/2008/09/22/pdfs/A38472-38582.pdf [Accessed May 15th 2012].

Chapter 10

Lessons learnt from analyses of the water footprint of tomatoes and olive oil in Spain

Gloria Salmoral & Daniel Chico
Water Observatory of the Botín Foundation;
CEIGRAM, Technical University of Madrid, Madrid, Spain

ABSTRACT: This chapter evaluates the water footprint (WF) of Spanish tomatoes and olive oil over the period 1997–2008. It analyses the three types of water: green, blue and grey. Water apparent productivity (WAP) and virtual water exports for tomatoes and olive oil have been studied. The ranges of the WF of tomatoes and olives per unit of product (m³/t) show that providing a unique value for a product WF may be a strong assumption because of the widely different climatic conditions, production systems, productivity levels and irrigation schedules across the country. The greenhouse tomato system presents the greatest WAP, which is influenced by the much higher price of off-season productions and larger crop yield. However, tomato production also shows a high grey WF, implying a pressure over water resources related to nitrogen pollution. The increase of groundwater consumption in the upstream Guadalquivir basin caused concerns about the sustainability of olive irrigation. Recently, the situation seems to be under control given the deficit irrigation practices and the constraints imposed by the sharp increase in energy prices. The virtual water related to olive oil exports illustrates still the importance of the green water footprint of rainfed olives amounting to about 77% of the total virtual water exports.

Keywords: water productivity, groundwater, irrigation, sustainability, virtual water

1 INTRODUCTION

Spain is the largest world producer and exporter of olive oil and table olives. In the 2009/2010 agricultural season, nearly 50% of the estimated olive oil world production was produced in Spain, with 56% of Spanish olive oil intended for export. In this season olive oil production amounted to 1.4 Mt [Mt = million tonnes = 10^9 kg] and almost 2,000 M€ in 2.3 Mha [Mha = million hectares = 10^6 ha] (IOC, 2012; MARM, 2012). Over the period 1995–2009, olive production comprised in economic terms about 13% of the gross national agricultural production (MARM, 2012). During the same period tomatoes had a yearly average production of about 4 Mt from 61,000 ha. In economic terms, tomato production represented approximately 5% of the gross national agricultural production (MARM, 2012). In the 2009/2010 season 15% of tomatoes produced (4.8 Mt) were exported, though the period 1998–2009 showed an average of 25% for exports (MARM, 2012; DataComex, 2012).

Because of the economic and social importance of olive and tomato production in Spain, this chapter analyses the water footprint (WF) of green water (stored in the soil profile), blue (freshwater) and grey water (an indicator of pollutant assimilation capacity, see Glossary) for tomatoes and olive oil. The analysis is set from the Extended Water Footprint (EWF) perspective, which combines the contribution of the WF relying on water resources accounting (Hoekstra *et al.*, 2011) with an economic perspective based primarily on determining the economic value of water (Garrido *et al.*, 2010; see Chapter 6).

The innovations of this chapter in relation to previous studies are:

- Three growing systems for tomatoes (irrigated open air/greenhouses and rainfed) were distinguished. In addition, four yearly seasons (spring short, spring-summer, short autumn and long periods) were considered in greenhouse tomato production.
- It was assumed that olive orchards do not meet their irrigation water requirements. Irrigation water was restricted yearly based on level of drought.
- The grey WF for both crops was calculated based on nitrogen surplus instead of the nitrogen doses rate per hectare times the leaching fraction.

The following sections synthesize the main findings of two studies elaborated by the Water Observatory of the Botín Foundation (Chico *et al.*, 2010; Salmoral *et al.*, 2011), where more detailed information on data, applied methodology and references can be found.

2 METHODOLOGY OF THE WATER FOOTPRINT FOR TOMATOES AND OLIVE OIL

The green, blue and grey water components of tomatoes and bottled olive oil were calculated in absolute (volume) and relative terms (volume/unit of product) for the time period 1997–2008. We used the CROPWAT model to estimate the green and blue WF of both tomatoes and olive fruits in terms of evapotranspiration. For tomatoes, a distinction was made between growing systems (open air/covered irrigated and rainfed) and production cycle throughout the year (spring short, spring-summer, short autumn and long periods). Under irrigated conditions we assumed that the tomato crop water demand is completely satisfied and that irrigation was applied regularly on regardless of rainfall.

For olive fruit, irrigated versus rainfed systems were analyzed. It was assumed that the crop water requirements were not met. In the Guadalquivir basin water allowances vary from 1,200 to 2,500 m³/ha in olives, the large figure corresponding to high density olive tree plantations; but this information is not from the public domain. As a result, we took a water allowance of 2,280 m³/ha for a normal climatic year that was reduced annually according to the level of drought in the basin. The WF of olive oil as a product was calculated by dividing the olives water consumption by a product fraction of 20%, which indicates the quantity of olive oil obtained per kilogram of olives. The WF of one litre of bottled olive oil also required assessment of the water embedded in the bottle, cap and label. However, we did not quantify the WF of the production process of olive oil since the amount required on this step is insignificant as previous studies have shown (Avraamides & Fatta, 2008).

The grey WF is defined as the volume of freshwater that is required to assimilate a load of pollutants based on the natural concentration of pollutants in a receiving water body and the existing ambient water quality standards (see Hoekstra *et al.*, 2011; Glossary). An ambient water quality standard of 50 mg/L NO_3^- was used following the European Nitrates and Groundwater Directives. The natural concentration of pollutants in the receiving water body was assumed to be negligible. The grey WF was calculated for nitrogen since it is a very dynamic element which can be the source of diffuse pollution caused by leaching (See Chapter 12). Improvements in this study are achieved since the grey WF is calculated based on nitrogen surplus from the Spanish Agricultural Nutrient Balances (MARM, 2008) instead of the fertilizer application rate per hectare times the leaching fraction, which previous studies have used (Chapagain & Hoekstra, 2011; Mekonnen & Hoekstra, 2010; 2011a).

The water apparent productivity (WAP) and virtual water exports calculations for both tomatoes and olive oil were also assessed. Further details on the applied methodology can be found in Chico *et al.* (2010) and Salmoral *et al.* (2011).

3 RESULTS

3.1 The water footprint of tomatoes

Over the period 1997–2008 the green water ranged from 15 to 25 hm³/year [hm³ = cubic hectometre = million m³ = 10^6 m³] and the blue water varied between 250 and 460 hm³/year. The lower green component in comparison to the blue one is because crop water requirements were fully meet. CROPWAT firstly evapotranspirates water from irrigation than water from rain when both events occur on the same day. The national grey WF varied from 470 to 710 hm³/year during the same time period (Figure 1). As a result, pollution is a greater concern at national level for tomato production compared with water consumption.

The average green (20 hm³) and blue (260 hm³) WFs for the time period 1997–2008 represent 0.12% and 2.3%, respectively, of their water colour components for the national crop production as presented on Chapter 6. The mean grey WF (550 hm³) during the same time period represents 6.6% of the national grey WF of crop production calculated by Mekonnen & Hoekstra (2011b).

There are sharp differences in the WF in relative terms (m³/t) [t = tonne = 10^3 kg] for tomato production (Figure 2). Rainfed production has by far the highest WF (970 m³/t); however the WF in absolute terms (hm³) is negligible for rainfed production of tomatoes in Spain. The grey WF of open air irrigated and greenhouse production systems are smaller, partly because their yields are much higher.

Figure 3 summarizes the average green, blue and grey tomato WF per unit of product by province, grouping them according to their production importance over the period of investigation. The large differences encountered are related to the different production system (open-air rainfed or irrigated *vs.* covered), crop yields and climate parameters prevailing in each province. The largest producing provinces show a significantly smaller WF (m³/t).

Differences among growing systems are significant, according to the proportion of water consumed or polluted in the main producing provinces (Table 1). The primary

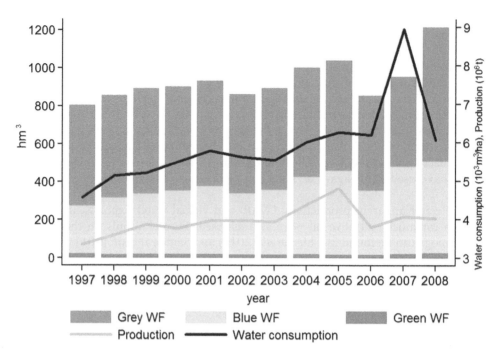

Figure 1 National green, blue and grey water footprint (WF) (hm³, left axis), national production (Mt, right axis) and water consumption (10³ m³/ha, right axis) for tomatoes. (Source: Chico *et al.* (2010)).

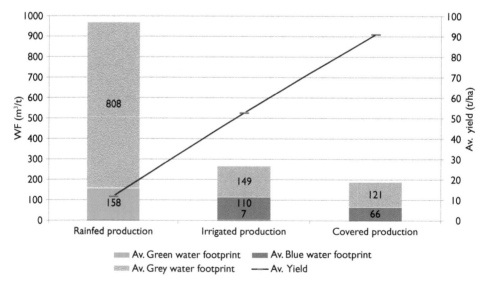

Figure 2 Mean green, blue and grey water footprint per unit of product (WF in m³/t) of open air (rainfed and irrigated) and greenhouse tomato production and associated average yields (t/ha). (Source: Chico *et al.* (2010)).

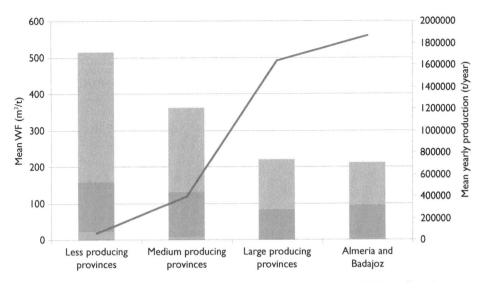

Figure 3 Mean green, blue and grey water footprint per unit of product (WF in m³/t, left axis) in the different Spanish provinces and their mean annual production (t/year, right axis). (Source: Chico *et al.* (2010)).

Table 1 Blue, green and grey share (%) of the total tomato WF (hm³) across the main producing provinces (percentages sum up by row).

| Province | Rainfed | | Open air | | | Greenhouse | | Total WF |
	Green (%)	Grey (%)	Green (%)	Blue (%)	Grey (%)	Blue (%)	Grey (%)	(hm³)
Badajoz			3	51	46			215
Almería			0.3	4	12	30	54	183
Murcia			1	11	26	20	42	80
Las Palmas	0.1	0.5	0	6	10	35	48	26
Granada			1	9	39	15	37	42
Caceres			2	51	47			45
Seville	0.2	1	2	64	32	0.1	1	25
Navarra			3	28	68	0.5	1	44

Source: Chico *et al.* (2010).

impact of the tomato production in Badajoz (also in Caceres or Seville), is the high volume of blue water consumed, whereas in Almería (but also Murcia, Navarra or Granada) the main impact resulting from tomato production is the grey water generated. The later indicates the volume of freshwater required for assimilating the nitrogen discharge into water bodies.

3.1.1   The water apparent productivity of tomatoes

The WAP is an indicator of the economic performance of water use. Over the study period, the WAP of tomatoes varied from 0.025 to 46 €/m³ taking into account all

(50) provinces of study, period and system of production. As shown in Table 2 and Table 3, the WAP depends on the production system, type of water (green or blue) and season of the year. On average, the WAP of tomatoes was 4.30 €/m³ during the period 1997–2008. Greenhouse production has much higher water productivity compared to irrigated open air (Table 2). In tomato production, the prices vary significantly depending on the time of the year, being a strong stimulus for off-season production (autumn and winter) (Table 3). Most of this off-season production takes place in greenhouses.

Tomato production in Almería was worth 641 M€ in 2007, approx. 35% of the total value generated by agriculture in the province (1,800 M€), and 4.9% of the regional Gross Domestic Product (GDP). In Badajoz, on the contrary, tomato production was valued at 649 M€ in 2007, approx. 62% of the value produced by agriculture in this region (1,048 M€), and 6% of the regional GDP.

Table 2 Average yearly production (t), share of greenhouse production (%) and average water apparent productivity (WAP in €/m³) over the period 1997–2008 in the main producing provinces and under different production systems.

Province	Average yearly production (t)	% of greenhouse production	WAP of rainfed systems (€/m³)	WAP of open air irrigation systems (€/m³)	WAP of greenhouses (€/m³)
Badajoz	1,067,555	0.4		3.1	0.03
Almería	801,324	92		3.9	7.1
Murcia	335,012	76	3.8	3.9	8.8
Las Palmas	179,242	92	18.1	4.6	9.3
Granada	163,314	56		7.3	7.2
Caceres	160,934	0.1		2.2	
Navarra	141,998	2.7		3.4	6.3
National average	**3,968,767**	**3.6**	**2.1**	**3.1**	**7.8**

Source: Chico et al. (2010).

Table 3 Percentage of green and blue WF in comparison to the total WF and average water apparent productivity (WAP in €/m³) over the period 1997–2008 for greenhouse production and main producing provinces in relation to the year season. Percentages sum up by row and year season.

	Early season			Middle season			Late season		
Province	Green (%)	Blue (%)	WAP (€/m³)	Green (%)	Blue (%)	WAP (€/m³)	Green (%)	Blue (%)	WAP (€/m³)
Badajoz	3	97	5.70	5	95	3.80	2	98	10.40
Almería	6	94	9.30	22	78	2.10	3	97	7.90
Murcia	28	72	9.20	24	76	3.40	24	76	10.40
Las Palmas	5	95	11.80	5	95	3.80	3	97	11.10
Granada				5	95	2.20			
Caceres	1	99	22.70	39	61	3.00	1	99	24.00
Navarra	20	80	7.50	24	76	2.70	15	85	9.50
National average	**3**	**97**	**5.70**	**5**	**95**	**3.80**	**2**	**98**	**10.40**

Source: Chico et al. (2010).

It is also noteworthy that the source of water is linked to the production system. Provinces using surface water (Badajoz, Cáceres and Navarra) produce around 98% of their production in open-air systems, while the three provinces using almost exclusively groundwater (Almería, Las Palmas and Tenerife) produce over 90% of their tomatoes in greenhouses. The average groundwater apparent productivity is notably higher in production systems that use groundwater (3.70–10.50 €/m³) compared to those using surface water (3–6.40 €/m³). This distinction of water productivity based on the origin of water is analyzed in more depth in Chapter 7.

3.1.2   *The virtual water exports of Spanish tomatoes*

A large proportion of Spanish-produced tomatoes are intended for export, especially those grown in the south-eastern Mediterranean provinces (Almería, Murcia and Granada). On average the annual amount of virtual water exported through tomatoes is 4, 88 and 134 hm³ of green, blue and grey water, respectively. 93% of these virtual water exports go to the European Union, mainly the UK, Germany and the Netherlands. These exports represent around 0.03% and 1.7%, respectively, of the national green and blue virtual water exports presented on Chapter 6 of this book, and 10% of the grey virtual water exports of Spanish crop products according to Mekonnen & Hoekstra (2011b). However, in economic terms tomatoes exports are more than 6 times larger (9.08 €/m³) than the national average exports of 1.34 €/m³.

3.2   The water footprint of olive fruit and olive oil

Over the period 1997–2008 Spanish olive fruit production consumed 5,340–9,720 hm³/year under rainfed conditions, 630–2,550 hm³/year of green water in irrigated systems, 460–890 hm³/year of blue water. Grey water ranges were 950–1,210 hm³/year (Figure 4). Variation of green water from year to year depends mainly on rainfall. Rainfed olives account for the largest green WF since they occupy from 3.5 (year 2008) to 7.4 (year 1997) times the irrigated area. The lowest annual rainfall in 2005 (with 430 mm) clearly reflected the decrease of the green WF both under rainfed and irrigated conditions. In comparison to the national WF crop production, olive fruit production accounts for 20% and 5% of the green and blue water respectively, and 13% of the Spanish grey water footprint for crop production (Mekonnen & Hoekstra 2011b).

Between 1997 and 2008 the olive orchard area increased from 2.2 to 2.4 Mha, although this increase was mainly due to the expansion of irrigated olive orchards in Andalusia, particularly in the Guadalquivir basin. Olive orchards grew in Andalusia from 230,200 to 507,400 ha. In fact, Andalusia consumed almost 90% of the national blue WF of olive fruit production (760 hm³) in the year 2008. The expansion of irrigated olives occurred mainly from groundwater sources in the Guadalquivir basin (see Chapter 8), which includes Jaén, Córdoba, Seville and a portion of Granada province (Figure 5). In the last decade the large increase in olive oil production has produced what could be called an *olive oil bubble*. After 2008 olive oil prices dropped, as they tend to approach the market fat world prices. The result is that the hardest hit will be those on rainfed olive groves and irrigated land with high intakes of energy (in some cases the water is raised up to 600 m). In addition, problems related to diversity losses

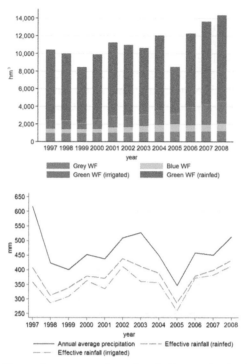

Figure 4 National green, blue and grey water footprint of olive production in hm³ (left) and annual average rainfall and effective rainfall in mm (right) for the period 1997–2008. (Source: Salmoral *et al.* (2011)). This figure has been extracted from the Span. J. Agric. Res. 9 (4): 1089–1104 (2011) with kind permission of INIA (National Institute of Agricultural Research).

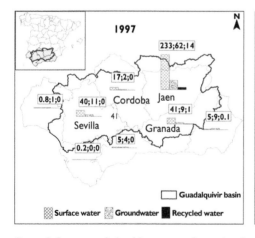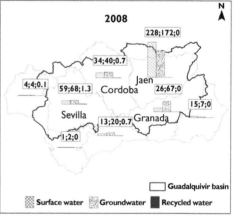

Figure 5 Sources of the blue water footprint (hm³) for olive production along provinces within Guadalquivir basin in 1997 (left) and 2008 (right). (Source: Salmoral *et al.* (2011)). This figure has been extracted from the Span. J. Agric. Res. 9 (4): 1089–1104 (2011) with kind permission of INIA.

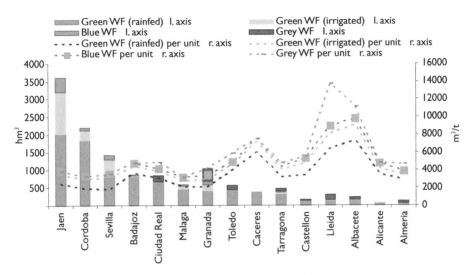

Figure 6 Total water footprint (green rainfed + green irrigated + blue + grey, in hm³) for olive production and the water footprint per unit of product (m³/t) in 2008. Only provinces that comprise
≥1% of the national olive production in 2008 are illustrated. (Source: Salmoral et al. (2011)).
This figure has been extracted from the Span. J. Agric. Res. 9 (4): 1089–1104 (2011) with kind
permission of INIA.

and environmental pressures arise with more intensive olive orchards (Scheidel &
Krausmann, 2011).

In 2008 Jaén, Córdoba and Seville jointly accounted for nearly 70% of the national
olive production and for 54% (7,260 hm³) of the national WF of olive fruit production (Figure 6). While their total WFs in hm³ are the largest, they are very efficient in
terms of WF per unit (m³/t). The provinces showing the highest nitrogen pollution per
tonne of crop produced are minor olive producers such as Lleida and Albacete.

The WF of the bottle, cap and label for one litre of bottled olive oil does not represent more than 0.5% of the total supply chain for each year and province of study. In
conclusion, most of the water used to produce olive oil can be directly associated with
olive fruit production in the field. Spain has the following annual ranges of the WF
per litre of olive oil produced: 6,300–11,760 L/L green WF (rainfed); 2,770–4,640 L/L
green WF (irrigated); 1,430–2,780 L/L blue WF (irrigated) and 710–1,510 L/L grey
WF (rainfed and irrigated). These ranges indicate that providing an unique value of
water consumed/polluted by litre of olive oil means giving an average of a broad
interval because of changing climate conditions, soil characteristics, water irrigation
applied and crop yield across provinces and time.

3.2.1 The apparent water productivity of olive oil

The WAP of olive oil varies in a similar way over the period of study in rainfed
and irrigated systems comparing two typical olive oil producing provinces (Jaén and
Toledo) located in different regions of Spain (Figure 7), due to the variation of olive
oil market prices. In rainfed systems the WAP in Jaén ranges from 0.20 to 0.62 €/m³
and from 0.07 to 0.36 €/m³ in Toledo. The WAP of irrigated systems has shown a

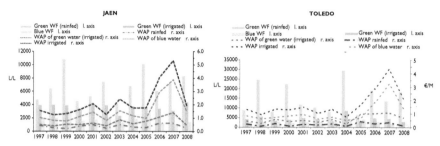

Figure 7 Olive oil water footprint (L/L) and water apparent productivities (€/m³) for both rainfed and irrigated production systems in Jaén (left) and Toledo (right) over the period 1997–2008. (Source: Salmoral *et al.* (2011)). This figure has been extracted from the Span. J. Agric. Res. 9 (4): 1089–1104 (2011) with kind permission of INIA.

relatively stable trend between 1997 and 2005 with values below 2.40 and 1.70 €/m³ in Jaén and Toledo, respectively. The peaks of WAPs in 2006 and 2007 are related to highest olive oil market prices that took place during both years.

3.2.2   *Virtual water exports of olive oil*

The mean annual amount of exported olive oil over the studied period consisted of 4,350; 1,180 and 570 hm³ of green, blue and grey water, respectively. Differences between years are very significant, with green water being the most unstable component and closely dependent on annual precipitation, whereas blue virtual water exports are much more stable. These values show that approximately 70% of the total virtual water exports are green water, which denotes the importance of the green component in the virtual water trade, as reported in previous studies (Aldaya *et al.*, 2010). 23% of olive oil exports correspond to irrigation water, which suggests that expanding groundwater irrigated olive orchards and olive oil exportation may add further pressure to the already stressed Guadalquivir basin. Compared to the national exports of crop products estimated by Mekonnen & Hoekstra (2011b), our results show that olive oil accounts for 32%, 15% and 43% of the green, blue and grey water of the Spanish crop exports. Rainfed olives therefore have an important role in virtual water exports, even if both the area of irrigated olive trees and the related blue water component have increased during the period of study. The volume of grey water exported as olive oil in relation to the total Spanish exports of crop products is relevant, although the largest producing provinces do not generate significant volumes of grey water.

4   CONCLUSIONS

Our study provides an overview of the three water types (green, blue and grey) ascribed to the production of tomatoes and oil olive in Spain, differentiating the various production systems found across provinces and the resulting variation for the water consumed/polluted in absolute terms (hm³) and per unit of product (m³/t). Tomato production shows relatively high blue water use efficiency, although it also shows a high grey WF, implying a pressure over water resources related to nitrogen pollution. WF evaluations that omit the grey component would lead to incomplete conclusions, as they may contribute to increased efficiency in water consumption but fail to consider the environmental quality aspects. In contrast to tomatoes, the total olive oil WF stands out because of the large portion of green component for both rainfed and irrigation conditions. This circumstance makes olive oil production largely dependent on the precipitation pattern.

To obtain a more comprehensive picture, not only water consumed or polluted was considered, but also the associated economic value (WAP in €/m³). We note important differences in the WAP between provinces, production systems and also throughout the year. The greenhouse tomato system shows the greatest WAP, which may be related to the much higher price of off-season productions and larger crop yield. In addition, groundwater production showed higher blue water productivity than that of open-air irrigated. While the provinces irrigating with surface water mainly produce tomatoes intended for the industry in open-air systems, those using groundwater produce foremost fresh tomatoes for export, which are more valuable. In the case of olive oil, water productivity also varies among provinces over the period depending on production system, climatic-soil conditions and market price variation.

The pattern in both products is that the largest producing places show high water use efficiency per product and WAP, but imply great pressure on the available water resources. The production of fresh vegetables in the southeast of Spain (Almería) shows signs of aquifer depletion. Reductions in leaching would lessen the grey WF per unit of product. Still, overuse would have to be dealt with. In the case of Badajoz province, if production continues increasing, rises in water use efficiency and leaching reductions may be needed. There doesn't seem to be a risk of over-use since significant quantities are available thanks to the presence of large reservoirs, but the impact on the water quality may need to be tackled.

Between 1997 and 2008 olive orchard area more than doubled in the major production centre (Andalusia). Olive oil production has increased and has led to what could be called an *olive oil bubble*. The drop of olive oil prices after 2008 has hit hardest the rainfed olive orchards and those irrigated with high energy costs. This growth took place due to increasing groundwater abstractions in the Upper Guadalquivir basin and has caused concerns about the sustainability of olive irrigation in those areas. This situation led to a water-stressed basin because of over-allocation of available water resources. 23% of olive oil exports rely on irrigation water, which suggests that expanding groundwater-irrigated olive orchards and olive oil exportation may add further demand to the already stressed Guadalquivir basin. However, recently the situation seems to be under control given the deficit irrigation practices with very limited amounts and the constraints imposed by the sharp increase in energy prices.

Another important conclusion of the study is the ranges given for tomatoes and olives WF per unit of product. They show the strong assumptions taken when providing a unique value of the WF because of the widely different climatic conditions, production systems, productivity levels and irrigation schedules across the country. As every water footprint calculation is an estimation, values given are the average of a broad interval.

Irrigation farmers' decisions depend on several factors, particularly for woody crops that have traditionally been grown under rainfed conditions (i.e. olives and vineyards). Firstly, energy costs might not allow farmers to irrigate as much as they wish, particularly in deep groundwater wells (see Chapters 14 and 19). They are also motivated to applied greater amount of irrigation water when market olive oil prices are high, since their cost would be then compensated. The scale of our study did not enable us to take into account farmers' decisions regarding precipitation during irrigation management, by assuming that rainfall is sufficient and reducing their irrigation schedules (García-Vila et al., 2008). All these climatic, agronomical and management aspects heavily influence the calculated value of the WF, and should be carefully taken into account for an accurate evaluation of the water use performance of crops.

The results of this study also confirm the importance of a detailed WF assessment of ingredients in the case of agriculture-based products since olives comprise more than 99.5% of the WF of one litre of bottled olive oil, emphasizing the olive fruit production in the field as key to improving water management. Recently, the water label has become a new way of certificating the efficient use of water resources for all water users (EC, 2011). Our results lead us to conclude that the water label would not be accurate for food products if only a single value is given regarding to water consumed/polluted.

Further local crop production studies, completed with a wider range of social, economic and environmental indicators are required for an appropriate sustainability assessment and informed decision making.

REFERENCES

Aldaya, M.M.; Allan, J.A.; Hoekstra, A.Y. (2010). Strategic importance of green water in international crop trade. *Ecological Economics*, 69: 887–894.

Avraamides, M. & Fatta, D. (2008). Resource consumption and emissions from olive oil production: a life cycle inventory case study in Cyprus. *Journal of Cleaner Production*, 16: 809–821.

Chapagain, A.K. & Hoekstra, A.Y. (2011). The blue, green and grey water footprint of rice from production and consumption perspectives. *Ecological Economics*, 70(4): 749–758.

Chico, D.; Salmoral, G.; Aldaya, M.M.; Garrido, A. & Llamas, M.R. (2010). *The Water Footprint and Virtual Water Exports of Spanish Tomatoes*. Papeles de Agua Virtual, no. 8. Botín Foundation, Santander, Spain.

DataComex (2012). *Estadísticas del Comercio Exterior Español*. [*Spanish International Trade Statistics*] Available from: http://datacomex.comercio.es/principal_comex_es.aspx [Accessed 25th January 2012].

EC (European Commission) (2011). *Assessment of the Efficiency of the Water Footprinting Approach, and of the Agricultural Products and Foodstuff Labelling and Certification Schemes. Final Report: Part B – Recommendations for Certification*. Directorate-General Environment, European Commission.

García-Vila, M.; Lorite, I.J.; Soriano, M.A. & Fereres, E. (2008). Management trends and responses to water scarcity in an irrigation scheme of Southern Spain. *Agricultural Water Management*, 95: 458–468.

Garrido, A., Llamas, M.R., Varela-Ortega, C., Novo, P., Rodríguez Casado, R., Aldaya, M.M. (2010). *Water footprint and virtual water trade in Spain. Policy implications*. Fundación Marcelino Botín. Springer. 153 p.

Hoekstra, A.Y.; Chapagain, A.K.; Aldaya, M.M. & Mekonnen, M.M. (2011). *The Water Footprint Assessment Manual: Setting the Global Standard*. Earthscan, London, UK.

IOC (2012). *World Olive Oil Figures*. Available from: http://www.internationaloliveoil.org/ [Accessed 25th January 2012].

MARM (2008). *Balance del Nitrógeno en la Agricultura Española. Año 2006*. [*Nitrogen Balance in Spanish Agriculture. Year 2006*] Dirección General de Recursos Agrícolas y Ganaderos. Secretaría General de Medio Rural. Ministerio de Medio Ambiente, y Medio Rural y Marino, Madrid. [Directorate General for Agricultural and Livestock Resources. General Secretariat for Rural Affairs. Ministry of Environment and Rural and Marine Affairs, Madrid].

MARM (2012). *Agricultural and Statistics Yearbooks*. Spanish Ministry of Environment, and Rural and Marine Affairs. [Online] Available from: http://www.marm.es/es/estadistica/temas/anuario-de-estadistica/ [Accessed 25th January 2012].

Mekonnen, M.M. & Hoekstra, A.Y. (2010). A global and high-resolution assessment of the green, blue and grey water footprint of wheat. *Hydrology and Earth System Sciences*, 14(7): 1259–1276.

Mekonnen, M.M. & Hoekstra, A.Y. (2011a). The green, blue and grey water footprint of crops and derived crop products. *Hydrology and Earth System Sciences*, 15(5): 1577–1600.

Mekonnen, M.M. & Hoekstra, A.Y. (2011b). *National Water Footprint Accounts: The Green, Blue and Grey Water Footprint of Production and Consumption*. Value of Water Research Report Series, no. 50. UNESCO-IHE.

Salmoral, G.; Aldaya, M.M.; Chico, D.; Garrido, A. & Llamas, M.R. (2011). The water footprint of olive oil in Spain. *Spanish Journal of Agricultural Research*, 9(4): 1089–1104.

Scheidel, A. & Krausmann, F. (2011). Diet, trade and land use: a socio-ecological analysis of the transformation of the olive oil system. *Land Use Policy*, 28: 47–56.

Part 3

Looking at the environment and sector uses

Linking land management to water planning: Estimating the water consumption of Spanish forests

Bárbara Willaarts
Water Observatory of the Botín Foundation;
CEIGRAM, Technical University of Madrid, Madrid, Spain

ABSTRACT: The role of vegetation in the partition of rainfall and the provision of water supply downstream have seldom been addressed in Spain's water planning, despite the fact that changes in the vegetation cover and land management have large hydrological implications. This chapter gives a first overview on what are the water requirements of major forest types across Spain, including woods, shrubs, agroecosystems and pastures. Likewise, it provides a preliminary accounting on how droughts and changes in the forest cover modify the evapotranspiration rates, and the likely impacts on streamflow availability. Our results suggest that forests consume the largest fraction of annual rainfall in Spain, above agriculture, and that changes in land use have a greater impact compared to droughts on runoff reduction downstream.

Keywords: Forest evapotranspiration; Forest expansion; Land abandonment; Integrated land and water resources management

1   INTRODUCTION

Land use and land cover play a crucial role in the partition of rainfall and the provision of blue water for downstream users in catchments. The number of studies addressing the hydrological implications linked to land conversions has grown significantly over the last years, particularly in humid regions (Gordon *et al.*, 2005; Jackson *et al.*, 2005; Scanlon *et al.*, 2007; Trabucco *et al.*, 2008). These driven changes on the water cycle have been argued to rival or even surpass those ascribed to climate change (Vörösmarty *et al.*, 2000).

In semi-arid countries like Spain research on the links between land use and water is still scarce (Gallart & Llorens, 2003; Gallart *et al.*, 2011; Willaarts *et al.*, 2012). This knowledge gap is paradoxical bearing in mind that forest area has increased approximately 1.5 Mha [Mha = million hectares = 10^6 ha] in the last two decades, and over 30% of this augmentation is concentrated in some of the most water-stressed provinces along the Mediterranean arc (MAGRAMA, 2012a; 2012b). This forest expansion largely responds to the increase of afforestation and rural abandonment processes taking place on previous agricultural fields (Hill *et al.*, 2008; Lasanta *et al.*, 2009).

Alongside with the changes in the forest area, a significant reduction in runoff has been observed in different catchments across Spain over the last decades (Beguería *et al.*, 2003; Otero *et al.*, 2010; Lorenzo-Lacruz *et al.*, 2012). Most of these studies agree that climate change, and particularly the reduction in precipitation observed during the second half of the 20th century, is a major driver of streamflow reduction. However, these studies also conclude that observed decreases cannot be explained by climatic factors alone.

The reasons why water planning in Spain has – so far – disregarded the potential impacts on water availability resulting from land conversions, and particularly from changes in the forest cover, might be diverse. On one hand, knowledge of forest water requirements is scarce, as oppose to the extensive research carried out within agriculture, probably because commercial forestry in Spain represents a small economic sector (less than 3% of the GDP) and it is mainly concentrated in water-abundant basins of northern Spain with few water conflicts. On the other hand, conventional approaches to water planning have mostly focused on managing the demand side, securing blue water availability by adopting structural approaches and constructing dams and large infrastructures. Only recently, the tight nexus existing between land and water is becoming more understood; and the fact that managing water in relation to land might not alleviate the high blue water demand among competing users, but it can contribute to optimizing the supply of blue water. Foremost, because the largest fraction of the annual rainfall (over 80%) in semiarid regions like Spain turns into soil moisture, and the way we manage this green water can make a difference in the fraction of available runoff downstream.

Managing water in connection with land requires a deep understanding about the relationship existing between landscape structure and rainfall partition. Usually, forests with a thick vegetation cover and well developed soils have a larger soil water holding capacity; thus, the largest fraction of annual rainfall is evapotranspired. On the contrary, short grass vegetation like pastures, have less access to soil moisture and generate larger flows of runoff. Nevertheless, rainfall partition varies greatly across landscapes, depending on the type of vegetation cover, land management, climate and soil characteristics.

The European Water Framework Directive (WFD) has made a huge effort in order to adopt a broader perspective in relation to water management, by trying to conciliate the use of blue water by different stakeholders with the protection of aquatic systems, in an attempt to reverse and prevent their further deterioration. However, the WFD still needs to further expand its approach, since it has poorly addressed the interrelations between land and water-related processes. Accordingly, the main goal of this chapter is to provide a first estimation of what are the water requirements (evapotranspiration rates) of major forest types, including woods, pastures, agroecosystems (mainly *dehesas*) and shrubs across the national territory. A further objective is to assess how droughts and changes in the forest cover could be altering forest evapotranspiration and consequently streamflow availability. Gaining understanding on this front can help to lay the foundations for moving towards an *Integrated Water and Land Resources Management* (IWLRM) approach (Falkenmark & Rockström, 2004) in Spain.

2 DATA AND METHODS

The water requirements of Spanish forests were estimated following Zhang *et al.* (2001), who argue that forests evapotranspiration is controlled by climate but also by the capacity of the different species to pump water from the unsaturated zone. This capacity depends, among other factors, on the physiological conditions of the plant species and their root depth. Using experimental data from over 300 studies, Zhang and colleagues found a semi-empirical relationship to estimate annual forest water consumption, expressed as:

$$\frac{ET_i}{PP_i} = \left(\frac{1 + w_i \times \frac{E_{oi}}{PP_i}}{(1 + w_i \times \frac{E_{oi}}{PP_i}) + (\frac{PP_i}{E_{oi}})} \right) \tag{1}$$

where ET represents the annual evapotranspiration rate of forest i (mm); PP and E_o are respectively, the annual rainfall and potential evapotranspiration of the area where forest i grows (mm); and w is an adimensional indicator called *water availability coefficient*, equivalent to the crop water coefficient, whose value varies from species to species depending on its physiology and plant architecture.

To estimate w for Spanish forests, we conducted a literature review on experimental studies where actual evapotranspiration rates were measured for different forest types across Spain and/or in areas with similar climate conditions (see Table 1). Based on this information, we used Equation (1) to estimate the w parameter for the different forest types present in Spain.

Information regarding Spanish forest area and species composition was obtained from the *Forest National Inventory* (FNI) (MAGRAMA, 2012a). Specifically, we used the data provided by the third FNI, which represents the most updated source of information on forest status in Spain (period 1996–2006), to quantify the actual rate of forest water evapotranspiration.

To account for the variations that changes in the forest cover over the last years could have had in the water balance of forests, we used the information of the second FNI to estimate the forest evapotranspiration for time period 1986–1996. Variations in forests' water consumption observed across both periods (period 1986–1996 against 1997–2007) were quantified and described at provincial scale.

Annual ET for each forest type was calculated using mean annual values of PP and E_o obtained from the Integrated Water Information System (*Sistema Integrado de Información del Agua*, SIA) for the time period 1980–2008. We used mean values instead of year to year data because we wanted to isolate at first the effect of climate variability on the forest water accounting. In doing so, changes in the water consumption observed across time will be due to changes in the coverage and area surface of forest and not due to the intrinsic variability of the Mediterranean climate. Additionally, in order to estimate the impacts of droughts on water availability downstream, we also estimated the actual forest's water consumption using the E_o and PP annual values of a particularly dry year (2005).

Table 1 Data sources used to estimate the water availability coefficient (w) for the most representative vegetation species found in Spain. Precipitation (PP), Potential evapotranspiration (Eo) and actual evapotranspiration (ET) represent mean annual values and (standard deviation) from the different observations recorded for each species.

Forest species	Number of observations	Year of measurement	PP (mm/year)	E_o (mm/year)	ET (mm/year)	Source	Region – Country
Fagus sylvatica	3	2000, 2001	817 (54)	528 (14)	509 (9)	Fernández et al. (2006); Dalsgaard et al. (2011)	Galicia (Spain); Denmark
Pinus halepensis	2	1997	454 (0.1)	1,160	332 (37)	Bellot et al. (2001)	Alicante (Spain)
Pinus nigra	1	1989	650	1,209	559	Domingo et al. (1994)	Almería (Spain)
Pinus pinea	2	1961–1990	796 (47)	894 (35)	451 (100)	Willaarts (2010)	Seville (Spain)
Pinus ssp	3	1961–1990	544 (131)	678 (231)	277 (149)	Willaarts (2010)	Seville (Spain)
Pinus sylvestris	2	2005	682 (188)	654 (63)	328 (8)	Vincke & Thery (2008)	Belgium
Quercus robur/Q. petraea	1	1989	965	665	588	Frank & Inouye (1994)	Central Europe
Quercus pyrenaical/Q. pubescens	2	1991–1993	932 (218)	968 (22)	516 (13)	Moreno et al. (1996)	Salamanca (Spain)
Quercus ilex	11	1961–1999	596 (155)	942 (54)	515 (79)	Piñol et al. (1999); Joffrey & Rambal (1993);Willaarts (2010)	Seville and Barcelona (Spain)
Quercus robur	3	1992–1994	914 (53)	510 (66)	447 (18)	Frank & Inouye (1994)	Denmark
Quercus suber/Q. ilex/Q. faginea	5	2000–2007	579 (178)	782 (17)	482 (45)	Pons (2009)	Valencia (Spain)
Abies Alba	16	2001–2002	946 (203)	607 (46)	560 (65)	Vilhar et al. (2005)	Slovenia
Dehesas with shrubs (Quercus ssp)	1	1961–1990	910 (86)	851 (51)	512 (136)	Willaarts (2010)	Seville (Spain)
Dehesas (Quercus ssp)	1	1961–1990	895 (108)	882 (60)	478 (102)	Willaarts (2010)	Seville (Spain)
Eucalyptus spp.	4	1961–1990	1,364 (405)	1,035 (141)	839 (177)	Van Lill et al. (1980); Bosch & Hewlett (1982); Fernández et al. (2006); Jiménez et al. (2007)	South Africa; Australia; Galicia and Seville (Spain)
Mediterranean shrubs (Quercus coccifera; Pistacia Lentiscus; Erica multiflora; Stipa tenacissima)	4	1961–2001	695 (246)	866 (194)	384 (159)	Lewis (1968); Bellot et al. (2001); Willaarts (2010)	USA;Alicante and Seville (Spain)
Pastures	3	1984–1986, 1997	699 (174)	966 (81)	430 (90)	Joffrey & Rambal (1993); Bellot et al. (2001); Willaarts (2010)	Seville and Alicante (Spain)

3 RESULTS

3.1 Forest water requirements

Table 2 summarizes the water availability coefficients (w) obtained for the major forest species found in Spain. Overall tree species have a greater w compared to shrubs, *dehesas* and pastures. This is mainly because trees and shrubs have greater amount of biomass stored per unit of space and larger root systems than pastures, which increases their water demand and access. The range of w values estimated for Spanish forests are in agreement with those obtained by Zhang *et al.* (2001) for forest and pastures elsewhere.

Figure 1 describes the implication different w coefficients have in terms of water consumption for three main typologies of forests types. In the most humid regions of Spain (aridity index close to 1), *Quercus* forests, *dehesas* and pastures have similar evapotranspiration rates (*ET/PP*). Foremost, because under these climatic conditions radiation as oppose to precipitation is the limiting factor influencing evapotranspiration. However, as we move in to the warmer and drier south of Spain, precipitation becomes the limiting factor. In these regions, the *ET/PP* ratio (covering other factors such as soil properties or topography constant) will be determined by the capacity each forest type has to access water (w). For instance, in areas where the aridity index is close to 2.5 (e.g. Valencia), *Quercus ilex* consumes approximately 90% of the total annual precipitation. However, under the same climate conditions, pastures evapotranspire only 60% of the annual rainfall.

Table 2 Water availability coefficients (w) obtained for different forest types and species found in Spain.

Vegetation type	Composition	w
Forest	*Fagus sylvatica*	2.4
	Abies alba	2.0
	Quercus suber/Quercus faginea	2.0
	Quercus ilex	1.9
	Quercus robur/Quercus petraea	1.8
	Quercus pyrenaica/Quercus pubescens	
	Eucalyptus spp.	1.5
	Pinus halepensis	1.2
	Pinus nigra	
	Pinus pinea	
	Pinus ssp	
	Pinus sylvestris	
Shrubs	Mediterranean (*Quercus coccifera; Pistacia Lentiscus; Erica multiflora; Stipa tenacissima*)	0.7
Dehesa (Agroecosystems)	*Quercus ssp* with annual pastures and Mediterranean shrubs	0.4
	Quercus ssp with annual pastures	0.3
Pastures	Annual	0.1

Source: Own elaboration.

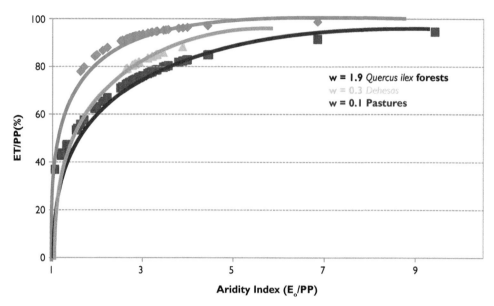

Figure I Mean annual evapotranspiration rates (expressed as % of annual rainfall) of three representative Spanish vegetation types. Observations refer to the *ET/PP* ratio calculated for the different Spanish provinces where each one of the three vegetation types grow. (Source: Own elaboration).

Table 3 summarizes the mean annual evapotranspiration rates of Spanish forests at present. Conifer, broadleaf, afforested and mixed forests consume together almost 75,000 hm³ [hm³ = cubic hectometre = million m³ = 10^6 m³], which represents 24% of the Spanish mean annual precipitation (approximately 318,000 hm³). Pastures are the second largest water consumers (32,000 hm³/year) as they occupy almost 9 Mha. Shrubs and *dehesas* occupy around 3.7 Mha and consume approximately 16,000 hm³, 13% of the national forest consumption.

Tree species have also different average water requirements (Table 3). Non-native species such as *Eucalyptus ssp*, frequently used in commercial and afforestation programs, evapotranspire larger amounts of water (over 5,400 m³/ha/year) compared to native *Quercus* or *Pinus* species, which consume less than 5,000 m³/ha/year. These results have large implications from a water planning perspective, especially in water-scarce areas within Spain, where several reforestation programs are being subsidized in previously used rain-fed agricultural areas as a climate change mitigation measure. Such afforestation programs have often been regarded as a cost-effective way to simultaneously sequester carbon, increase wood and paper supplies and diversify rural incomes, especially in the developing world (Vertessy, 2001). However, in the Spanish context these reforestation programs might contribute to sequester CO_2 emissions, but large scale programs undertaken with non-native species might have important trade-offs in terms of water availability downstream.

Climate is a major driver of forest's distribution across Spain. Forest species extending along the Mediterranean arc of Spain (e.g. *Quercus ilex*, *Q. faginea*, *Q. suber* or *Pinus halepensis*) are well adapted to dry conditions and therefore have smaller aver-

Table 3 National accounting of water consumed by Spanish forest types. Species composition and actual surface area was obtained from the third FNI (MAGRAMA, 2012a).

Vegetation type	Composition	Mha	m³/ha/year	SD[1]	hm³/year
Conifer forest	Pinus halepensis	1.9	4,370	756	7,949
	Pinus pinaster	1.2	4,581	761	5,708
	Pinus sylvestris	1.0	4,728	759	4,726
	Pinus nigra	0.7	4,461	801	2,916
	Pinus pinea	0.4	4,497	473	1,624
	Pinus ssp	1.3	4,685	873	6,012
	Abies & Juniperus ssp	0.4	4,846	690	1,879
Afforestation	Pinus ssp	0.1	4,451	1,218	238
	Eucaliptus ssp	0.6	5,480	750	3,155
Broadleaf forest	Quercus ilex	2.6	4,847	738	12,683
	Quercus pyrenaica	0.7	5,076	753	3,564
	Fagus sylvatica	0.4	6,044	718	2,747
	Quercus suber	0.3	5,602	448	1,688
	Quercus robur	0.2	5,870	620	1,080
	Quercus faginea	0.2	4,721	787	687
	Other broadleaf species	1.1	4,938	957	6,119
Dehesas (Agroecosystems)	Quercus ssp	2.4	4,175	479	10.225
Mixed forest	Quercus & Pinus ssp	0.7	4,695	883	3,338
	Pinus & broadleaf ssp	0.4	5,218	1,131	2,093
	Mix of broadleaf ssp	<0.1	5,569	676	207
Shrubs	Shrubs	1.3	4,307	728	5,495
Pastures	Pastures	8.9	3,693	604	32,105
TOTAL		**27.5**			**122,604**

Source: Own elaboration.

[1] SD = standard deviation; [hm³ = cubic hectometre = million m³ = 10⁶ m³].

age evapotranspiration rates (~4,800 m³/ha/year) compared to typical Atlantic species (e.g. *Fagus sylvatica, Quercus robur, Q. pyrenaica* or *Abies ssp*), which consume over 5,400 m³/ha/year. *Dehesas*, pastures and shrubs have lower water requirements as their biomass content is smaller (≤4,000 m³/ha/year).

From a hydrological perspective, the differences found across species water requirements have important implications. The presence of Atlantic forests with high water requirements in the northern basins does not necessarily have negative implications from a water availability perspective. These basins are water abundant and reducing their forest cover could have rather worse than positive impacts. Foremost, because these forests play a critical role regulating the baseflow and surface runoff within catchments and the removal of forest cover will disrupt this regulating process, increasing the risk of floods. Mediterranean forests perform similar ecological functions, but they extend in basins where competition for water is much higher. In this part of Spain as in other Mediterranean areas of southern Europe there is a growing problem of land abandonment (Lasanta *et al.*, 2009). Official statistics have reported a decrease in the agricultural area of 1.5 Mha since the mid 1990s (MAGRAMA, 2012b).

This reduction is mainly related to the low productivity of much of this agricultural land, which has either been converted into other uses (e.g. urban) or abandoned. This process of land abandonment entails important hydrological implications as it encompasses an increase in the vegetation cover (mainly serial shrubs and forest regrowth), which augments the vegetation water demand and reduces water availability downstream (Otero et al., 2010; Willaarts et al., 2012). From an ecological perspective no real positive effects result from the abandonment and homogenization of the agrarian Mediterranean landscape, as it increases the risks of fires, reduces the biological diversity and lowers the efficiency of forests as carbon sinks as they age, particularly in *dehesas*.

As Table 3 shows, Spanish forests currently evapotranspire approximately 122,000 hm³/year. Agriculture alone consumes around 30,000–33,000 hm³ (see Figure 1, Chapter 6), evidencing that forests consume the largest amount of annual rainfall in Spain. This is not to say that forest cover should be reduced to increase water availability downstream. On the contrary, forests play a critical role regulating the water cycle in catchments and supplying multiple ecosystem services, e.g. by preventing floods and regulating the micro and meso-climate. However, these results provide evidence of the importance of placing forest and land management at the core of hydrological planning. Even though much of the water conflicts in water stress basins are frequently caused by poor demand management (e.g. over-allocation of water rights, low efficiency, etc.), managing forests (e.g. by preventing shrub encroachment and forest aging, and maintaining a proper forest cover) can be part of the solution in water stress basins, since it might contribute to optimize the supply of water for downstream uses. The former results also raise an important issue: the existing trade-offs between climate change mitigation options (e.g. afforestation programmes) and water supply optimization options in catchments.

3.2   Changes in forest water consumption under dry conditions

Table 4 summarizes the estimated changes in forest water consumption in a dry year. Under dry conditions like those experienced in Spain in 2005, forest evapotranspiration at national level was estimated to be 20,000 hm³ lower than under mean climatic conditions. However, the ratio of forest water evapotranspiration to precipitation increased by 3%, which means an equivalent reduction in runoff availability. This cutback mainly occurs because most of the scarce rainfall infiltrated in soils beneath forests is used for the most part to satisfy the demand of the stressed vegetation cover and only a small fraction during the wetter months exceeds the soil's field capacity, feeding rivers and aquifers.

From the water planning perspective the previous results have great implications for securing blue water availability (see Figure 2). Under dry conditions the largest reduction in blue water availability resulting from greater forest water consumption occurs in arid and semi-arid regions with an aridity index closer to 2 (e.g. Extremadura, Castille-La Mancha and Andalusia). These results evidenced that impacts on water availability under dry conditions without proper forest management are greater in the naturally water-scarcer regions.

Table 4 Annual national forest water consumption (hm³) under dry and mean average climatic conditions.

	Mean condition (period 1980–2008)	Dry year (2005)
Forest evapotranspiration (ET)	122,604	99,251
Precipitation (PP)	318,124	237,156
ET/PP (%)	39	42

Source: Own elaboration.

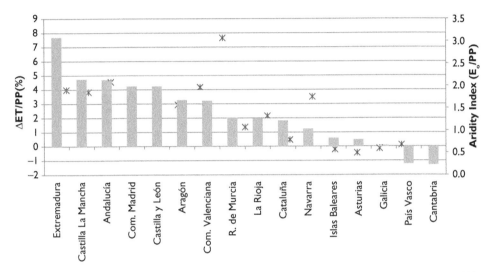

Figure 2 Changes in the ratio of annual forest evapotranspiration (ET/PP) comparing mean (1980–2008) and dry annual conditions (year 2005) (grey bars). The aridity index (red dots) represents the mean annual ratio of E_o to PP for the time period 1980–2000. (Source: Own elaboration).

3.3 Changes in forest water consumption due to land cover changes

According to the data provided by the second and third FNI, between 1986 and 2006 the forest area in Spain increased approximately by 1.5 Mha (Table 5). Such forest expansion has augmented forest's annual ET by almost 10,500 hm³. If mean annual rainfall in Spain is around 318,000 hm³, the percentage of precipitation annually consumed by Spanish forests has increased from 35 to 39%.

The largest increases in forest cover and water consumption have occurred mainly in the southern half of the country, in the regions of Extremadura, Andalusia and Castille-La Mancha (see Figure 3). Important increases of the forest water consumption have occurred also in the northern regions of Asturias and Galicia, where large commercial forest programmes are well established. Overall, it is noteworthy that almost 50% of the observed increase in the water consumed by forests has taken place in arid and semi-arid provinces (aridity index > 2.5).

Table 5 Changes in forest water consumption between 1996 and 2007.

Vegetation type	Area 1996 (Mha)	Area 2007 (Mha)	Water requirements (m³/ha/year)	ET 1996 (hm³/year)	ET 2006 (hm³/year)
Forests	9.8	14.9	5,074	49,882	74,779
Dehesas	2.3	2.4	4,175	9,511	10,225
Shrubs	2.1	1.3	4,307	9,047	5,495
Pastures	11.8	8.9	3,693	43,482	32,105
Total	**26.0**	**27.5**		**111,922**	**122,604**

Source: Own elaboration.

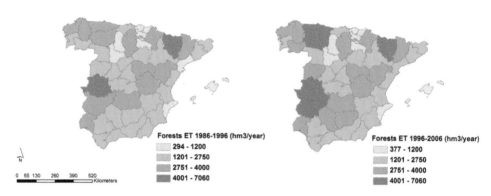

Figure 3 Trends in annual forest's ET (hm³) in Spain during the last three decades. (Source: Own elaboration).

Despite the importance of these figures, the former results need to be interpreted with caution as the uncertainty surrounding the changes in forest surface area in Spain during the last 30 years is very high. Other official land use data sources like the Corine Land Cover Project (EEA, 1990; 2005) or the Crops and Land Use map created by the Ministry of Agriculture (*Mapa de Cultivos y Aprovechamientos*) (MAGRAMA, 2012b) report different land use and land cover trends compared to the FNI datasets. A detailed and contrasted assessment between the different land use sources needs to be carried out to obtain a better and clearer picture of what has really happened within the Spanish territory in terms of land use and land cover flows over recent decades.

4   CONCLUSIONS

Forests consume the largest amount of annual rainfall in Spain, above agriculture, and this highlights the crucial role they play in the water balance. Currently, Spanish forests evapotranspire 39% of the annual incoming rainfall, although during droughts this ratio can rise up to 42%. This increase in water consumption implies an equivalent reduction in runoff availability downstream. But above climate, land use changes and particularly the increase of forest cover linked to land abandonment and afforestation programmes seem to have a greater impact on water resources in Spain, reducing blue water availability by up to 4% annually. The magnitude of these reductions is especially relevant in the most water-stressed regions. Nevertheless, a contrasted assessment is needed to ascertain the real scale of land use and land cover changes across Spain to accurately judge the impacts of changes in forests area on water resources. Even though efforts to use water more efficiently need to be placed on managing the demand better, integrating land and forest management into water planning might be a cost-effective solution to optimize the supply of water in catchments. Such realization questions the potential negative *co-effects* of afforestation programs, currently being undertaken as a climate change mitigation option, from a water perspective.

REFERENCES

Beguería, S.; López-Moreno, J.I.; Lorente, A.; Seeger, M. & García-Ruiz, J.M. (2003). Assessing the effect of climate oscillations and land-use changes on streamflow in the central Spanish Pyrenees. *Ambio*, 32: 283–286.

Bellot, J.; Bonet, A.; Sánchez, J.R. & Chirino, E. (2001). Likely effects of land use changes on the runoff and aquifer recharge in a semiarid landscape using a hydrological model. *Landscape and Urban Planning*, 778: 1–13.

Bosch, J.M. & Hewlett, J.D. (1982). A review of catchment experiments to determine the effect of vegetation changes on water yield and evapotranspiration. *Journal of Hydrology*, 55: 3–23.

Dalsgaard, L.; Mikkelsen, T.N. & Bastrup-Birk, A. (2011). Sap flow for beech (*Fagus sylvatica*) in a natural and managed forest – effect of spatial heterogeneity. *Journal of Plant Ecology*, 4: 23–35.

Domingo, F.; Puigdefabregas, J.; Moro, M.J. & Bellot, J. (1994). Role of vegetation cover in the biochemical balances of a small afforested catchment in southeastern Spain. *Journal of Hydrology*, 159: 275–289.

EEA (European Environment Agency) (1990). Corine Land Cover maps of Spain, year 1990. Available from: http://centrodedescargas.cnig.es/CentroDescargas/. [Accessed 24th February 2012].

EEA (European Environment Agency) (2005). Corine Land Cover maps of Spain, year 2005. Available from: http://centrodedescargas.cnig.es/CentroDescargas/. [Accessed 24th February 2012].

Falkenmark, M. & Rockström, J. (2004). *Balancing Water for Humans and Nature: The New Approach in Eco-hydrology*. EarthScan, London, UK.

Fernández, C.; Vega, J.A.; Gras, J.M. & Fonturbel, T. (2006). Changes in water yield after a sequence of perturbations and forest management practices in an *Eucalyptus globulus* Labill. Watershed in northern Spain. *Forest Ecology and Management*, 234: 275–281.

Frank, D.A. & Inouye, R.S. (1994). Temporal variation in actual evapotranspiration of terrestrial ecosystems: patterns and ecological implications. *Journal of Biogeography*, 21: 401–411.

Hill, J.; Stellmes, M.; Udelhoven, Th.; Röder, A. & Sommer, S. (2008). Mediterranean desertification and land degradation. Mapping related land use change syndromes based on satellite observations. *Global and Planetary Change*, 64: 146–157.

Gallart, F. & Llorens, P. (2003). Catchment management under environmental change: Impacts of land cover change on water resources. *Water International*, 28: 334–340.

Gallart, F.; Delgado, J.; Beatson, S.J.V. *et al.* (2011). Analysing the effect of global change on the historical trends of water resources in the headwaters of the Llobregat and Ter river basins (Catalonia, Spain). *Physics and Chemistry of the Earth*, A/B/C, 36: 655–661.

Gordon, L.J.; Steffen, W.; Jonsson, B.F. *et al.* (2005). Human modification of global water vapor flows from the land surface. *Proceedings of the National Academy of Sciences USA*, 102: 7612–7617.

Jackson, R.B.; Jobbágy, E.B.; Avissar, R. *et al.* (2005). Trading water for carbon and with biological carbon sequestration. *Science*, 310: 1944–1947.

Jiménez, E.; Vega, J.A.; Pérez Gorostiaga, P. *et al.* (2007). Evaluación de la transpiración de *E. globulus* mediante la densidad de flujo de savia y su relación con variables meteorológicas y dendrométricas [Transpiration assessment of *E. globulus* through sap flow density analysis and its relation with meteorological and dendometric variables]. *Boletín del CIDEU [CIDEU Bulletin]*, 3: 119–138.

Joffrey, R. & Rambal, S. (1993). How tree cover influences the water balance of Mediterranean rangelands. *Ecology*, 74: 570–582.

Lasanta, T.; Arnáez, J.; Errea, M.P.; Ortigosa, L. & Ruiz-Flaño, P. (2009). Mountain pastures, environmental degradation and landscape remediation, the example of a Mediterranean Policy initiative. *Applied Geography*, 29: 308–319.

Lewis, D.C. (1968). Annual hydrologic response to watershed conversion from oak woodland to annual grassland. *Water Resources Reserarch*, 4: 59–72.

Lorenzo-Lacruz, J.; Vicente-Serrano, S.M.; López-Moreno, J.I.; Morán-Tejeda, E. & Zabalza, J. (2012). Recent trends in Iberian streamflows (1945–2005). *Journal of Hydrology*, 414–415: 463–475.

MAGRAMA (Ministerio de Agricultura, Alimentación y Medio Ambiente) (2012a). *Datos de superficies del II y III Inventario Forestal Nacional* [Second and third Forest National Inventory datasets]. Available from: http://www.magrama.gob.es/es/biodiversidad/temas/inventarios-nacionales/inventario-forestal-nacional/ [Accessed 24th February 2012].

MAGRAMA (Ministerio de Agricultura, Alimentación y Medio Ambiente) (2012b). *Mapa de Cultivos y Aprovechamientos de España escala 1:50.000. Periodo 1980–1990 y 2000–2010.* [Crops and Land Use map of Spain, scale 1:50,000, time period 1980–1990; 2000–2010]. Available from: http://www.magrama.gob.es/es/cartografia-y-sig/temas/sistema-de-informacion-geografico-de-datos-agrarios-siga-/mca.aspx [Accessed 24th February 2012].

Moreno, G.; Gallardo, J.F.; Schneider, K. & Ingelmo, F. (1996). Water and bioelement fluxes in four Quercus pyrenaica forest along a pluviometric gradient. *Annals of Forest Science*, 53: 625–639.

Otero, I.; Boada, M.; Badía, A.; Pla, E. *et al.* (2010). Loss of water availability and stream biodiversity under land abandonment and climate change in a Mediterranean catchment (Olzinelles, NE Spain). *Land Use Policy*, 28: 207–218.

Piñol, J.; Ávila, A. & Escarré, A. (1999). Water balance in catchments. In: Rodà, F.; Retana, J.; Gracia, C.A. & Bellot, J. (eds.), *Ecology of Mediterranean evergreen oak forests*. Springer-Verlag, Berlin, Germany: 273–282.

Pons, J. (2009). *Cork Oak Regeneration: An approach based on species interactions at landscape scale*. PhD Thesis, Universidad de Alicante, Spain.

Scanlon, B.R.; Jolly, I.; Sophocleous, M. & Zhang, L. (2007). Global impacts of conversions from natural to agricultural ecosystems on water resources: quantity versus quality. *Water Resources Research*, 43.

Trabucco, A.; Zomer, R.J.; Bossio, D.A.; van Straaten, O. & Verchot, L.V. (2008). Climate change mitigation through afforestation/reforestation: a global analysis of hydrologic impacts with four case studies. *Agriculture, Ecosystems and Environment*, 126: 81–97.

Van Lill, W.S.; Kruger, F.J. & Van Wyk, D.B. (1980). The effect of afforestation with *Eucalyptus grandis* Hill ex Maiden and *Pinus patula* Schlecht. et Cham. on streamflow from experimental catchments at Mokobulaan, Transvaal. *Journal of Hydrology*, 48: 107–118.

Vertessy, R.A. (2001). Impacts of plantation forestry on catchment runoff. In: Nambiar. E.K.S. & Brown, A.G. (eds.), *Plantations, Farm Forestry, and Water*. Proceedings of a National Workshop, Melbourne, 21–22 July 2000. Rural Industries Research and Development Corporation Report No. 01/20: 9–19.

Vilhar, U.; Starr, M. & Urbancic, M. (2005). Gap evapotranspiration and drainage fluxes in a managed and a virgin dinaric silver fir-beech forest in Slovenia: a modelling study. *European Journal of Forest Research*, 124: 165–175.

Vincke, C. & Thery, Y. (2008). Water table is a relevant source for water uptake by a Scots pine (*Pinus sylvestris* L.) stand: evidence from continuous evapotranspiration and water table monitoring. *Agricultural and Forest Meteorology*, 148: 1419–1432.

Vörösmarty, C.J.; Green, P.; Salisbury, J. & Lammers, R.B. (2000). Global water resources: vulnerability from climate change and population growth. *Science*, 289: 284–288.

Willaarts, B.A. (2010). *Dinámica del paisaje en Sierra Norte de Sevilla. Cambios funcionales e implicaciones en el suministro de servicios de los ecosistemas. [Land Use Trends in Sierra Norte de Sevilla (Spain). Functional changes and ecosystem services trade-offs]* PhD Thesis, Universidad de Almería, Spain.

Willaarts, B.A.; Volk, M. & Aguilera, P. (2012). Assessing the ecosystem services supplied by freshwater flows in Mediterranean agroecosystems. *Agricultural Water Management,* 105: 21–31.

Zhang, L.; Dawes, W.R. & Walter, G.R. (2001). Response of mean annual evapotranspiration to vegetation changes at catchment scale. *Water Resources Research*, 37: 701–708.

Chapter 12

The challenges of agricultural diffuse pollution

Emilio Custodio[1], Alberto Garrido[2],
Carmen Coleto[3] & Gloria Salmoral[2]
[1] Department of Geotechnical Engineering and Geo-Sciences,
  Technical University of Catalonia, Barcelona, Spain
[2] Water Observatory of the Botín Foundation;
  CEIGRAM, Technical University of Madrid, Madrid, Spain
[3] General Directorate for Water, Ministry of Agriculture,
  Food and Environment, Madrid, Spain

ABSTRACT: Agricultural diffuse pollution is of concern in Spain, as it is in most European countries. The mountainous nature of the Iberian Peninsula and the Spanish islands explain the coexistence of areas little affected by nitrate with others in which this pollution is intense, especially where there is intensive cropping and livestock production, mostly pig farms. The use of chemicals in agriculture and intensive livestock production do not differ from other European countries, although the different climate and soil conditions explain differences in fertilizers and pesticides application rate and chemicals used. The enforcement of different European Directives poses specific technical and also administrative coordination problems due to the different authorities responsible for water resources and the environment at national and regional level. Many aquifers and the related springs and surface watercourses will not reach the good quantitative and chemical status by 2015 required by the European Water Framework Directive. This means that new terms have to be negotiated, with action that needs to be well defined in technical, economic, social and administrative terms. Most of the work carried out concerns nitrate pollution, vulnerable zones and good agricultural practices, but there is also saline pollution due to the evapoconcentration effect when irrigation water has originally high solute contents, something that is not rare in Spain. Pesticides and emergent pollutants in groundwater and the related surface water are being studied, but results are still in the early stages; their presence may become a serious future challenge for water supply and care of the environment.

Keywords: agricultural diffuse pollution, nitrate, European Water Framework Directive, good groundwater chemical and quantitative status, vulnerable zones

1 INTRODUCTION

The total agricultural land surface area in Spain is about 20 Mha [Mha = million hectares = 10^6 ha] (40% of the territory), of which close to 3.4 Mha are irrigated. About 70% to 75% of the total water consumption is for crop irrigation (see Chapter 6).

Water planning has traditionally focused on quantity issues, with reduced attention to water quality, a serious affair in some rivers and aquifers. Pollutants reach water bodies from point and non-point sources. Non-point sources are mainly due to agriculture, livestock and the related airborne contaminants, which cause diffuse pollution, mostly nitrate excess from fertilizers and in some cases salinity increase from return irrigation flows. This pollution has environmental and social consequences and often a significant associated cost to users.

The effect of agriculture and animal breeding on water quality is the subject of the European Union (EU) Nitrate Directive (ND) (OJEU, 1991). After previous efforts from water and environmental administrations in several European countries, a final decisive push to consider water quality and the environmental implications of pollution came from the Water Framework Directive (WFD) (OJEU, 2000), which is developed and complemented in what refers to groundwater by the *Daughter* Groundwater Directive (GWD) (OJEU, 2006). These Directives have already been transposed into the Spanish water law.

Spain is politically and administratively divided into 17 Autonomous Communities, here called Regions. They are responsible for their territorial resources and environment. Besides, for water management, 16 River Basins Districts have been defined, of which 7 are inside a particular region and depend on each region's own Administration, and the other 9 comprise territories over two or more regions and are under the responsibility of the Government of Spain for water planning, although the regional Governments have their shared responsibility inside their territory in relation to water quality and the environment.

Spain has a varied geology (which includes soluble salt formations), is mountainous, has a long coastline, numerous aquifers (many of them of small size, although thick and yielding), and climatic conditions vary from sub-humid to semiarid and even arid. Thus very different recharge and natural quality conditions can be found in the aquifers. In general terms, groundwater, and consequently springs and river base flow, have moderate to good quality for most uses, although naturally brackish or saline water can be found, and in some cases inconvenient solutes may appear, such as F excess and the presence of Fe and Mn, rarely of As and B. Large areas of river headwaters and aquifer recharge are in the highlands, which are little affected by agriculture and feedstock breeding.

This chapter presents an overview of non-point (diffuse) source pollution in Spain, with emphasis on agrochemicals. The nitrate monitoring network and recommended codes of best management practices are considered. Finally, the current economic, social and management issues relating to non-point pollution are briefly commented. Only general references are given. Some statements come from the direct experience of the authors.

2   USE OF AGROCHEMICALS IN SPAIN

Agricultural fertilizers are widely used in Spain, in both rainfed and irrigated areas. The common nutrient elements (N, P and K) are applied in different forms. Since 1998 the total fertilizer consumption shows a decreasing trend (Figure 1), as well as nutrient consumption per unit of crop production value (Figure 2). Improved technology, the increasing cost of fertilizers and in their application, and more strict farming rules help in reducing fertilizer application, especially when plots are large or there is cooperative action. Soil correctors are applied in sandy soils. The production of manure from intensive livestock is about 75 Mt/year [Mt = million tonnes = 10^9 kg], although only a fraction is used as fertilizer.

Average N application as manure and in inorganic form over the total agricultural area of Spain is about 40 to 60 kg/ha/year. In rainfed areas average values are about 10 kg N/ha/year, up to 60 kg N/ha/year. N application is much larger in irrigated fields: average values increased from about 60 kg N/ha/year in 1950 to a maximum of 360 kg N/ha/year around 1985, and in year 2005 decreased to about 250 kg N/ha/year. According to the Spanish nitrogen balance in the year 2008 (MARM, 2010), average doses in kg N/ha/year were 350 for flowers and citrus, 280 for vegetables and potatoes, 220 for industrial crops, 90 for cereals and leguminous plants, and 50 for vineyards and olives, although actual values vary over a large range.

Pesticides are widely used. The largest consumption over the period 1995–2009 took place between 1999 and 2005 (MARM, 2011). The basic components are variable and diverse, and change from period to period and according to the area. Methyl bromide was widely applied in sandy soils before plantation of some intensive crops but since the 1990s it has been banned.

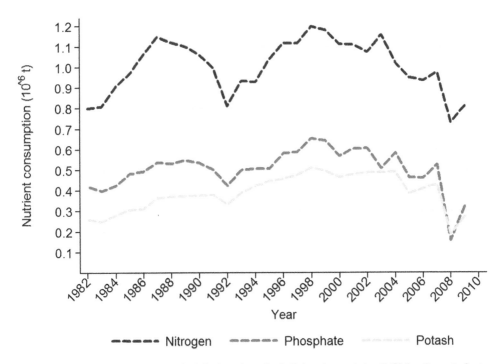

Figure 1 Consumption of nitrogen (as N), phosphate (as P_2O_5) and potash (as K_2O) fertilizers in Spain. Period 1982–2009. (Source: IFA (2011)).

Figure 2 Ratio between nutrient consumption and crop production (kg/€). (Source: Own elaboration based MARM (2011) and IFA (2011)).

3   AGRICULTURAL DIFFUSE POLLUTION IN SPAIN

Diffuse agricultural pollution in Spain refers mostly to nitrate. In some areas irrigation return flows may be brackish and even saline due to application of relative high salinity irrigation water, especially in coastal areas. Diffuse pollution affects primarily groundwater, which is mainly considered here, but also surface and coastal waters through groundwater discharge and agricultural drainage. Besides, intensive aquifer use for irrigation produces hydrodynamic changes in the aquifer and in the related surface waters, which may enhance quasi-diffuse effects in some areas due to ground salt dissolution, enhanced mixing with saline or poor quality groundwater, large-scale seawater intrusion, and in some of the new irrigated arid areas the downward displacement of saline waters held in the unsaturated zone.

NO_3 is easily leached to groundwater, while ammonia, urea-based fertilizers and manure-N are less leachable. However they may be partly volatilized in relatively warm, acidic soils, and later on deposited into the soil over a wider area than the fertilized one. This has been noticed around the Doñana National Park, in south-eastern Spain.

Nitrate leaching is a concern in Spain, as it is in many other European countries. Although average application intensity is lower, nitrate is more unevenly distributed due to the varied climate, soils, depth to the water table, land elevation, crops, and fertilization practices. The average agricultural N excess (surplus) in Spain is estimated to be 20–25 kg/ha/year, varying from less than 25 to more than 150 kg/ha/ year (MARM, 2010). However local studies may show higher values. For example, in Valencia (eastern Spain) studies show leaching rates of 230–250 kg N/ha/year for open-air cultivated vegetables, whereas for orange trees they are 90–160 kg N/ha/ year. In many cases the N actually applied may be greater since nitrate in irrigation water is generally not accounted in Spain. In the European Union (EU) N average excess values are about 50 kg/ha/year, up to 125 kg/ha/year in Germany and 225 kg/ ha/year in the Netherlands (Cuevas Navas, 2008).

In arid areas, even low fertilization rates may lead to high NO_3 concentration in recharge; for example a recharge of 10 mm/year and a low N application of 10 kg/ ha/year, used by plants at 50% efficiency, leads to a concentration of about 280 mg/L NO_3 in recharge water. In some hot spots in Spain, more than 500 mg/L NO_3 are found in groundwater; they often coincide with areas intensively irrigated with local groundwater where the NO_3 has been progressively accumulating in the aquifer due to recycling (input exceeding output over a long time), often evolving towards a high steady state value.

Under common scenarios, nitrate behaves as a non-retarded, conservative solute, since the underground ambient is generally oxidizing, even for many of the thick carbonate aquifers in Spain. Nitrate reduction in the ground is a rare circumstance in most Spanish aquifers.

In some of the semiarid and arid areas, the combination of low recharge and deep water table means long transfer times of conservative solutes from the surface to the aquifer, from decades to centuries, even longer for solutes affected by retardation, as is the case for many pesticides. In some areas nitrate and especially pesticide pollution is still in the unsaturated zone, above the water table. Besides, deep wells in thick aquifers produce mixed groundwater, so arriving nitrate is diluted in the abstracted water and this delays the early recognition of diffuse pollution severity. However nitrate pollution is already a serious problem in some areas, forcing to look for alternative water sources (a difficult task in small islands like Menorca), or to mixing with other waters, or even membrane treatment, as in the area of Denia, in Valencia Region.

Little information is available on agricultural groundwater pollution by phosphorus. Most of the reported surface water contamination is due to household chemicals. Phosphates are highly retained in the widespread carbonate-rich soils and aquifers of Spain, so they do not tend to appear in springs and river baseflow in many areas. This explains the relative good trophic conditions of some groundwater-dependent lakes, reservoirs and wetlands, in spite of the high NO_3 content of agricultural origin in the groundwater that feeds them, when surface runoff contribution is low and livestock is sparse in the area. This is the case of the Lagunas de Ruidera Natural Park (lagoons in central-southern Spain) or S'Albufera wetland in north-eastern Mallorca Island.

The behaviour of pesticides and similar agrochemicals in soils and groundwater bodies is little known. This is due to patchy studies, variable behaviour in soils, changing active principles and poorly documented field application. The European Directive 2008/105/EC (OJEU, 2008) on environmental quality standards of priority substances gives limits on pesticide concentrations between 0.04 and 1.8 µg/L in inland surface water, depending on the substance, but they do not legally apply to groundwater. Some regions and river basin districts report that their waters are pesticide-free, but this refers mostly to surface waters. Modern herbicides have been reported in groundwater, in spite of their assumed high retention and fast decay in soil, such as glyphosphate. In recent years there have been events in which the pesticide content in reservoirs and aquifers have exceeded the quality standards and have caused closing down sources supplying water to many towns in Andalusia; residues of herbicides used in olive groves led to temporary cuts in supply of drinking water between 2001 and 2004 to 135 municipalities with a total population of about 2 million inhabitants. Many supply systems have had to install expensive activated carbon filters to address these pollution episodes (J. Corominas, pers. comm., 20th January 2012). Pharmaceuticals used for feedstock have been identified in some intensive pig raising areas of Catalonia (northeastern Spain), as sulphonamides in high NO_3 content groundwater. The importance of these organic compounds can become a major water quality problem in the future since they are difficult to monitor because of the frequent substitution and introduction of new compounds, and the slow transit time in the ground.

4   NITRATE MONITORING AND VULNERABLE ZONES

The water sampling monitoring networks provide data on NO_3 concentration in natural waters all over Spain. Results are available in the four-year reports requested by the Nitrate Directive (ND). Except in the north-western area, the 50 mg/L NO_3 is exceeded in about 30% of groundwater sampling points, or about 17% of the surface area, with a slight increasing trend. Surface water tends to be stable with respect to average and winter values, although maximum values tend to increase slightly (MARM, 2008). This is small compared with the average 40–45% of points above limits in the sampling points of the EU, but the values are similar when inhabited, flat areas are compared. The most deteriorated aquifers are along the Mediterranean coast, in Andalusia (southern Spain) and in the Canarian and Balearic archipelagos.

After the ND, Vulnerable Zones to Nitrate Pollution (VZN) due to agricultural activities have been officially declared in Spain, based on available information. This includes the identification of the affected waters, the formulation of codes of good agricultural practices, the designation of the vulnerable zones, the development of action programmes, and the analytical monitoring to be carried out. In total 57 VZN have been declared, which cover about 13% of the national territory. This will be commented on further in Section 6.

5   CODES OF GOOD AGRICULTURAL PRACTICES

In pursuance of the ND, each EU Member State has to produce and enforce codes on good agricultural practices, both for agriculture and livestock, in order to reduce nitrate losses to aquifers, especially in the VZN, and consequently into inland and coastal surface waters. This is a very sensitive issue due to the economic, social and political implications. In Spain this is the responsibility of the Regions, which are in charge of producing, enforcing and controlling the compliance. The preparation and publication of the codes has been slow.

These codes should take into account local circumstances and consider 1) a common upper limit of N application (from 150 to 170 kg/ha/year), 2) periods and areas in which fertilization is forbidden or subject to more strict limitations, 3) the convenience of maintaining a plant cover on the fields during the cold months, 4) limits to total use of fertilizers and the mode of application, 5) rules for a. calculating fertilization rates and methods, including fractionating their application; b. use of alternating crops; and c. performing soil analysis before fertilization; this allows for calculating maximum acceptable yearly irrigation depths and methods, 6) maximum animal density in extensive feedstock breeding, 7) required conditions against leakage and storage tank capacity to hold manure and liquid animal wastes, 8) obligation to keep a book on fertilizer use, and on production, storage, and distribution of animal wastes. Economic studies to help in fertilizer use and application are not included, except preliminary in Catalonia.

In the codes there are no explicit provisions and methods to account for the NO_3 already in the irrigation water, which may be significant in the case of groundwater in some areas. Taking this NO_3 into account will contribute to reducing nitrate losses and their accumulation by recycling, but this is not an easy task since the right timing for fertilization and water application usually do not coincide. The codes of La Rioja region (central-northern Spain), which has easy access to surface water, recommend abandoning the use of groundwater with high nitrate content and irrigate with surface water, which contains lower concentrations of NO_3. This measure is intended to alleviate the NO_3 concentrations in local aquifers and springs, as well as to save energy. What seems sound from the Region's point of view, is not so for the whole river basin, due to less good quality and the different time distribution of NO_3 available in downstream surface water.

High application of irrigation water may be beneficial when it comes from external sources since this helps to sustain aquifer recharge in intensively exploited aquifers, and may reduce the use of herbicides. This is the case of the fruit tree plantations in the Lower Llobregat river alluvial plain (Barcelona, north-eastern Spain). However, an integrated water balance is needed.

In some cases nitrate pollution is not strictly dependent on good agricultural practices, but on the convenience to apply out-of-season irrigation to limit frost damage to crops in critical days, but this leaches NO_3 already in the soil. This is the case of the Inca-Sa Pobla plain in eastern Mallorca Island. To reduce this leaching, the application of water as mist is advisable but this increases the operational cost to the farmers.

The application of the codes has encountered some practical pitfalls: 1) high cost and complex inspection and control, 2) difficulty in enforcing the guidelines when agricultural plots are small, or the plot owner is not the person who is cultivating it, or the application of the rules is rejected by farmers, 3) complaints by farmers on unfair treatment and over-charges, especially those inside the VZN with respect to neighbours just outside it, 4) poorly coordinated action of involved authorities and political foot-dragging.

6 SOCIAL, ECONOMIC AND MANAGEMENT ISSUES

Spain is responsible to the EU Commission for the enforcement of the Directives. The River Basin Districts produce 4-yearly reports on nitrate issues and yearly reports on groundwater. The EU monitors the application of the Directives and issue procedures of infringement to the Member States when regulations are not satisfactorily accomplished.

In relation to nitrate pollution, the designation of VZN is a complex task due not only to poor knowledge and data scarcity but also to the high economic and social costs involved, the reluctance of farmers, sometimes plain rejection. So, the number and surface area of VZN initially declared in Spain were short in some regions. In 2002 the European Commission filed an infringement procedure against Spain for breaching the Directive on Nitrates, which has been pending for almost nine years. It was required to increase nitrate monitoring due to agricultural sources and to review the number and boundaries of the VZN, as well as some of the programmes of action, including the small storage capacity of animal wastes and their excessive application. Through coordination between the Central Government and the regional administrations, in particular in Catalonia where there are many large pig farms, new vulnerable zones were designated and their programmes of action were modified. The result was accepted by the European Commission.

The slow evolution of groundwater makes difficult attaining the WFD good quantitative and chemical status of aquifers by the year 2015. To cope with this, the WFD allows for agreed delays, following the 6-year period between successive water plans (2021 and 2027), if supported by studies and accompanied by proposals for action. Several EU Members have already applied for such delays, but not yet Spain, although this is currently considered in some of the recent River Basin Management Plan drafts.

Not every Spanish River Basin Authority or responsible institution has enough knowledge, staff, economic and technical means for due action since this is costly and often limits their actual possibilities. In some cases this is worsened by excessive political interference and giving the responsibility to politically appointed, technically unskilled persons, since water is a very sensitive political and social issue (See Chapter 4). Water mobilizes people and especially farmers' lobbies, which are able to create high media pressure and may influence the polls, even if jobs in agriculture affect only quite a small fraction of the population. Currently the main issue is water quantity, although the concern on water quality is increasing.

As has been already mentioned, agricultural pollution appears long-delayed in wells, springs, streams and coastal waters. For this reason, the recovery of aquifer quality after a diffuse pollution source ceases or slows down often takes decades. This is well known by specialists but not by decision-makers, water administrators and the public in general, who often expect prompt and visible responses. Enforcing actions that are economically, socially and politically costly becomes difficult when the expected improvements will be noticed much later and sometimes with further deterioration over some time. This is already mentioned in some of the River Basin District Water Plan drafts of Spain.

The social cost of agricultural pollution is not known. Studies are rare and incomplete. In some areas, getting drinking water is difficult and suppliers have to mix local groundwater with waters of other origins, often imported from long distances, or they have to treat the NO_3 excess by means of expensive ion exchange or reverse osmosis plants. There is research on effective and cheap new treatments, but applicable results are not yet at hand. The related cost increase is fully transferred to the population, which is not directly responsible for the water quality deterioration. Some unpublished preliminary data tend to show that treatment costs are possibly one order of magnitude higher than the cost to reduce nitrate losses.

The involvement of water users and stakeholders is very important for water quality management, particularly for agriculture, and this seems the best way to get positive results. Experience on users associations is just beginning, slow to build, and limited, although there is some encouraging and pioneering progress in Spain, mostly as a civil society initiative. About 20 groundwater users' associations already exist, as well as a national association linking them. They have a legal status recognized in the Water Act. Most current action refers to water quantity since farmers are still reluctant to consider quality issues.

Implementing the codes of good agricultural practices is not an easy task since this means an additional cost to farmers, who have to sell their production in highly competitive markets. The increased cost is not accompanied by an immediate benefit for them, apart from possible but unproved savings in agrochemicals and water use. This implementation is also poorly attractive for politicians since in many cases results will only appear after a much longer time than normal political life, and thus enforcement is not palatable for them. Giving subsidies for pollution abatement looks better, but is expensive and, if continued, may often lead to perverse results. This is a general issue that in some areas of Spain may be serious. However, since farmers have a wide margin of error in the rate of N application, some room for N savings are possible with little change in profitability. Some studies under way on the economics of nitrate pollution reduction and environmental improvement show that efficient irrigation and fertirrigation (incorporation of fertilizers to irritation water) may produce good results at low or no cost, and even with some benefits, but field-scale experience is needed. Reducing fertilizer use through taxes is rarely considered. Most efforts refer to irrigated agriculture, but rainfed farming, although less polluting, also produce significant diffuse pollution over a large territory in the semi-arid areas of Spain, but this is poorly documented.

For feedstock wastes, improved collection and storage is at a cost that greatly depends on local circumstances. There is some experience in the EU, with costs crudely evaluated at a few €/kg of N leaching reduction. The disposal of wastes from feedstock is a growing serious problem in Spain if nitrate load to groundwater has to be limited. The control by water authorities is poorly feasible in practice since they lack staff, resources and means. Then, important progress is needed to get the participation and co-responsibility of farmers, groundwater users and water stakeholders in the control and monitoring, and also to help water authorities to find solutions to existing problems and how to comply with legislation, but this needs time and continuous backing and help, and this is just starting. In the meanwhile a mix of enforcement of rules and compensations has to be applied in Spain. Policy makers and people should be aware that what is not compensated by those causing the problem, means passing the cost to third parties, such as neighbours, urban supply and future generations. How measures are applied affect the EU economy as a whole and the balance among Member States, and should not be at the expenses of socially inacceptable loss of natural resources, ecological services and inheritance. In Spain this message is still poorly internalized and politicians and media avoid dealing with it.

In difficult-to-solve situations there is the possibility of declaring some aquifers as too costly to be brought into good condition, and using them only for low water quality uses, but this should not cause damage to other water resources, coastal waters and the environment, nor maintain comparative temporary economic privileges with respect other areas. Good quality water has to be guaranteed for demanding uses, especially for drinking purposes, looking for other water sources or treating carefully low quality groundwater before distribution, as is currently done in the area of Barcelona. But this vision may be incomplete since the environment and the economic contribution provided by nature are not duly considered, nor is there equity in the distribution of the economic costs of pollution. The principle of *the polluter pays* needs to be reformulated and adapted to diffuse pollution by agriculture, according to actual circumstances and taking into account inherited situations. This is an ethical and moral issue still not adequately addressed in Spain.

7   CONCLUSIONS

Agricultural diffuse pollution in Spain is variable according to the diverse characteristics of the mainland and the islands. From aquifers, pollutants are transferred to springs and river base flow, as well as to littoral water. Most study and monitoring efforts refer to nitrate, which is of concern in many areas where the concentration limits are exceeded, though pesticides and some substances added to them are the next challenge to be faced. According to the European Directives many aquifers will not attain the good quantitative and chemical status by 2015. Thus delays are needed and exceptions have to be negotiated taking into account not only the slow turnover time of groundwater, but also the role of pollutants in the soil and in the unsaturated zone, and the delaying effect of sorption. Action is rather delayed. Spain has specific problems that have to be faced not only by means of regulations and good agricultural practices, but also considering its specific circumstances, and the economic, social and environmental impacts. Improvements come slowly but can be achieved. A combined action of water, environmental and land use authorities on one hand, and water users' associations on the other, are needed as a key issue for governance and water planning. Steps are being implemented.

REFERENCES

Cuevas Navas, R. (2008). *Zonas vulnerables en Andalucia* [Vulnerable zones in Andalusia]. Agencia Andaluza del Agua. Sevilla, Spain. Available from: www.ruena.csic.es/pdf/presentacion_cuevas_medio_ambiente.pdf [Accessed 17th July 2011].

IFA (2011). International Fertilizer Industry Association. *IFADATA* Available from: http://www.fertilizer.org/ [Accessed 15th September 2011].

MARM (2008). *Estado y tendencias del medio acuático y las prácticas agrarias, cuatrienio 2004–2007* [State and trends of water environment and agrarian practices, 2004–2007]. Ministerio de Medio Ambiente y Medio Rural y Marino. Madrid, Spain: 1–200.

MARM (2010). *Balance del nitrógeno en la agricultura española (año 2008)* [Nitrogen balance of Spanish agriculture, year 2008]. Ministerio de Medio Ambiente, y Medio Rural y Marino. Madrid, Spain.

MARM (2011). *Anuario de estadística agroalimentaria 2009* [Statistical agro-food yearbook 2009]. Estadísticas del Medio Rural. Medios de Producción. Fitosanitarios. Ministerio de

Medio Ambiente y Medio Rural y Marino. Available from: http://www.marm.es [Accessed 15th September 2011].

OJEU (1991). *Directive 91/676/EEC of the Council of the European Communities of 12 December 1991 concerning the protection of waters against pollution caused by nitrates from agricultural sources.* Official Journal of the European Union.

OJEU (2000). *Directive 2000/60/EC of the European Parliament and of the Council of 23 October 2000 establishing a framework for Community action in the field of water policy.* Official Journal of the European Union.

OJEU (2006). *Directive 2006/118/EC of the European Parliament and of the Council of 12 December 2006 on the protection of groundwater against pollution and deterioration.* Official Journal of the European Union.

OJEU (2008). *Directive 2008/105/EC of the European Parliament and of the Council of 16 December 2008 on environmental quality standards in the field of water policy, amending and subsequently repealing Council Directives 82/176/EEC, 83/513/EEC, 84/491/EEC, 86/280/EEC and amending Directive 2000/60/EC of the European Parliament and the Council.* Official Journal of the European Union.

Chapter 13

Urban and industrial water use challenges

Enrique Cabrera
Departamento de Ingeniería Hidráulica y Medio Ambiente,
Universitat Politècnia de València, Valencia, Spain

ABSTRACT: In countries in which agricultural dominates water uses, like Spain, urban and industrial water demand is, from the quantity point of view, a minor issue. Nevertheless, in the last decades this use is gaining relevance, a fact that explains an elemental risk analysis: although even in dry periods the probability of a supply failure is low, the costs associated to that fail are extremely high. In any case, problems linked to agricultural uses are, generally speaking, different of those related to urban and industrial uses. In very few occasions both uses compete, and then the problems are rather decoupled. However, as they share the framework, most of the challenges these uses faces (i.e., to adapt old policies to the current scenario), are common. And because of that, those who are interested in agricultural uses can find in this chapter some interesting ideas on the challenges that urban water faces and on the guidelines to walk towards a more sustainable water management.

Keywords: urban water, urbanization, water prices, population growth

1   INTRODUCTION

The passing of the centuries has not changed human beings' ability to create. Yet as a result of technology, today's customs and lifestyles bear no resemblance to those of just fifty years ago. The solid foundations of knowledge established by our predecessors began to bear fruit in the industrial revolution, in the second half of the 18th century, but this was nothing compared to the technological advances of recent decades which have led to the globalization that currently holds sway. Everything has happened so fast that our present way of life does not resemble that of even a few decades ago. Contrast this with people who lived in the 10th century. Had they been born two centuries earlier, they would have noticed little change.

These advances have taken place in all engineering fields, although a distinction should be made between those with a short, as opposed to a long, history. The former (telecommunications or aeronautics) have evolved in lockstep with their also young socio-economic frameworks. But this is not the case for water engineering. As a result, the technological advance of 20th century water engineering was obliged to coexist with an almost ancestral culture and a rigid and consolidated legal system of rights that has led to significant dysfunctions. To dovetail rapid technological and social change with ancient laws and rights is a complex task, because a rigid framework makes adapting water policy to the contemporary scenario difficult, especially in countries where the history of water is of great importance. However, rectifying disparities is easier in the case of new engineering fields since progress and frameworks move virtually in tandem.

At present, water administration, training of decision-makers and cultural attitudes of consumers are inadequate to deal with the collateral damage of progress. The required reforms need to be undertaken urgently, because problems are growing with time, and, if the complexity of the reforms continues to frighten those who have to carry them out, the end is already known; a major crisis arrives, forcing them to be implemented painfully.

The current imbalance began with the development of civil, hydraulic and electromechanical engineering in the 20th century. Until then, everything had taken place much more gradually with problems and their solutions going hand-in-hand. However, huge dams and pumps changed the dimensions of hydraulic engineering as it became possible to store large volumes of water and transport it over hundreds of kilometres. This development enabled the achievement of goals that previous generations could only dream of, with millions of hectares of dry land being irrigated and hitherto uninhabitable places (such as Las Vegas) being occupied.

Water engineering had achieved its greatest standing and arrived at its zenith by the end of the 20th century. It was in 1987 when the Brundtland Commission, concerned about the environmental deterioration worldwide, called for more sustainable policies. Since then, the only water policy that humans had implemented (making more resources available) has found its counterweight in a demand management more committed to efficiency. But, as this latter goal has been timidly pursued, the mismatch between supply and demand remains. Inertia hinders progress because the solution chosen first by humans, who tend to be resistant to change, is *business as usual*. Only enhanced environmental education for society can counteract the weight of history, social stress and recession.

2 THE NEW CONTEXT OF URBAN WATER IN THE 21ST CENTURY

Supplying the world's growing population with water of quality is today a complex challenge. Some facts explain why this is so. In order to highlight that the causes of the main problems are rather the same, these facts are presented first in general and afterwards particularised to Spain. Later, as problems are shared, the same diagnosis and the guidelines to follow (see Sections 3 and 4) apply. After that, no other specific mention of Spain is required.

2.1 Extraordinary and asymmetric population growth

Over the past six decades the world's population has nearly tripled (Figure 1a). While in 1950 the Earth had 2,500 million inhabitants, today there are 7,000 million of us, an impressive increase given that in the previous nine centuries the rise was *only* 1,400 million (from 300 million at the start of the second millennium up to 1,700 million at the beginning of the 20th century). And although the population growth slope

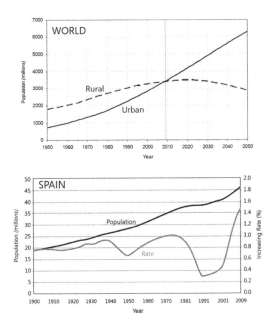

Figure 1 a) World urbanisation prospects (UN, 2010); and b) Spanish population growth in recent decades (BBVA, 2010).

has lessened, the ordinate is increasing so that by 2050 the world's population will reach 9,000 million people.

Yet the figures are even more striking if comparisons are made in terms of urban population. In 1950 only one third of the inhabitants of the Earth (approximately 700 million people) lived in cities. By mid-2009 there were already 3,400 million urban dwellers, accounting for 50% of the world's population, and Spain shows similar trends. Figure 1b depicts both the population growth and the increasing rate, while Table 1 shows the asymmetry.

2.2 Water needs are increasing in an uncertain scenario dominated by climate change

At present in the world, agriculture uses 70% of water resources although forecasts indicate that, in absolute terms, this will decrease slightly (Figure 2), while domestic and manufacturing uses (both urban uses) will increase. In any case it is important to underline, from the food crisis of 2008, that it is believed that food production will have to increase substantially, and that water supply for irrigation will also rise moderately in 2050, adding more pressure to this valuable resource. Spain does not have the official forecasts that should be a substantial part of the new river basin management plans. They still are pending despite the stipulated requirements (*River Basin Management Plans shall be published at the latest nine years after the date of entry into force of this Directive*) in Article 13 of the Water Framework Directive (WFD) published in

Table 1 Distribution of the population in urban and rural areas of Spain (BBVA, 2010).

Year	Population (total)	People P, living in cities P ≥ 50,000	People P, living in cities 50,000 > P ≥ 1,000	People P, living in rural areas (P < 1,000)
1900	18,830,649	13.73%	74.16%	12.11%
2001	40,847,371	50.63%	45.53%	3.84%
2009	46,745,807	52.47%	44.31%	3.22%

Baseline
World water use by sector, 2000–2050

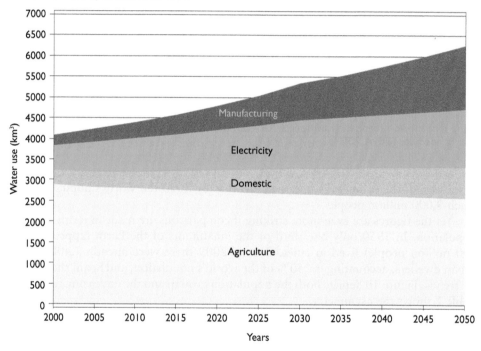

Figure 2 World water use by sector prospects. (Source: OECD Environmental Outlook Baseline 2007).

December 2000. In any case, percentages, over a total water demand of 30,000 hm³/year [hm³ = cubic hectometre = million m³ = 10^6 m³], are similar to those depicted in Figure 2. In this respect, Spain is rather average and in this context, water resources will suffer a growing stress, mainly taking into account the threat of a global climate change that will (according IPCC models) dramatically affect the Mediterranean countries. As population will follow an increasing trend, the solution must come hand in hand with efficiency.

2.3   The area of land altered and urbanised by humans is growing

Land use changes due to urbanisation affect the hydrological cycle. The creation of impervious surfaces increases runoff, reduces aquifer recharge and gives rise to urban flooding. Its growing frequency and the high economic and, occasionally human, damage it entails, have made this issue one of the greatest concerns for some cities. Then, the urban development of cities, which obviously grows in tandem with population numbers, has a direct impact and increases the complexity of sustainable water management in the world. *It can be stated that a land use decision is a water policy decision as well.*

From this point of view Spain is not on the world's average. The pace of land change during the last decade (1.95%) has been three times higher than the average (0.68%) of the 23 countries that have participated in the European Union (EU) Corine Land Cover Project 2000, which had as its objective to quantify land uses changes between 1990 and 2000. Although the recent burst of the *housing bubble* has provisionally stopped the trend, the problem is already on the floor (Figure 3) and so the study of strategies to minimise its adverse impacts is a crucial issue.

2.4   Water pollution is steadily increasing

Another dramatic change over recent decades has been the deterioration of water quality. In fact, each passing day more expensive and sophisticated treatments are required to make water potable. While in the past irrigation did not bring with it toxic substances, the massive use of fertilisers and pesticides has contaminated many of the aquifers that supply urban areas. These aggressive practices began in the second half of the 20th century and the reaction came in Europe in 1991 when the Nitrates Directive was enacted to become one of the first pieces of environmental legislation in the EU. It marked a turning point and has helped to slightly improve the situation. Yet, it remains one of the issues that cause most environmental concern.

Nevertheless urban (including storm water collection) and industrial uses, due to the pollutants they bring with them, have the highest impact on water quality. Furthermore, the quantity of water polluted and its accumulation in the aquifers is continually increasing due to population and urbanization growth and to the average rise in unit consumption (litres per person per day). Thus, between 1950 and 1990 water used tripled while the population only doubled. Restoring water to its initial quality calls for major investments that many developing countries are unable to afford. Also, what is worse, a long time is required to restore what has been changed (see Chapter 12). Finally, in some cases, once the treatment plants have been built, municipalities cannot afford the operational and energy costs they require.

As all over the world, during recent decades Spanish water bodies have received many sources of pollutants (urban, industrial and agricultural), and (although people only react if the pollution is clearly perceived, such as when it affects public spaces like beaches) society is sensitive to contamination, so the response has been effective to some extent, although with two weak points. First, most of the money devoted to these investments came from Brussels (EU). But this time is gone. Second, most of the small cities and rural settlements do not have waste water treatment plants. In Catalonia, for instance, 34.57%

Figure 3 Land use change. The city of Benidorm in the early 1960s and today.

of the municipalities (hosting just 4.59% of the population), do not have such a facility, and as water is subsidised, the main body responsible for its management (the Catalan Water Agency) supports a deficit of 2,500 M€. In any case, it must be underlined that Catalonia is not the worst region of Spain. It is in a better position than others.

2.5   The current economic crisis

We are living in turbulent times with few developed countries, if any, not affected by the global crisis. One of the biggest problems is the result of excessive borrowing by governments that now must reduce their deficits. That means that in forthcoming years they will manage tight budgets. This will hamper what has hitherto been standard practice in many countries: paying for this infrastructure with public money. To put it another way, subsidies will gradually be withdrawn. Although banned since 2010 in Europe by the WFD, many countries have so far been ignoring it.

The probable end of subsidies, in addition to leading to an increase in prices, will make it necessary to seek alternative funding and to reduce costs. The need to attract private capital will reopen the perennial debate about public-private management which, irrespective of their pros and cons, is a subject that should be discussed from a pragmatic standpoint far away from the political arena. In addition, the need to reduce costs will foster efficiency and a search for economies of scale. In short, the current economic crisis will, in all probability, mark a turning point from subsidies to full urban water cost recovery. Spain, with unemployment over 21%, an external debt over the 160% of the Gross National Product (GNP) and a public deficit of around 9%, should start to think about it. Otherwise, this *hydric bubble* will burst soon.

2.6   Investment has been huge – the challenge now is to maintain infrastructures and renew them

The 20th century, particularly in its early decades, witnessed the building of large water structures, mainly in semiarid countries or regions, like Spain or California. As part of a water policy geared towards mobilizing more water resources, water engineering structures were an excellent driving force for an economy which had to grow after the Great Depression and World War II. Furthermore, at that time governments were not burdened by current debt levels, and these were popular projects that enhanced quality of life and were therefore political vote winners. And since their environmental impacts were not well known, no one, or few persons, objected to them.

But that is now over, especially in developed countries where water management needs are much more necessary than water development. The water engineering structures and systems of the past now have to be preserved, if not replaced, including millions of kilometres of urban pipelines. Yet this is not attractive to the public at large who assign little value to those policies, unless they are obviously necessary. Nor is it attractive to politicians (renewing pipelines means *burying* money). However, since the need is obvious so as not to further jeopardise future generations, it cannot be delayed. Civil society must react and act.

In Spain, urban water infrastructure assets can be summarised in very few (although impressive) figures: some 200,000 km of water supply pipes, around 70,000 km of sewer pipes and roughly 5,000 water treatment plants. An important part of these assets is, or will soon become, old. But current urban water tariffs do not fully allow for maintaining or, when necessary, replacing them. Hopefully, still there are some few years of margin to react. Most of the water treatment plants have been built in the last decade and they still show acceptable performances. But as the return period of these infrastructures is shorter than most of the other urban water assets (pipes, reservoirs, tanks, etc.) the time to react is rather short.

3   ON THE NEED TO ADAPT DECISION-MAKERS' TRAINING TO THE NEW SCENARIO

The problems to be tackled are consequently formidable, Mexico City being a catalogue of the most significant ones. With over 20 million inhabitants, its urban water comes mostly from its aquifers because the surface resources in its surroundings are scarce. However, the city's unremitting growth and the degradation of its aquifers, severely affected by urban activity, made it necessary to find resources in neighbouring basins, and this generated serious social problems for the farmers who, until then, had been using this water. There is, indeed, a high leakage level that can only be remedied by huge investment, which will in turn require raising prices with all the social problems that this generates for the poorest people. In short, what had previously been only a one-dimensional engineering problem is now multidimensional. Technology, although important, is playing today a secondary role.

Profound changes have happened in just a few decades that require adapting both the training of decision-makers and the mindset of the citizens to this new scenario. Solutions will only be acceptable when they harmoniously combine all standpoints on the problem. *Integration* should therefore govern the decisions made about future water policies, especially in urban areas with serious social implications. Water is a very transversal issue and sustainable solutions require integration of all interests and points of view. This calls for a cultural change, particularly in terms of the major players, ranging from decision-makers to all stakeholders. In short, there is a need to map out new curricula which both bring the training of those who will be the leaders of the future into closer contact with reality and also take into account the three areas (political, managerial and technical) in which decisions are made. The training will prepare the future leaders for the role they are to play, and all will have a common knowledge of the issues to be tackled in order to be familiar with the viewpoint of the other positions. Only through integration can sustainable solutions be found.

Yet, training for those who make the decisions is not enough on its own. The administration must be as professional as possible in order to avoid political interference in technical decisions. Last, but not least, the general public has to be educated about the environment as there will always be a conflict between what is good in the short term and what is good in the medium-to-long term. This is because while politicians make their decisions with an eye on the next election, solutions are sustainable only if they take into account the interests of future generations. Yet, political interests and sustainability can be made compatible by educating the public. If most voters support an unpopular decision because they understand the need for it, their political representatives will only become visible in the medium-to-long term.

The need to deal with problems using an interdisciplinary approach is obvious, and perhaps as a result it has become increasingly prominent. Experts are now beginning to take the idea on board, and it is already present in international forums. This interdisciplinary approach has not yet been put into practice, and this is one of the major outstanding issues. Hence, unless the training of decision-makers and society is adapted to the new scenario, it will be virtually impossible to ensure in the near future that most of the people living on the planet (hopefully all of them) will have enough water in their homes and drinking water of quality at a reasonable price.

Once more Spain is a paradigmatic example because it is crucial to adapt the water culture of the citizens and the administrative structure to the new scenario. When framework changes and new challenges arise, there appear inconsistencies between the traditional institutions and the emerging needs. In Spain, ballasted by historical water rights, that becomes evidence of the need for change.

4   THE BIG CHALLENGES

Today the major challenge facing urban water is to achieve the Millennium Development Goals set at the 2002 Johannesburg Summit, one of which is to provide access by 2015 to water and sanitation for billions of people in developing countries who at that time did not have it, and to reduce by half the percentage of people without. But this formidable task (its solution requires a Herculean effort) is out of the scope of this chapter. Attention is paid now to those general issues, arising from the rapid changes in recent decades, discussed above, which call for new approaches, a process that has begun in some countries.

This can be highlighted by the differences in the price of 200 m^3 (the volume corresponding to the yearly demand of a family) of tap water (excluding urban drainage and sanitation) around the world. The wide divergence between some of the world's largest cities is hard to understand. According the last International Water Association report on this matter, the highest price is US\$ 765 in Copenhagen while the lowest one, in Milan at US\$ 33, is not even 5% of the Copenhagen rate. This disparity can only be explained qualitatively by historical reasons and quantitatively by costly structures and quality of delivered water. Hence analysing which terms of the water costs are taken into account leads to an understanding of what at first sight seems very difficult to justify. Yet while part of the bill is subsidised and some costs, especially environmental ones, are ignored, the differences will continue to exist. However, because water is of great social interest, decision-makers are not prepared to break with ancient customs.

In the light of the foregoing, a natural transition demands both educating decision-makers and users. Thus, the former will learn how much is at stake and will not hesitate to drive change, while the latter will bear the sacrifices they are asked to make because they understand the need for the decisions. This will be a gradual change based on the following five points:

i   A cross-cutting approach and breadth of vision. Actually water issues are located in a multidimensional space. Decisions should consider and balance at least the three key axes of that space (economic, social and environmental).

ii Integration with other environmental policies. Water decisions influence and are influenced by other environmental policies. As noted above, water/land is the most important but by no means the only nexus. For instance the energy/water pairing has come much more to the fore over the last ten years.

iii Joining science and politics. At present there is a lack of common ground in all matters relating to the environment because while science follows the pace of events in real time, politics usually thinks only in the short term.

iv Administration, the gradual driving force behind progressive change, must be adapted to the current context. That includes competent and free public servants, far away from the political arena. If water policy reforms are not gradually implemented, a water crisis will finally force the changes abruptly. Society's awareness will make these changes easier.

v   Enhancing demand management as the best option to mitigate water stress and to be sustainable. This entails improving efficiency, encouraging saving and promoting reuse. With growing demand for water in a scenario of climate change, the future asks for a more efficient water use. Four key actions are necessary:

   a   Passing on all the costs of water services to users encourages saving and efficiency. Some countries (e.g. Germany) have been passing costs on into the bills of users for over a decade. This is done by dividing the drainage fee into two blocks: one, the stormwater fee (proportional to the water drained by the property, a function of the paved land) and another for wastewater treatment (water use dependent). As rainwater drainage requires high investments, this *stormwater utility fee* has become significant and independent of the blackwater fee (see Table 2). As a matter of fact, in Berlin it has risen up to 1.90 €/m²/year [http://www.bwb.de].

       The full cost recovery principle is questioned on the grounds that water is a universal right, which justifies subsidies. Here it should be noted that progressive rates make it possible to subsidise the poorest without having to do so with the whole service. An analogous case is the taxation in a country. It can be designed, or not, to protect low rents. However, no one disputes that a country needs to balance its income and expenditure. The question is how to do so.

       Finally, it must be ensured that all the money users pay is invested in maintaining and improving these services.

   b   Renewing infrastructures, some of which are now very old.

   c   Commitment to provide a quality supply. The relentless rise in the consumption of bottled water, with all its environmental drawbacks, needs to be halted

Table 2 Urban drainage rates in Germany in 1999 (BUNR, 2001).

	Drainage fee (divided into two blocks)		Drainage fee, €/m³
	Blackwater (€/m³)	Rainwater (€/m²/year)	(one block)
Germany	1.79	0.77	2.28
Old West Germany	1.72	0.78	2.23
Old East Germany	2.39	0.59	2.54

soon. As a matter of fact, global bottled water sales have increased dramatically over the past several decades. Total revenues of bottled water are in Spain higher than those of all water utilities. Its unitary cost is hundreds, even thousands, of times higher than that of tap water, and nobody argues that this commodity is too expensive, while it is socially very complex to adapt the price of public water to its real costs.

d   Improving knowledge to continue moving forward. Research should take into account the concerns of industry professionals.

In this list of challenges the technical ones have not been included (e.g. water for human use should be of the highest quality, and the importance of promoting grey water reuse or water harvesting). They all can arrive in a natural way, once the *structural barriers*, to which we have paid all our attention, have been removed.

5   CONCLUSIONS

Addressing the supply of quality water to a growing population in a world that is changing at breakneck speed is one of the biggest challenges facing society in the 21st century. And although their magnitude very much depends on the starting point of each country, on their water awareness and, for sure, on the specific socio-economic characteristics of each country, the main guidelines are rather the same all around the world. Spain is, indeed, a representative case of this general picture.

REFERENCES

BBVA (Banco Bilbao Vizcaya Argentaria) (2010). *Población 51: La población en España: 1900–2009* [Spain's Population: 1900–2009] Fundación BBVA, Madrid, Spain.
BUNR (Bundesminiterium für Umwelt, Naturschutz und Reaktorsicherheit) (2001). *Water Resources Management in Germany*. Federal Ministry for the Environment, Bonn, Germany.
OECD (Organization for Economic Cooperation and Development) (2007). *OECD Environmental Baseline 2007*. OECD, Paris, France.
UN (United Nations) (2010). World *Urbanization Prospects. The 2009 Revision. Highlights*. Report ESA/P/WP/215. United Nations, New York, USA.

Chapter 14

Challenges and opportunities related to the Spanish water-energy nexus

Laurent Hardy[1] & Alberto Garrido[2]
[1] *Department of Geodynamics, Complutense University of Madrid, Madrid, Spain*
[2] *Water Observatory of the Botín Foundation; CEIGRAM, Technical University of Madrid, Madrid, Spain*

ABSTRACT: Water-energy nexus outcomes are progressively going beyond the frontiers of academic institutions. The recent Bonn 2011 Conference gathered water, energy, and food perspectives and brought the issues to the policy arena. The private sector sees the water-energy nexus as a source of challenges but also identifies business opportunities for the near future. This chapter presents some striking outcomes of the consideration of the water-energy nexus for both public and private levels of action. The first of the two connections is the *energy for water*. We emphasize the energy needs of irrigated agriculture, a challenge for the coming decades (today, 67% of the withdrawn water in Spain) and the trade-offs between water conservation and energy demands of alternative supplies for the agricultural sector. The second connection is the *water for energy*. Around 8,500 million m³ per annum of water are withdrawn per year in Spain to cool electricity generation plants. According to future electricity demand scenarios and the types of technology being installed, this volume could double, adding further pressure to already stressed Spanish basins. We conclude by highlighting the need to integrate both energy and water issues jointly in all decisions related to energy generation decisions and water use and conservation issues.

Keywords: water-energy nexus, energy scenarios, water demand, irrigation system modernization, Spain

I UNDERSTANDING THE WATER-ENERGY NEXUS

Nexus has been the word used to emphasize the intimate connection between water, energy, and the recently added factor, food (Hoff, 2011). Water and energy still top the list of priority issues in sustainability assessments. The realization that they are closely linked and that both should not be treated independently is a source of complexity, but is the only way to make progress towards more sustainable water and energy management.

Within the water-energy nexus, we define the connection energy for water as the energy consumption in the integral water-use cycle. Numerous studies have already shown the existence of the energy consumption in the water use cycle in several countries and sectors (CEC, 2005; Pate *et al.*, 2007; Water Environment Federation, 2009;

Cabrera *et al.*, 2010). Hardy & Garrido (2010) and Hardy *et al.* (2012) have shown that the Spanish integral water-use cycle accounted for 6% of the electricity demand in 2008 (see Table 1). 64% of the 16,500 GWh consumed in the water use cycle was required for extraction and water treatment; of which, 25% was required to supply water for food production.

Table 1 does not include final usage of water, such as heating domestic water. It has been estimated (IDAE, 2010) that 21% of the primary energy bill (electricity, gas, and oil) of a household is allocated to heating domestic water. In addition, according to the same source, from the average household electricity bill of 4,000 kWh/year in Spain, 3% is necessary for domestic hot water. Annual electricity consumption for domestic hot water in the urban sector would be 2,260 GWh.

The *water for energy* connection accounts for the water required to procure the raw material to produce one unit of energy (the fuel) and the water used in the power plant cooling systems. Each energy generation technology has different energy needs. Based on data from Rio Carrillo & Frei (2009) and Linares & Sáenz de Miera (2009), fossil and renewable energies show significantly different volumes of water use (18,000 m³/GWh *vs.* 29,000 m³/GWh). Hardy *et al.* (2012) have shown that

Table 1 Water-related energy use in Spain in 2008.

Stages	Water volume (hm³)	Electricity	
		Consumption (GWh)	Percentage (%)
Extraction and Water Treatment	34,940	10,418	64
Urban	4,343	5,457	33
(from desalination)	(694)	(2,275)	(14)
Agriculture	20,360	4,141	25
Energy	8,683	521	3
Industry	1,554	299	2
Distribution/Water Use	25,587	3,374	21
Residential	2,540	440	3
Commercial	833	144	0.9
Municipalities and Other	359	62	0.4
Industrial	286	49	0.3
Agricultural	20,360	2,469	15
Non registered water	1,210	210	1.3
Wastewater Treatment	2,842	2,530	15
Wastewater collection	3,788	189	1.2
Wastewater treatment	2,842	1,454	9
Recycled water (treatment and distribution)	1,510	887	5.4
Total	34,940	16,323	100
Total Spain electricity use		279,392	
Percentage		5.8%	

Source: Hardy *et al.* (2012).

Note: The water volume column gives the volume of water for each stage of the water use cycle. "Total" is the total volume of extracted water in Spain. Not all the water extracted is distributed nor treated because of own extraction and treatment systems (agricultural sector, energy sector, and industrial sector).

depending on how we deal with a 153% increase in electricity demand for the year 2030 (following UNESA, 2007), there could be an impact in terms of required water withdrawal (see Section 3). According to the same authors, the energy sector withdrew 25% of the 35,000 hm³ in Spain in 2008 (mainly for cooling purposes) [hm³ = cubic hectometre = million m³ = 10^6 m³]. Severe consequences on water availability should be expected according to some of the climate change scenarios.

The water-energy nexus has planning and economic implications for water management. Sustainability assessments have to integrate both resources and establish technical recommendations.

2   ENERGY FOR WATER

2.1   Energy use in irrigation communities

The Spanish irrigation system has experienced profound transformations in the last decade. Up until 2000, flood irrigation systems were common and did not involve major energy consumption (0.02–0.15 kWh/m³). Since 2002, the modernization of the sector has led to the replacement of superficial irrigation systems by sprinkler and drip irrigation systems, which are much more energy intensive (0.28–0.68 kWh/m³). Between 2002 and 2008 (Figure 1), while drip irrigation systems increased by 40%, electricity needs increased by 10% during the same period (MARM, 2008a; 2009), showing that, to some extent, water savings had been achieved at the expense of a higher energy consumption. At the same time, the price of electricity went up by 30–70% in 2007–2008 (Ederra & Murugarren, 2010; see Figure 2b). As we shall see below, modernization requires investment costs that might not justify the water savings considered, especially when there is an alternative source of water available like regenerated water or desalination. There is the risk that better irrigation technologies may end up increasing water consumption (Cots, 2011; Ward & Pulido, 2008).

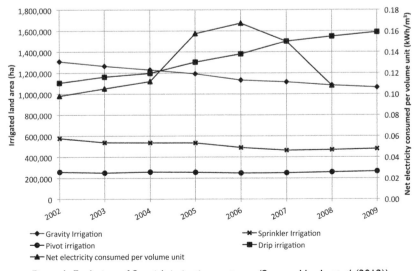

Figure 1 Evolution of Spanish irrigation systems. (Source: Hardy et al. (2012)).

Even if the evolution shows that improvements have been made in the use of energy, it is clear today that water savings come second because energy consumption has become the real issue. Abadía *et al.* (2010), and Carrillo-Cobo *et al.* (2010) show the importance of making energy audits in water users' associations because energy savings could be achieved through reorganization of irrigation periods and irrigation district management.

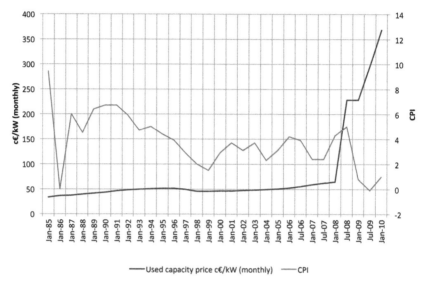

Figure 2a Evolution of contracted capacity price and the Consumer Price Index (CPI). (Source: Own elaboration with Ederra & Murugarren (2010)).

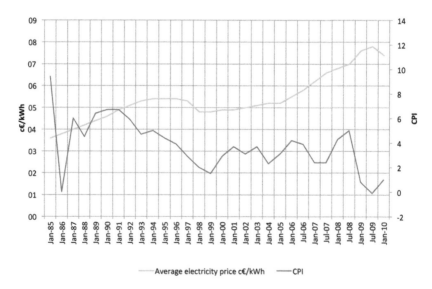

Figure 2b Evolution of the price of electricity consumed and the Consumer Price Index (CPI). (Source: Own elaboration with Ederra & Murugarren (2010)).

2.2   Further energy savings in irrigation districts

In Spain, the relation between water consumed and water used is already close to 0.8 ± 0.27 according to Krinner *et al.* (1994) and 0.8 according to Corominas (2009). Without getting into deficit irrigation (i.e. to reduce the amount of water available for irrigation without reducing the yield, quality and production), it becomes very difficult to further increase the ratio. A method to assess modernization of irrigation systems *versus* using alternative sources of water (like regenerated water or desalination) has been described in Hardy & Garrido (2010) and is briefly presented in Box 1. It is important to notice that, although other solutions for modernization exist,

---

Box 1   Numerical model to assess modernization against the use of alternative sources of water

In a situation of water scarcity and water use efficiency (Y_1: water consumed/water used), we want to assess the best way of expanding the resource base. There are two possibilities: to modernize the irrigation system to save water at the expense of higher energy consumption or to get the water from an alternative source (like regenerated water or desalination). Equations 1 and 2 below show the situation in which the change is neutral.

$$E_2 \cdot \frac{Y_1}{Y_2} = E_1 + E_a \cdot \left(1 - \frac{Y_1}{Y_2}\right) \tag{1}$$

which is equivalent to

$$\frac{Y_2 - Y_1}{Y_1} = \frac{E_2 - E_1}{E_a + E_1} \cong \frac{E_2 - E_1}{E_a} \tag{2}$$

where E_1 and E_2 are the energy consumption for initial and modern irrigation systems. E_a is the energy consumption of the alternative water source. Both are in kWh/m³.

The decision rule is based on the value of Y_2, which is the expected water use efficiency in a modernized irrigation system that justifies this option. If such efficiency is not realistic, then getting water from the alternative source might be the best option. The same analysis is carried out for the associated investment cost (see Equation 3).

$$\frac{Y_2 - Y_1}{Y_1} = \frac{C_2 - C_1}{C_a + C_1} \cong \frac{C_2 - C_1}{C_a} \tag{3}$$

where

$$C_{[\text{€}/m^3]} = \frac{C_{[\text{€}/ha]} \cdot d_{[year^{-1}]} \cdot Y}{ET_{[m^3/ha/year]}} \tag{4}$$

This decision tool can be used for instance to compare two irrigation techniques (drip irrigation and sprinkler), two alternative water sources (desalination and regenerated water) and all possible values for the evapotranspiration (ET: water loss by the plant or plant water needs, in m^3/ha/year). Desalination requires E_a = 2.70 kWh/m^3 and C_a=0.40 €/m^3 (Torres Corral, 2005) whereas regenerated water supposes E_a=0.60 kWh/m^3 and C_a = 0.06 €/m^3 (Mujeriego, 2006). Damping factor (d) has been set to 0.12/year. C [€/ha] is the investment cost per hectare for the new irrigation system. The decision criterion requires the evaluation of two criteria (the energy needed and the cost performance) to assess the usefulness of modernization of irrigation systems instead of using alternative sources of water. The reference for acceptable water use efficiency is set to 0.80 (higher efficiencies would require deficient irrigation). In Tables 2 and 3, the value of Y_2 has been coloured so that green indicates that Y_2 < 0.8, yellow indicates that 0.8 < Y_2 < 1 and red indicates that Y_2 > 1.

the analysis here focuses on what generally happened in Spain, i.e. the modernization of irrigation systems through pressurized systems of irrigation. Irrigation management can also produce water conservation even without making large investments (see Pérez Pastor, 2010).

Graphic results of the model presented in the Box 1 are shown below. Energy criteria of Table 2 (left hand side) shows that strong modernization (maximum energy increment) is worthwhile even with initial water use efficiencies Y_1 up to 0.75, but only when the alternative water source is desalination. Regenerated water has lower energy consumption per m^3, therefore at a higher level of modernization the best option for high initial water use efficiencies is not modernization but – if available – regenerated water (Table 3). If irrigation techniques with higher energy consumption were used (sprinkler instead of drip irrigation), the expected water use efficiency Y_2 could be so high that modernization would not be justified; therefore, the best option might be to use alternative sources of water.

The investment cost (right-hand side of Tables 2 and 3) is a strong limiting factor and a final decision will be based on the combination of both energy criteria and investment cost. Investment cost is site-specific, i.e. the cost is determined by the importance of the installations required to modernize the irrigation system. Where modernization is justified from an energy point of view, we can see from Tables 2 and 3 that only irrigation areas with very low initial water use efficiencies (Y_1) would be worth modernizing, otherwise the best option is to rely on the alternative source of water (regenerated water or desalination).

In addition, we observe that crops need of water per hectare (ET) affect the minimum final water use efficiency (Y_2) that would justify modernization. If water use efficiency remains constant, as ET increases (i.e. plant water necessity increases), modernization becomes a better option over alternative sources of water. This is because as more water is needed for the whole irrigation system, the investment cost per m^3 decreases (see Equation 4).

Modernization of irrigation systems entails an increase in energy demand. Consequently, the energy embodied in the food product will increase and so will the CO_2 footprint of the agricultural stage of the product. In Table 4 some examples of food products cultivated in Spain are given. We assume the modernization of an irrigated land from gravity irrigation to drip irrigation system and desalination as alternative source of water.

Table 2 Appreciation of the usefulness of modernizing irrigation systems. Irrigation system: drip irrigation; alternative water source: desalination; ET = 4,100 m³/ha/year.

YI (Water consumed/Water used)	Increase in energy consumption E2-E1 (kWh/m³)									
	0.00	0.02	0.04	0.06	0.08	0.10	0.12	0.14	0.16	0.18
1.00	1.00	1.01	1.01	1.02	1.03	1.04	1.04	1.05	1.06	1.07
0.95	0.95	0.96	0.96	0.97	0.98	0.98	0.99	1.00	1.01	1.01
0.90	0.90	0.91	0.91	0.92	0.93	0.93	0.94	0.95	0.95	0.96
0.85	0.85	0.86	0.86	0.87	0.87	0.88	0.89	0.89	0.90	0.91
0.80	0.80	0.81	0.81	0.82	0.82	0.83	0.83	0.84	0.85	0.85
0.75	0.75	0.76	0.76	0.77	0.77	0.78	0.78	0.79	0.79	0.80
0.70	0.70	0.70	0.71	0.71	0.72	0.73	0.73	0.74	0.74	0.75
0.65	0.65	0.65	0.66	0.66	0.67	0.67	0.68	0.68	0.69	0.69
0.60	0.60	0.60	0.61	0.61	0.62	0.62	0.63	0.63	0.64	0.64
0.55	0.55	0.55	0.56	0.56	0.57	0.57	0.57	0.58	0.58	0.59
0.50	0.50	0.50	0.51	0.51	0.51	0.52	0.52	0.53	0.53	0.53
0.45	0.45	0.45	0.46	0.46	0.46	0.47	0.47	0.47	0.48	0.48
0.40	0.40	0.40	0.41	0.41	0.41	0.41	0.42	0.42	0.42	0.43
0.35	0.35	0.35	0.35	0.36	0.36	0.36	0.37	0.37	0.37	0.37
0.30	0.30	0.30	0.30	0.31	0.31	0.31	0.31	0.32	0.32	0.32
0.25	0.25	0.25	0.25	0.26	0.26	0.26	0.26	0.26	0.26	0.27
0.20	0.20	0.20	0.20	0.20	0.21	0.21	0.21	0.21	0.21	0.21
0.15	0.15	0.15	0.15	0.15	0.15	0.16	0.16	0.16	0.16	0.16
0.10	0.10	0.10	0.10	0.10	0.10	0.10	0.10	0.11	0.11	0.11
0.05	0.05	0.05	0.05	0.05	0.05	0.05	0.05	0.05	0.05	0.05

YI (Water consumed/Water used)	Cost of water (€/m³)									
	0.03	0.06	0.08	0.11	0.13	0.16	0.19	0.21	0.24	0.27
	Investment cost (k€/ha)									
	1000	1900	2800	3700	4600	5500	6400	7300	8200	9100
1.00	1.00	1.06	1.12	1.18	1.25	1.31	1.37	1.43	1.49	1.55
0.95	0.95	1.01	1.07	1.12	1.18	1.24	1.30	1.36	1.42	1.47
0.90	0.90	0.96	1.01	1.07	1.12	1.18	1.23	1.29	1.34	1.40
0.85	0.85	0.90	0.95	1.01	1.06	1.11	1.16	1.22	1.27	1.32
0.80	0.80	0.85	0.90	0.95	1.00	1.05	1.09	1.14	1.19	1.24
0.75	0.75	0.80	0.84	0.89	0.93	0.98	1.03	1.07	1.12	1.16
0.70	0.70	0.74	0.79	0.83	0.87	0.91	0.96	1.00	1.04	1.09
0.65	0.65	0.69	0.73	0.77	0.81	0.85	0.89	0.93	0.97	1.01
0.60	0.60	0.64	0.67	0.71	0.75	0.78	0.82	0.86	0.89	0.93
0.55	0.55	0.58	0.62	0.65	0.69	0.72	0.75	0.79	0.82	0.85
0.50	0.50	0.53	0.56	0.59	0.62	0.65	0.68	0.71	0.75	0.78
0.45	0.45	0.48	0.51	0.53	0.56	0.59	0.62	0.64	0.67	0.70
0.40	0.40	0.42	0.45	0.47	0.50	0.52	0.55	0.57	0.60	0.62
0.35	0.35	0.37	0.39	0.41	0.44	0.46	0.48	0.50	0.52	0.54
0.30	0.30	0.32	0.34	0.36	0.37	0.39	0.41	0.43	0.45	0.47
0.25	0.25	0.27	0.28	0.30	0.31	0.33	0.34	0.36	0.37	0.39
0.20	0.20	0.21	0.22	0.24	0.25	0.26	0.27	0.29	0.30	0.31
0.15	0.15	0.16	0.17	0.18	0.19	0.20	0.21	0.21	0.22	0.23
0.10	0.10	0.11	0.11	0.12	0.12	0.13	0.14	0.14	0.15	0.16
0.05	0.05	0.05	0.06	0.06	0.06	0.07	0.07	0.07	0.07	0.08

Source: Own elaboration with Hardy & Garrido (2010).

Table 3 Appreciation of the usefulness of modernizing irrigation systems. Irrigation system: drip irrigation; alternative water source: regenerated water; ET = 4,100 m³/ha/year.

YI (Water consumed/Water used)	Increase in energy consumption E2-E1 (kWh/m³)									
	0.00	0.02	0.04	0.06	0.08	0.10	0.12	0.14	0.16	0.18
1.00	1.00	1.03	1.05	1.08	1.11	1.15	1.18	1.22	1.26	1.30
0.95	0.95	0.98	1.00	1.03	1.06	1.09	1.12	1.16	1.20	1.24
0.90	0.90	0.92	0.95	0.98	1.00	1.03	1.06	1.10	1.13	1.17
0.85	0.85	0.87	0.90	0.92	0.95	0.98	1.00	1.04	1.07	1.11
0.80	0.80	0.82	0.84	0.87	0.89	0.92	0.95	0.98	1.01	1.04
0.75	0.75	0.77	0.79	0.81	0.84	0.86	0.89	0.91	0.94	0.98
0.70	0.70	0.72	0.74	0.76	0.78	0.80	0.83	0.85	0.88	0.91
0.65	0.65	0.67	0.69	0.70	0.72	0.75	0.77	0.79	0.82	0.85
0.60	0.60	0.62	0.63	0.65	0.67	0.69	0.71	0.73	0.75	0.78
0.55	0.55	0.56	0.58	0.60	0.61	0.63	0.65	0.67	0.69	0.72
0.50	0.50	0.51	0.53	0.54	0.56	0.57	0.59	0.61	0.63	0.65
0.45	0.45	0.46	0.47	0.49	0.50	0.52	0.53	0.55	0.57	0.59
0.40	0.40	0.41	0.42	0.43	0.45	0.46	0.47	0.49	0.50	0.52
0.35	0.35	0.36	0.37	0.38	0.39	0.40	0.41	0.43	0.44	0.45
0.30	0.30	0.31	0.32	0.32	0.33	0.34	0.35	0.37	0.38	0.39
0.25	0.25	0.26	0.26	0.27	0.28	0.29	0.30	0.30	0.31	0.32
0.20	0.20	0.21	0.21	0.22	0.22	0.23	0.24	0.24	0.25	0.26
0.15	0.15	0.15	0.16	0.16	0.17	0.17	0.18	0.18	0.19	0.19
0.10	0.10	0.10	0.11	0.11	0.11	0.11	0.12	0.12	0.13	0.13
0.05	0.05	0.05	0.05	0.05	0.06	0.06	0.06	0.06	0.06	0.06

YI (Water consumed/Water used)	Cost of water (€/m³)									
	0.03	0.06	0.08	0.11	0.13	0.16	0.19	0.21	0.24	0.27
	Investment cost (k€/ha)									
	1000	1900	2800	3700	4600	5500	6400	7300	8200	9100
1.00	1.00	1.30	1.59	1.89	2.18	2.48	2.77	3.07	3.36	3.66
0.95	0.95	1.23	1.51	1.79	2.07	2.35	2.63	2.91	3.19	3.47
0.90	0.90	1.17	1.43	1.70	1.96	2.23	2.49	2.76	3.02	3.29
0.85	0.85	1.10	1.35	1.60	1.85	2.10	2.35	2.61	2.86	3.11
0.80	0.80	1.04	1.27	1.51	1.74	1.98	2.22	2.45	2.69	2.92
0.75	0.75	0.97	1.19	1.41	1.64	1.86	2.08	2.30	2.52	2.74
0.70	0.70	0.91	1.11	1.32	1.53	1.73	1.94	2.15	2.35	2.56
0.65	0.65	0.84	1.03	1.23	1.42	1.61	1.80	1.99	2.18	2.38
0.60	0.60	0.78	0.95	1.13	1.31	1.49	1.66	1.84	2.02	2.19
0.55	0.55	0.71	0.87	1.04	1.20	1.36	1.52	1.69	1.85	2.01
0.50	0.50	0.65	0.80	0.94	1.09	1.24	1.39	1.53	1.68	1.83
0.45	0.45	0.58	0.72	0.85	0.98	1.11	1.25	1.38	1.51	1.65
0.40	0.40	0.52	0.64	0.75	0.87	0.99	1.11	1.23	1.34	1.46
0.35	0.35	0.45	0.56	0.66	0.76	0.87	0.97	1.07	1.18	1.28
0.30	0.30	0.39	0.48	0.57	0.65	0.74	0.83	0.92	1.01	1.10
0.25	0.25	0.32	0.40	0.47	0.55	0.62	0.69	0.77	0.84	0.91
0.20	0.20	0.26	0.32	0.38	0.44	0.50	0.55	0.61	0.67	0.73
0.15	0.15	0.19	0.24	0.28	0.33	0.37	0.42	0.46	0.50	0.55
0.10	0.10	0.13	0.16	0.19	0.22	0.25	0.28	0.31	0.34	0.37
0.05	0.05	0.06	0.08	0.09	0.11	0.12	0.14	0.15	0.17	0.18

Source: Own elaboration with Hardy & Garrido (2010).

As we can observe from Table 4, modernization of irrigation systems certainly has to undergo an in-depth analysis to be justified from both technical and economical points of view. That way, from the energy standpoint, improvements in irrigated areas in Spain would be optimally made: either modernization or alternative source of water. The use of energy would be more efficient and sustainability of agriculture could be improved.

Table 4 Estimation of energy increase and increase in CO_2 footprint for some products grown in Spain.

Product – Agricultural system	Energy increase (kWh/tonne)	CO_2 footprint increase (g CO_2/tonne)	Percentage of agricultural CO_2 footprint (%)
Wine – Organic	51.9	12,814	5.4
Wine – Conventional	51.9	12,814	14.4
Tomato cherry	18.3	4,509	4.0
Olive Oil – Organic	86.0	21,246	2.3
Olive Oil – Conventional	86.0	21,246	4.3
Apple	33.5	8,273	23.6

Source: Hardy & Garrido (2010), Junta de Andalucía (2010), AQUAVIR (2005), EPEA (2009), MARM (2008b) and MITYC (2011).

2.3 Economic considerations of energy use in irrigated agriculture

The price of electricity for farmers has been increasing since July 2008, since Spain's electricity sector entered the free market, especially because of the contracted capacity. Ederra & Murugarren (2010) estimate that between 2005 and 2009, the electricity bill increased by 82%. Figure 2a shows how the contracted capacity price skyrocketed since January 2008 (increased by 470%). Figure 2b shows that the price of electricity consumed also increased from January 2008 (augmented 9%).

A possibility proposed by the National Federation of Irrigation Communities of Spain (FENACORE) is to turn water users associations (WUA) into green electricity producers (solar photovoltaic, solar thermoelectric, wind or hydropower) and hence generate their own electricity. The irrigation period in Spain usually begins in March and ends in October. The electricity that they would produce should be enough to supply their own needs during the irrigation period and could represent an extra income during the rest of the year. WUAs are demanding their status as energy generators to be granted (currently not possible due to legal barriers). In the meanwhile, WUAs are finding ways either to save energy or money, for example developing collective agreements with private companies to negotiate better electricity supply contracts.

2.4 Long term perspectives for regenerated water

Successive national water and wastewater treatment plans (PNSD) were implemented in order to enforce the 91/271/CE Directive from the European Commission on urban wastewater treatment. The 1995–2005 PNSD ended up with 77% of the cities and villages in conformity with the 91/271/CE Directive, which aimed at getting wastewater treatment in all cities and villages of at least 2,000 inhabitants equivalent. In 2008, although 92% of the population is connected to a wastewater treatment system, only 51% are connected to a tertiary wastewater treatment system (EuroStat, 2008). Energy consumption for primary and secondary wastewater treatment in Spain is estimated to be 0.53 kWh/m^3 (Hardy & Garrido, 2010) and adding a tertiary treatment stage, a supplementary 0.13 kWh/m³ (Water Environment Federation, 2009). Although the Spanish wastewater treatment system is well developed, with 83% of the wastewater generated from all the sources being treated (EuroStat, 2008), improvements still have to be made. Table 5 provides a complete breakdown of the wastewater treatment sector according to the most recent data (2008) where 17% of the wastewater generated in Spain is not treated at all. Regarding urban wastewater treatment system, almost all the population is connected to a primary treatment system (96%) but percentages for secondary and tertiary treatment are lower (respectively 37% and 51%). In addition, we give an estimation of the supplementary energy consumption for the full treatment (tertiary treatment) of the 650 hm³ that are not treated.

Compared to the energy consumption of the total water use cycle of Spain (see Table 1), the extension of the wastewater treatment system to a 100% tertiary treatment system would increase the water-related electricity consumption by 3%.

2.5   Water-related greenhouse gas emissions

The water-related electricity consumption in the Spanish water use cycle was 16,500 GWh for 2008. The CO_2 emitted due to electricity generation is about 4.3 million tonnes of CO_2. Over a total of 406 million tonnes of CO_2 emitted in 2008 (EuroStat, 2008), the Spanish water use cycle accounts for 1% of total Spanish CO_2 emissions. This does not include CO_2 emissions from the energy required for final usages of water such as domestic hot water.

As shown earlier in this chapter, if 100% of the generated wastewater were to be treated and recycled, i.e. all the wastewater generated undergoes tertiary treatment, the related energy consumption would be close to 430 GWh/year. The question (still open) is whether the CO_2 emissions of the wastewater saved due to the treatment (air contamination avoided) compensate for the CO_2 emissions due to the production of energy necessary for their treatment. An integral wastewater treatment system in Spain would suppose 106,000 tonnes of supplementary CO_2 emitted due to electricity generation, or 0.03% of total Spanish CO_2 emissions.

In the UK, it has been estimated that nearly 6% of the greenhouse gas emissions relate to water use and 90% of the water-related greenhouse gas emissions result from

Table 5 Estimated energy consumption (GWh) for a 100% tertiary wastewater treatment.

	2006	2008
Wastewater generated by all sources	3,962	3,788
from industry sector	905	828
from urban sector	3,057	2,960
Treated discharges of wastewater treatment plants	85%	83%
Total wastewater not connected to urban wastewater collecting system	585	649
Energy consumption for primary, secondary and tertiary treatment (GWh)	**396**	**429**

Source: Own estimation with EuroStat (2008).

Note: All values are in hm³ unless specified. The energy consumption (GWh) are calculated from standard energy consumption of a 190,000 L/day treatment plant (see Water Environment Federation, 2009).

water usage inside the house, i.e. final usage of water is amongst all the stages of the water use cycle the most greenhouse gas emissions intensive (Clarke *et al.*, 2009).

Although no study has been carried out for Spain yet, Cabrera *et al.* (2010) showed that the interest in carrying out energy audits in water networks relies not only in making energy savings, but also in making CO_2 credit savings. According to MITYC (2011), Spanish electricity production in 2009 emitted 0.306 kg CO_2/kWh, but the evolution of the Spanish technology mix toward a cleaner production system brought this figure to 0.247 kg CO_2/kWh in 2010.

3 WATER NEEDS FOR FUTURE ENERGY GENERATION

Energy demand is expected to increase in the next 20 years in Spain like in the rest of the world. Scenarios for future electricity generation propose different demands and different technology mixes (IIT, 2005; UNESA, 2007; PwC, 2010). The water-energy nexus is relevant in determining the best technology mix. Rio Carrillo & Frei (2009) have shown for Spain that renewable energy systems are less water-intensive (in terms of withdrawn water) than fossil fuel energy systems (18,000 m³/GWh *vs.* 29,000 m³/GWh). Nuclear energy is the most water-intensive technology with 75,362 m³/GWh. Geographic location is also important when planning a new power plant. For example, thermo solar power plants usually are constructed in arid regions where the access to water will be a limiting factor and demands have to be managed properly.

The energy sector is a water withdrawal sector, not a primary water consumptive sector. In Spain, it needs around 8,600 hm³/year (around 25% of the water extracted annually in Spain). If electricity demand were going to increase, it would be prudent to include the water needs comparing the different proposed scenarios and their respective technology mix. Figure 3 presents seven scenarios with technology mix from two different institutions (UNESA, 2007; PwC, 2010) for electricity production in the year

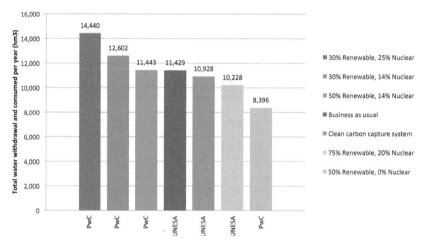

Figure 3 Water withdrawal and consumption for 7 scenarios of electricity production for the year 2030 in Spain. (Source: Rio Carrillo & Frei (2009), Linares & Sáenz de Miera (2009), UNESA (2007) and PwC (2010)).

2030. Scenarios have been regrouped as a function of the percentage of renewable energy technologies and the percentage of nuclear energy used in the technology mix. UNESA and PricewaterhouseCoopers have evaluated the electricity demand for the year 2030 to 428,773 GWh and 461,580 GWh respectively. The results presented in Figure 3 are volumes of water used for the power plant cooling systems (the *business as usual* scenario refers to the Spanish technology mix in 2007).

This suggests that water savings are higher when more renewable energy systems and less nuclear energy are present in the technology mix. Interestingly, we also observe that the reduction in total water required is higher if less nuclear plants are used than if more renewable energy technologies are used. Therefore and according to these scenarios, all else remaining the same, the larger savings in water for the electricity generation are made by taking nuclear technology out of the electricity production system, and to a lesser extent by introducing renewable energy technologies.

4 CONCLUSIONS

In Spain, two of the main conclusions after analysing the main aspects of the water-energy nexus are the importance of the irrigation sector and future energy generation. Spain has developed an intensive process of modernization of irrigated areas since 2002, with the aim of saving water (see Chapter 19). However, an undesirable consequence has been the increase in energy consumption. Hence, we developed a tool to appreciate the usefulness of modernizing irrigation systems that considers all the available options to face a situation of water scarcity before getting into the process of modernization. We find that from the water-energy perspective, unless there is a low initial water-use efficiency (around 50%), modernization of irrigation systems might not be the best option. It would be better instead to consider alternative sources of water such as desalination or regenerated water. Investment costs are always a limiting factor to modernization. As a water savings generation strategy, modernization generally performs worse than using desalinated water and much worse than using regenerated water.

Apart from becoming an alternative source of water, regenerated water production is an important challenge in the European Union. Considering the water-energy nexus, it would be an option to lower our dependency on resources, with the added advantage of providing an economical use for raw material like wastewater, which has no significant usage now. To extend wastewater treatment systems to include tertiary treatment all over Spain would account for 3% of the water-related electricity consumption (i.e. 0.2% of the Spanish electricity demand).

Spain still lacks a comprehensive analysis of final usages of water (in terms of both water and energy). However, it is estimated that the water use cycle (without considering final usage), could account for 1% of the total CO_2 emissions of Spain. If final usage (such as water heating) were taken into account, we surmise that the energy consumed per m^3 will grow, and hence, the CO_2 footprint would go above the said 1%.

One important conclusion in terms of energy planning from the study is that electricity consumption is likely to increase in the future, therefore several scenarios of technology mix and electricity demands exist. By relating the production of electricity to water needs, we show that more water is saved if nuclear power is removed than if more renewable energy systems are built.

REFERENCES

Abadía, R.; Rocamora, M.C.; Córcoles, J.I.; Ruiz-Canales, A.; Martínez-Romero, A.; Moreno, M.A. (2010). Comparative analysis of energy efficiency in water users associations. *Spanish Journal of Agricultural Research*, 8(2): 134–142.

AQUAVIR (2005). Superficie de los Cultivos de Regadío y sus Necesidades de Riego, en la Demarcación de la Confederación Hidrográfica del Guadalquivir [Irrigated Agriculture Area and Water Needs in the Guadalquivir River Basin District]. Final report.

Cabrera, E.; Pardo, M.A.; Cobacho, R. & Cabrera, E. Jr. (2010). Energy Audit of Water Networks. *Journal of Water Resources Planning and Management*, 136(6): 669–677.

Carrillo-Cobo, M.T.; Rodríguez-Díaz, J.A. & Camacho-Poyato, E. (2010). The role of energy audits in irrigated areas. The case of *Fuente Palmera* irrigation district (Spain). *Spanish Journal of Agricultural Research*, 8(S2): S152–S161.

CEC (California Energy Commission) (2005). California's Water-Energy Relationship. Final Staff Report. California Energy Commission, State of California, USA.

Clarke, A.; Grant, N. & Thornton, J. (2009). *Quantifying the energy and carbon effects of water saving*. Environment Agency, Energy Saving Trust, UK.

Corominas, J. (2009). Agua y energía en el riego, en la época de la sostenibilidad [Water and Energy in Irrigated Agriculture in Sustainable Times]. *Ingeniería del Agua*, 17(3): 221–233.

Cots, Ll. (2011). *Desarrollo y calibración de un modelo de simulación de recursos hídricos aplicado a la cuenca del río Corb dentro de la zona regable de los canales de Urgell (Lleida)* [Design and Calibration of a Water Resources Simulation Model Applied to Corb River Basin Located Within the Irrigation District of Urgell Canals (Lleida)]. Barragán, J. (dir.), PhD Thesis, Escola Tècnica Superior d'Enginyeria Agrària, Universitat de Lleida, Spain.

Ederra, I. & Murugarren, N. (2010). *La nueva tarifa eléctrica, la escalada de precios del agua de riego* [The New Electricity Tariff, the Escalation in Water Prices for Irrigation]. Navarra Agraria. March–April 2010, pp. 45–50. Available from: [http://www.riegosdenavarra.com/publica/TarifasElectricas2010 NavarraAgraria.pdf].

EPEA (Asociación de Empresas con Productos Ecológicos de Andalucía) (2009). *Dinámica de desarrollo de la agricultura ecológica e innovación*, Loriol-sur-Drôme [Development Dynamic of Organic Agriculture and Innovation]. Available from: http://www.agencebio.org/upload/pagesEdito/fichiers/JMLuque090909.pdf [Accessed 3rd March 2011].

EuroStat (2008). Water Statistics – Annual data, EuroStat Database. Available from: http://epp.eurostat.ec.europa.eu/portal/page/portal/statistics/search_database [Accessed 3rd March 2011].

Hardy, L. & Garrido, A. (2010). *Análisis y evaluación de las relaciones entre el Agua y la Energía en España* [Analysis and Evaluation of Spain's Water–Energy Relationship]. Botín Foundation, Santander, Spain.

Hardy, L.; Garrido, A. & Juana, L. (2012). Evaluation of Spain's Water-Energy Nexus. *International Journal of Water Resources Development*, 28(1), Special Issue: Water Policy and Management in Spain. Taylor & Francis, Leiden, the Netherlands.

Hoff, H. (2011). *Understanding the Nexus*. Background Paper for the Bonn 2011 Conference: The Water, Energy and Food Security Nexus. Stockholm Environment Institute, Sweden.

IDAE (Instituto para la Diversificación y Ahorro de la Energía) (2010). *Guía Práctica de la Energía, Consumo Eficiente y Responsible* [Practical Guide to Energy, Efficiency and Responsible Consumption]. Madrid, Spain.

IIT (Instituto de Investigación Tecnológica) (2005). *Renovables 2050 – Un informe sobre el potencial de las energías renovables en la España peninsular* [Renewables 2050 – Report on Renewable Energies Potential for Peninsular Spain]. Greenpeace.

Junta de Andalucía (2010). *Inventario de regadíos y su evolución en la última década* [Irrigation Districts Inventory and Evolution from the Last Decade]. Department of Agriculture and Fishing, Committee of Andalucía, Seville, Spain.

Krinner, W.; García, A. & Estrada, F. (1994). Method for Estimation of Efficiency in Spanish Irrigation Systems. *Journal of Irrigation and Drainage Engineering*, ASCE, 120(5): 979–986.

Linares, P. & Sáenz de Miera, G. (2009). *Implications for water of the world energy scenarios.* Working paper of the Instituto de Investigación Tecnológica, Universidad Pontificia Comillas, Madrid, Spain.

MARM (Ministerio de Medio Ambiente, Rural y Marino) (2008a). *Encuesta sobre superficies y rendimientos de cultivos* [Crop Areas and Yields Survey] (ESYRCE). Available from: http://www.magrama.gob.es/ministerio/pags/Biblioteca/Revistas/pdf%5FESRC%2FESRC%5F2008%2Epdf [Accessed 4th March 2011].

MARM (Ministerio de Medio Ambiente, Rural y Marino) (2008b). Anuario de estadística agraria 2009 [Agricultural Statistics Yearbook 2009]. Available from: http://www.magrama. gob.es/es/estadistica/temas/estad-publicaciones/anuario-de-estadistica/default.aspx#para3 [Accessed 4th March 2011].

MARM (Ministerio de Medio Ambiente, Rural y Marino) (2009). *Encuesta sobre superficies y rendimientos de cultivos* [Crop Areas and Yields Survey] (ESYRCE). Available from: http://www.magrama.gob.es/ministerio/pags/Biblioteca/Revistas/pdf%5FESRC%2FESRC%5F2009%2Epdf [Accessed 4th March 2011].

MITYC (Ministerio de Industria, Turismo y Comercio) (2011). *Balance Energético 2010 [Energy Balance 2010]*. Press release. Madrid, Spain.

Mujeriego, R. (2006). *La reutilización planificada del agua* [Managed Water Reuse]. Sabadell Universitat. Available from: http://www.canagua.com/es/pdf/reutilizacion.pdf [Accessed 4th July 2011].

Pate, R.; Hightower, M.; Cameron, C. & Einfeld, W. (2007). *Overview of Energy-Water interdependencies and the emerging energy demands on water resources.* Sandia National Laboratories, Albuquerque, New Mexico, USA.

Pérez Pastor, A. (2010) (Coord). Special issue (S2). Solutions to the water deficit in agriculture. *Spanish Journal of Agricultural Research* 8(S2), S3

PwC (PricewaterhouseCoopers) (2010). *El modelo eléctrico español en 2030, Escenarios y alternativas* [The Spanish Electricity Model in 2030, Scenarios and Alternatives]. PricewaterhouseCoopers.

Rio Carrillo, A.M. & Frei, C. (2009). Water: A key resource in energy production. *Energy Policy*, 37: 4303–4312, DOI: 10.1016/j.enpol.2009.05.074.

Torres Corral, M. (2005). Desalación y planificación hidrológica hoy [Desalination and Water Management today]. *Ingeniería y Territorio*, 72. Colegio de Ingenieros de Caminos, Canales y Puertos, Barcelona, Spain.

UNESA (2007). *Prospectiva de Generación Eléctrica 2030* [Prospective of Electricity Generation in 2030]. Madrid, Spain.

Ward, F.A. & Pulido, M. (2008). Water conservation in irrigation can increase water use. *Proceedings of the National Academy of Sciences of the USA* (PNAS), 105(47): 18215–18220.

Water Environment Federation (2009). *Energy Conservation in Water and Wastewater Facilities. Manual of practice No. 32*. WEF Press, Virginia, USA.

Chapter 15

Considerations on climate variability and change in Spain

Alberto Garrido[1], Bárbara Willaarts[1],
Elena López-Gunn[2] & Dolores Rey[3]
[1] *Water Observatory of the Botín Foundation;*
  CEIGRAM, Technical University of Madrid, Madrid, Spain
[2] *Water Observatory of the Botín Foundation; Department of*
  Geodynamics, Complutense University of Madrid, Madrid, Spain
[3] *CEIGRAM, Technical University of Madrid, Madrid, Spain*

ABSTRACT: This chapter summarizes the existing knowledge about climate change in Spain and its potential impacts on water resources and demands. Increasing evidence of climate changing conditions has prompted Spanish water agencies and governments to take into account possible water scenarios, which mostly indicate a reduction of runoff and the increased likelihood of extreme events, into basin management plans. Yet, uncertainties are still high and predictions have considerable interval ranges. This chapter argues that water institutions and adaptive management are important to anticipate no regret measures to tackle the worsening of hydrological regimes. Also greater efforts need to be placed in searching for mitigation measures.

Keywords: climate change, runoff decrease, adaptation, extreme events, evapotranspiration

1   INTRODUCTION

The latest published report of the Intergovernmental Panel on Climate Change concluded that between 1970 and 2004 global CO_2 emissions had increased 80% (IPCC, 2007). As a result mean global temperature has augmented significantly, more so in the Northern Hemisphere where annual temperature has risen between 0.2 and 2°C. Predictions to 2030 forecast a possible increase in global temperature ranging from 1.8 to 4°C, depending on the emission scenario used. Southern Europe is expected to be particularly vulnerable to climate change (CC), at least in all environmental and social aspects that depend on water resources (e.g. reduction of water availability, hydropower potential or agricultural productivity and increasing risk of wildfires) (Giorgi & Lionello, 2008). This chapter updates the CC projections for Spain and the possible impacts on water resources, by looking at both projected changes in runoff and expected variations in water demand in three different domains: agriculture, forests and urban areas.

2   PROJECTED CHANGES IN PRECIPITATION AND TEMPERATURE

The Mediterranean region is likely to suffer more severe climate change impacts than other EU countries (Bates *et al.*, 2008). Droughts could become more intense and frequent, and rivers' run-off may decrease (Fischer *et al.*, 2007; Lorenzo-Lacruz *et al.*, 2012). Despite these projections, Spain lacked a detailed spatial assessment of potential changes in temperature and precipitations induced by an increase in CO_2. This assessment is needed to forecast as far as feasible potential on-site hydrological changes and identify possible adaptation measures. Most national assessments performed so far (de Castro *et al.*, 2005; Iglesias *et al.*, 2005) relied on calculations obtained from broad-scale assessments from Atmosphere-Ocean coupled General Circulation Models (AOGCMs), which have low spatial resolution and a high degree of uncertainty. In response to this information gap, the Spanish Climate Change Office (*Oficina Española de Cambio Climático*, OECC), coordinator of the National Adaptation Plan to Climate Change (*Plan Nacional de Adaptación al Cambio Climático*, PNACC) issued in 2006, commissioned the elaboration of a regional assessment on the likely impacts of Climate Change on water resources in Spain. The first output has been the publication of the so-called *Assessment of the effects of CC on natural water resources* (CEDEX, 2011), which represents the most up to date report on future climate scenarios at the regional scale for Spain (1 km² resolution). According to this report, the mean annual temperature in Spain is expected to increase progressively along the 21st century, +0.065°C/year under A2 scenario and +0.048°C/year under scenario B2. This means that by 2040 mean annual temperature in Spain could increase between +1.4 to +1.9°C.

According to the CEDEX report, the mean annual precipitation is likely to decrease by up to −0.88 mm/year under A2 scenario, and −0.18 mm/year under a B2. This implies that annual precipitation could decrease between 5 and 6% by 2040. Yet, there is still a high degree of uncertainty linked to future rainfall trends in Spain, since most regional models show prediction errors above 15% when estimating annual variations. The uncertainty in forecasts is even higher across seasons, with mean errors ranging between −33% to +30%.

Despite the uncertainties surrounding future precipitation trends, it is pertinent to learn what may happen under different scenarios across different Spanish basins. According to the CEDEX report (2011), precipitation will decrease especially in the Canary Islands and the Southern Peninsular basins between 7–14%, depending on the scenario by 2040 (see Figure 1). The Eastern and Northern basins are not expected to experience large changes in rainfall patterns. Meanwhile a rise in temperature is expected in all basins, with inland catchments experiencing a larger increase (between +1.4 and +1.6°C). Coastal Atlantic and Mediterranean basins will be less affected by the increase in temperature.

If the above projections materialise, the Southern Mediterranean region of Spain will move towards its *aridification* while the Northern part of the country will experience a *Mediterranization* process. From a hydrological perspective, the projected increase in temperature together with the likely decrease in annual rainfall, might lead to an overall reduction in water availability in most basins. Figure 2 shows the projected regional decrease in runoff and recharge between 2010 and 2040 for the

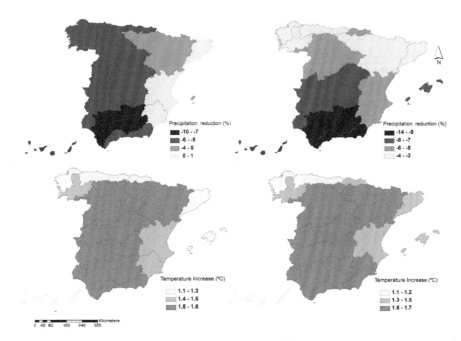

Figure 1 Forecasted changes in mean annual precipitation (%) and mean annual temperature (°C) between 2010 and 2040. Projections represent the mean value obtained from the regionalization of three AOGCMs (HadCM3, ECHAM4 & CGCM2). (Source: Own elaboration based on CEDEX (2011)).

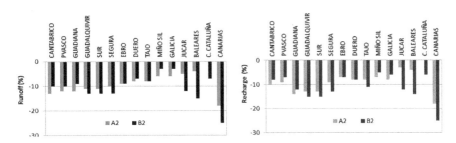

Figure 2 Mean annual reductions in runoff and recharge (%) in the different Spanish River basins between 2010–2040 under A2 and B2 scenarios. Projections represent the mean annual value obtained from the regionalization of three AOGCMs (HadCM3, ECHAM4 & CGCM2). (Source: Own elaboration based on CEDEX (2011)).

different basins. The Canary Islands would suffer the largest reduction in runoff and recharge (up to 25%). Within the Iberian Peninsula, the Southern basins of Guadalquivir, Sur, Guadiana and Segura would experience a significant reduction in surface and groundwater resources (up to 13% of runoff and 15% of groundwater recharge). Mediterranean basins like the Ebro, Catalonia Inland basins and Jucar would suffer a smaller reduction in water availability (below 10%), as the temperature increase along the coast is not expected to rise as much as inland. However, some detailed studies conducted in the Ebro Basin predict greater reductions. For instance, Quiroga *et al.* (2011) project a runoff reduction up to 29% under B2 scenario and even greater under scenario A2 (−46%). According to CEDEX (2011) Northern basins will also experience a similar runoff reduction, but the impact of water shortages might be smaller, since most of them are water abundant basins.

The water resource scenarios shown in Figure 2 have nevertheless a high level of uncertainty. Foremost, because future climate trends remain unclear and this is compounded by the inherent variability of the Mediterranean climate itself and the complex Spanish geography, which adds important challenges and unknowns when trying to make plausible climate change predictions.

3   INCREASED CLIMATE VARIABILITY AND MORE EXTREME EVENTS

The IPCC report on climate change and water (Bates *et al.*, 2008) forecasts that for mid-latitudes regions like Spain, the frequency of extreme rainfall events would increase and drought periods would be longer and more frequent. Yet, no clear increase in the frequency of floods has been observed during the 20th century, although the rate in short-term droughts has increased slightly since the 1950s (CEDEX, 2011).. These trends seem to also be supported by regional studies like Valencia *et al.* (2010), who found that in the Ebro basin the precipitation regime is now more homogenous than thirty years ago, although the rate of droughts has augmented. Vicente-Serrano & Cuadrat-Prats (2007) also demonstrated that from 1951 to 2000 there has been an increase in the severity of droughts in the Middle Ebro Valley, although with wide spatial variability.

Lorenzo-Lacruz *et al.* (2012) analyzed the evolution of the streamflow in the main rivers of the Iberian Peninsula during the last half of the 20th century. His study evidences a downward trend in annual, winter and spring streamflows, especially pronounced in the Central and Southern basins. The reduction in winter and spring streamflow is attributed to several causes, including changes in the seasonal rainfall pattern. Other important non-climatic factors such as reforestation, an increase in water demand and current water management strategies all play an important role in the observed streamflow evolution.

4   AGRICULTURAL WATER DEMAND

The uncertainty of climate change's projections makes difficult to develop and implement adaptation strategies. Small changes in agricultural water use could have significant economic and hydrological impacts. Water policy faces the dilemma of ensuring the sustainability of water resources in the future, while maintaining the strategic targets of agriculture, society and environment. Improving access reliability and meeting all users' expected availability are potential opposing goals for water management, which require compromises and adaptive capacity.

Crops' water demand is especially sensitive to changes in precipitation and runoff patterns, increase in temperature and to high levels of CO_2 concentration (Frederick & Major, 1997). If temperature rises, photosynthesis activity could increase and stomatal conductance could be lower. Therefore, crops' water use efficiency could be higher. Changes in crop's water needs will depend on the thermal requirements of each crop, and the period of the year in which the crop grows. It may be necessary to replace high water-demanding crops like rice, maize in some areas, and to stop the irrigation of inadequate soils (Iglesias *et al.*, 2005).

Crops' evapotranspiration rate could increase due to higher temperatures, and this could lead to greater water needs (Moratiel *et al.*, 2010). However, even if traditional varieties and sowing dates are maintained, the crop cycle could be shortened because of higher temperatures, and this would have the opposite effect on crops' total water needs. Controversy exists about whether improved yields under drought conditions must come at the expense of yields in the seasons when the rainfall is favourable. However, research is making improvements on those areas, and most likely varieties will emphasize one or another trait to offer farmers best or more suitable options to their specific climate circumstances.

Climate change is likely to have a wide range of impacts on agriculture, but there is a great deal of uncertainty in the implications that this might have for water management and water policy. In Spain, irrigation is the main water consumer, accounting for about 65% of total water demand (See Chapter 6). Changes in water demand will affect irrigated crops' profitability if more water is needed for irrigation. Also,

---

Box I   Climate change and the case of maize in Spain

Maize's evapotranspiration and irrigation requirements are expected to decrease in all sites studied in the Iberian Peninsula under A2 climate change scenario (Rey *et al.*, 2011) (Figure 3). The decrease in maize's evapotranspiration could be caused by decreases in the number of growing days and in Leaf Area Index due to higher temperatures, and a lower transpiration due to stomata closure caused by a higher concentration of CO_2. Maize's yield could be lower, because it is a very sensitive crop to high temperatures.

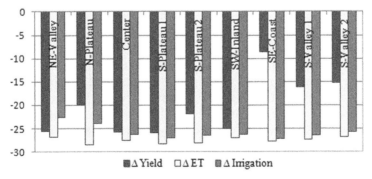

Figure 3   Variations (in %) of yield, evapotranspiration (ET) and irrigation needs of maize between control period (1961–1990) and future climate (2071–2100), due to CC in each site under study (A2 emissions scenario). Projections for current maize's varieties. (Source: Rey *et al.* (2011)).

As shown in Figure 3, ET decreases would be above 25% in most cases and almost 28% in the Plateaus. Reduction in maize's irrigation needs is lower. Yield decrease is less homogeneous, but significant in all sites. Therefore climate change could have very negative impacts on maize's yield in Spain, with decreases potentially exceeding 25% in the Central Plateau. However, maize's water needs due to new climatic conditions could also be lower than under current conditions. Ultimately, for each production area, water and maize prices will determine whether climate change makes the crops' profit larger or smaller. It largely depends on the price of maize, irrigation and energy prices. Probably, new maize's varieties better adapted to the new climate conditions will be developed in the next decades. If this was the case, the potential impacts reported here would not be so pronounced.

if precipitation decreases water will be scarcer, prices are expected to rise, increasing farmers' costs. Understanding how climate change could affect Spanish agriculture is the first step to mitigating the potential negative impacts of new climatic conditions. Moratiel *et al.* (2011) found that the expected climate change in the Duero Basin will cause an increase in reference evapotranspiration between 5% and 11% in the next 50 years compared to the current situation.

5　FOREST'S WATER DEMAND UNDER CLIMATE CHANGE

Forests, shrubs and natural pastures occupy approximately 54% of the Spanish territory (MARM, 2011), and consume on average 42% of the annual rainfall (see Chapter 11). Thus, if climate change predictions are plausible, these will have important effects on forest's ecology and its water balance.

From a water management perspective, gaining insight on the potential impacts of climate change on forests is crucial since changes in temperature and precipitation might alter forests' evapotranspiration, and consequently available water resources downstream (Otero *et al.*, 2011). Likewise, changes in forests' productivity due to water shortages might have important economic consequences for the Spanish forest sector, which annually generates over 1,000 M€ (AE, 2010).

Gracia *et al.* (2005) summarized the impacts of climate change on Spanish forest dynamics in three main aspects. First, if temperature and CO_2 increase, this will accelerate the leaf phenology in broadleaf forests and the renewal leaf capacity in evergreen species. As a result, leaf litter could increase and the overall carbon budget would become negative, implying that Spanish forests would become net sources of CO_2. Second, warmer conditions could increase the risk of pests in forests. Lastly, if the climate becomes drier, soil moisture could decrease and competition for water among trees could be higher, making trees potentially more vulnerable to droughts and extreme climatic events. Accordingly, CC is likely to reduce forest cover, and so will the forest's evapotranspiration. Also, there is a high chance that the remaining forest stands could shift from carbon sinks to net sources of CO_2.

Despite this general trend, the impacts of climate change on forests' water and carbon budgets would vary from region to region. The *fertilizer effect* caused by an increase in temperature and CO_2 concentrations could enhance primary productivity in the Atlantic region. In this area precipitation exceeds potential evapotranspiration, thus net primary productivity is likely to remain positive, implying a positive balance in carbon sequestration and a higher evapotranspiration. However, in the Mediterranean region where water is the most important limiting factor for forest growth, the likely impacts might differ. In these regions, and especially during spring and summertime when potential evapotranspiration is much greater than precipitations, primary productivity is likely to decrease due to a lack of water. Consequently, forest cover and evapotranspiration would decrease just in the Southern Mediterranean.

Several management measures have been studied to increase forest resilience and adaptation to CC. These measures include modifying the periods of forest intervention and the intensity of forest clearance. Different studies conducted in Mediterranean forests of Northern Spain show that lowering the intensity of forest intervention, i.e. maintaining a greater forest basal area will increase the capacity of forest as carbon sinks, both in the aerial part and in the soils. However, this increase in forest cover is likely to rise the water demand and reduce runoff. On the contrary, an increase in the intensity of forest clearance will reduce the carbon sequestration capacity of forest but it will increase soil water availability. According to Gracia *et al.* (2005) modifying the period of forest intervention seems to have a smaller effect on the carbon and water budget of forests. The EU project *Silvicultural Response Strategies to Climatic Change in the Management of European Forests* (SilviStrat) (Kellomäki & Leinonen, 2005) assessed the influence of different forest management regimes to cope with climate change in Europe. The report concludes that in forests subjected to extreme conditions – either due to low temperatures (boreal forests) or due to lack of sufficient water availability (Mediterranean forests) – none of the management measures mentioned above would remarkably increase the adapting capacity of forests to CC.

An important adaptation measure to cope with CC includes the improvement of land use planning by preventing forest ageing; and promoting the use of native species within afforestation programs which are best adapted to droughts. This is particularly important in the Mediterranean arc, since large afforestation programs might have negative consequences from the water availability perspective, due to the associated increase in water demand by forests (see Chapter 11). Further measures include the control of shrub encroachment linked to the abandonment agricultural fields.

6 URBAN DEMAND AND ADAPTATION MEASURES

Cities cover only 2% of the land surface, yet are responsible for almost 3/4 of CO_2 emissions concentrating more than half the world's population. In Spain, Madrid and Barcelona represent examples of these global metropolitan areas (see Box 2). The importance of cities in the context of climate variability and change is because often these metropolitan areas generate a large part of the country's Gross Domestic Product (GDP), e.g. the capital city of Madrid contributes 12% to the national GDP. Cities and urban areas are key areas for mitigation and adaptation for at least three reasons: first, in terms of mitigation, because of the large ecological footprint of cities, compared to its land area, with processes like water treatment which are energy intensive; second, because the urban water system is highly vulnerable to climate change impacts without foresight and planning (Loftus *et al.*, 2011); and third, due to the potential interaction of climate change with the urban heat island effect, the urban climate is hotter than surrounding areas, with implications into peak energy and water demands.

One of the difficulties in relation to water and CC is the upfront costs of adaptation, since these are immediate, whereas the potential benefits are uncertain, with pay back into the future and where there is an onus on no regret strategies, based on the precautionary principle. Infrastructure like water storage and distribution, water treatment plants, etc. normally requires large upfront investment, and key questions centre on the ideal timing of investment. Once investments are made, these infrastructure systems become locked in, which can perpetuate inefficiencies. There are therefore large associated issues related to investment risk, in e.g. water supply and drainage, flood management, or issues like how to retrofit existing infrastructure and how to plan for new or replacement infrastructure as this age or become obsolete. Normally, water and wastewater systems correspond to an important percentage of total infrastructure costs in an urban system. A key question therefore when considering both impacts of climate change and potential adaptation and mitigation measures are related to equity, i.e. the distributive effects of potential impacts of climate change and solutions like efficiency policies to invest on improved drainage and water infrastructure, or issues related to social cohesion, city competitiveness and the distributional effects of costs and benefits and who bears the costs of adaptation (OECD, 2008).

In terms of adaptive responses there are a number of options, from the technological, behavioural, economic instruments like markets or price signals, managerial and policy design (e.g. regulatory changes, incentives), or adequate insurance schemes (Fankhauser *et al.*, 2010; Garrido *et al.*, 2012). Some of these measures are softer, based on developing capacity or looking at institutional mechanisms like water rights and allocation, or water markets while others focus more on *hard infrastructure* like improved drainage.

In relation to integrated water (and land use) resources management, the implications and potential for regional planning are often overlooked, i.e. the consequences of a low carbon city and low carbon growth for spatial planning and establishing the potential implications of e.g. urbanization and urban development models in relation to CC. The potential for using spatial planning as an adaptation measure to CC, particularly for cities framed within their surrounding catchment, are largely unexplored in order to identify existing spatial climate variations, microclimates and the potential vulnerability to extreme events. Finally, on a more general note, a Climate Change National Adaptation Plan is being developed in Spain. This establishes a general reference framework to evaluate CC impacts, vulnerability and adaptation. According to Estrela (pers. comm.) climate change impacts are being taken into account in water balances of the upcoming River Basin Management Plans in Spain. This represents an opportunity to take into account climate change effects in water decision-making policies. This Plan includes an assessment of the management and capacity of the Spanish hydrological systems under different water resource scenarios, a second assessment of potential climate change effects on irrigation and a third one of climate change impacts on the ecological status of water.

Box 2   Climate change and water resources in the city of Madrid

The case of Madrid in some ways reflects general trends applicable to Spain as a climate variability and change hotspot. The Spanish capital experienced the highest temperature increases amongst 16 European capitals between 1970 and 2005. Moreover, the *Ayuntamiento de Madrid* (2008) on the basis of information from the Tagus Water Authority and scenarios from the National Adaptation Plan to Climate Change, projects that temperatures will increase significantly between 4°C and 7°C in the summer and between 2°C and 4°C in the winter in the last quarter of the 21st century, compared with temperature records between 1960 to 1990. Meanwhile precipitations are expected to decrease, especially during the summer and spring seasons. In addition, evapotranspiration is expected to decline between 40% and 60% during the summer, between 20% and 40% during the autumn, and less than 20% during the spring. In winter an increase in evapotranspiration of less than 20% is also predicted. Water availability is also expected to be negatively affected by CC. Predictions anticipate a reduction in reservoir water inflow and available water resources of 7%, whilst demand for irrigation from 2027 to 2050 is expected to increase by 10%, thus adding pressure on diminished water resources, with increased variability. Water resources have already decreased by 30% in the last 30 years in the so called short climatic cycle (from the 1970s), and this is already being incorporated into the current river basin planning cycle. As regards extreme weather events, uncertainty is large, but, overall, more frequent heat waves are expected, as well as more floods in areas close to the river Manzanares, which crosses the city of Madrid and is very close to residential housing. Finally, severe droughts are a further possible consequence of CC. Since Madrid is located in a continental Mediterranean area, with semi-arid conditions, reductions in water availability may lead to having to ban certain non-essential activities in the city of Madrid (watering parks and gardens, washing cars, etc.). Sometimes there will be trade-offs between mitigation and adaptation in other cases, some policy choices will be able to tackle both. In the case of Madrid, an example of autonomous adaptation for example is how the water supply company of Madrid is incorporating in current planning the probability that drought events might become more frequent and possibly longer, thus in equity terms the new customers and existing customers would be bearing this additional cost for potential events into the future. Reducing energy consumption and greenhouse gas emissions (GHG) derived from water pumping and final uses (heating and pressurising) could hence be fostered by regular awareness raising campaigns that may complement the use of water saving devices for showers or for example the installation of drip irrigation systems. In addition, the water supply company of Madrid has been pro-active in developing Manuals for Droughts (Cubillo & Ibáñez Carranza, 2003) to guarantee access to water resources, as well as a guide for decision-making under scarcity conditions. There are also plans for water trading with farmers in the Madrid catchment.

(Source: Lázaro-Touza & López-Gunn, 2011).

7   CONCLUSIONS

Scientific observations are showing ongoing climate change processes taking place in Spain. Future predictions indicate that CC may aggravate water scarcity in Spain, by increasing evapotranspiration, reducing precipitation and increasing the likelihood of extreme events. However these projections are still subject to considerable uncertainty. Difficult trade-offs will have to be faced in the future: if the reliability of water supplies is to be maintained, then demands might have to be curtailed. This will require proactive, flexible and adaptive management practices. Adoption of technological innovations will certainly help, more capital investment will be needed in urban and agricultural supply networks. A combination of more decentralization and liberalization, combined with targeted public intervention could be a win-win strategy. A consensus exists about the potential of water markets to add some degree of allocation flexibility, however these need to be better regulated (Chapter 16). As Chapter 6 shows, importing virtual water in the form of agricultural commodities is probably the cheapest adaptive mechanism Spain has to cope with drought cycles. However, an excess of imports reliance also generates large environmental externalities elsewhere (e.g. deforestation), which paradoxically might accelerate climate change. Overall, a combination of these measures will certainly contribute to adapt under eventual possible climate change scenarios. However, if we consider climate change as an effect rather than a cause, any major strategy to mitigate should place land use planning, including the agricultural and urban sectors, at the centre.

REFERENCES

AE (*Anuario de Estadística*) (2010). *Avance de los resultados del Anuario de Estadística 2010*. Ministerio de Medio Ambiente y Medio Rural y Marino. [Advance of the results of the Statistical Yearbook 2010-Ministry of Environment, and Rural and Marine Affairs]. Available online from: [http://www.marm.es/es/estadistica/temas/anuario-de-estadistica/] [Access: October 2011].

Ayuntamiento de Madrid (2008). Plan de Uso Sostenible de la Energía y Prevención del Cambio Climático de la Ciudad de Madrid 2008–2012 [Plan for the Sustainable Use of Energy and Climate Change Mitigation in the City of Madrid 2008–2012 - Madrid Government], Available online at: [http://www.madrid.es/UnidadWeb/Contenidos/Publicaciones/TemaMedioAmbiente/PlanEnergia/Planenergiasostenible.pdf].

Bates, B.C.; Kundzewicz, Z.W.; Wu, S. & Palutikof, J.P. (2008). *Climate Change and Water*. Technical Paper of the Inter-Governmental Panel on Climate Change. IPCC Secretariat, Geneva, Switzerland, 210.

CEDEX (2011). *Evaluación del impacto del cambio climático en los recursos hídricos en régimen natural* [Evaluation of the Impact of Climate Change in Water Resources under Natural Regime]. CEDEX, Madrid, Spain.

Cubillo, F. & Ibáñez Carranza, J.C. (2003). *Manual de Abastecimiento del Canal de Isabel II*. [Water Supply Manual of the Canal de Isabel II]. Canal de Isabel II, Madrid, Spain.

De Castro, M.; Martín-Vide, J. & Alonso, S. (2005). El clima en España: pasado, presente y escenarios de clima para el siglo XXI. [Climate in Spain: Past, present and scenarios of climate for the 21st Century] In: MMA (ed.), *Evaluación preliminar de los impactos en España por efecto del cambio climático*. [Preliminary Evaluation of Climate Change Impacts in Spain]. Ministerio de Medio Ambiente, Madrid, Spain: 1–65.

Fankhauser, S.; Hepburn, C. & Park, J. (2010). *Combining multiple climate policy instruments: how not to do it*. Centre for Climate Change Economics and Policy and Grantham Research Institute on Climate Change and the Environment.Working Paper no. 48.

Fischer, G.; Tubiello, F.N.; Van Velthuizen, H. & Wiberg, D.A. (2007). Climate change impacts on irrigation water requirements: Effects of mitigation, 1990–2080. *Technol. Forecast Soc.*74(7): 1083–1107.

Frederick, K.D. & Major, D.C. (1997). Climate change and water resources. *Climatic Change*, 37: 7–23.

Garrido, A.; Bielza, M.; Rey, D.; Mínguez, M.I. & Ruiz-Ramos, M. (2012). Insurance as an adaptation to Climate Variability in Agriculture. In: Dinar, A. & Mendelsohn, R. (eds.), *Handbook on Climate Change and Agriculture*. Edward Elgar Publishing Inc., Williston, Vermont, USA.

Giorgi, F. & Lionello, P. (2008). Climate change projections for the Mediterranean region. *Global Planet Change*, 63: 90–104.

Gracia, C.; Gil, L. & Montero, G. (2005). *Impactos sobre el sector forestal. Impactos del cambio climático en España* [Impacts on the Forest Sector. Climate Change Impacts in Spain]. Ministerio de Medio Ambiente, Madrid, Spain.

Iglesias, A.; Estrela, T. & Gallart, F. (2005). *Impactos sobre los recursos hídricos. Impactos del cambio climático en España* [Impacts on Water Resources. Climate Change Impacts in Spain]. Ministerio de Medio Ambiente, Madrid, Spain.

IPCC (Intergovernmental Panel on Climate Change) (2007). *Climate Change 2007: Synthesis Report*. Pachauri, R.K. & Reisinger, A.J. (eds.). Geneva, Switzerland.

Kellomäki, S. & Leinonen, S. (2005). *Final Report EU project Silvicultural Response Strategies to Climatic Change in Management of European Forests (SilviStrat)*. Contract EVK2-2000-00723. Available from: http://www.efi.int/portal/completed_projects/silvistrat/final_report/ [Access: 15th September 2011].

Lázaro-Touza, L.E. & López-Gunn, E. (2011). Climate change policies: mitigation and adaptation at the local level. The case of the city of Madrid (Spain). In: Tortora, M. (ed.), *Sustainable Systems and Energy Management at the Regional Level: Comparative Approaches*. IGI Global Publishers, USA.

Loftus, A.; Anton, B. & Philip, R. (2011). Adapting urban water systems to climate change: A handbook for decision makers at the local level. ICLEI; IWA; UNESCO-IHE. Bonn, Germany.

Lorenzo-Lacruz, J.; Vicente-Serrano, S.M.; López-Moreno, J.I.; Morán-Tejeda, E. & Zabalza, J. (2012). Recent trends in Iberian streamflows (1945–2005). *Journal of Hydrology*, 414–415: 463–75.

MARM (Ministerio de Medio Ambiente y Medio Rural y Marino) (2011). *Tercer Inventario Forestal Nacional (1997–2007)*. Available from: http://www.marm.es/es/biodiversidad/servicios/banco-de-datos-biodiversidad/informacion-disponible/ifn2.aspx [Accessed 30th November 2011].

Moratiel, R.; Snyder, R.L.; Durán, J.M. & Tarquis, A.M. (2011). Trends in climatic variables and future reference evapotranspiration in Duero Valley (Spain). *Natural Hazards and Earth System Sciences*, 11: 1–11.

Moratiel, R.; Durán, J.M. & Snyder, R.L. (2010). Responses of reference evapotranspiration to changes in atmospheric humidity and air temperature in Spain. *Climate Research*, 44: 27–40.

OECD (Organisation for Economic Cooperation and Development) (2008). *Competitive Cities and Climate Change*. 2nd annual meeting of the OECD roundtable strategy for urban development. OECD, Milan, Italy.

Otero, I.; Boada, M.; Badia, A.; Pla, E.; Vayreda, J.; Sabaté, S.; Gracia, C.A. & Peñuelas, J. (2011). Loss of water availability and stream biodiversity under land abandonment and climate change in a Mediterranean catchment (Olzinelles, NE Spain). *Land Use Policy*, 28(1): 207–218.

Quiroga, S.; Garrote, L.; Iglesias, A.; Fernández-Haddad, Z.; Schlickenrieder, J.; de Lama, C. & Sánchez-Arcilla, A. (2011). The economic value of drought information for water management under climate change: a case study in the Ebro basin. *Natural Hazards and Earth System Sciences*, 11: 643–657.

Rey, D.; Garrido, A.; Mínguez, M.I. & Ruiz-Ramos, M. (2011). Impact of climate change on maize's water needs, yields and profitability under various water prices in Spain. *Spanish Journal of Agricultural Research*, 9(4): 1047–1058.

Valencia, J.L.; Saa Requejo, A.; Gascó, J.M. & Tarquis, A.M. (2010). A universal multifractal description applied to precipitation patterns of the Ebro River Basin, Spain. *Climate Research*, 44: 17–25.

Vicente-Serrano, S.M. & Cuadrat-Prats, J.M. (2007). Trends in drought intensity and variability in the middle Ebro valley (NE Spain) during the second half of the twentieth century. *Theoretical and Applied Climatology*, 88: 247–258.

Part 4

Possible mechanisms and enabling conditions

Chapter 16

Water trading in Spain

Alberto Garrido[1], Dolores Rey[2] & Javier Calatrava[3]
[1] *Water Observatory of the Botín Foundation;*
  CEIGRAM, Technical University of Madrid, Madrid, Spain
[2] *CEIGRAM, Technical University of Madrid, Madrid, Spain*
[3] *Department of Business Economics,*
  Technical University of Cartagena, Cartagena, Spain

ABSTRACT: This chapter reviews Spanish water market regulation established in the Water Law Reform of 1999. It also provides an overlook of the type of exchanges that took place between 2004 and 2008, when market exchanges were more frequent. While exchanged amounts were not very significant in absolute terms, those that involved inter-basin transfers raised the most concerns and significance. The chapter describes in detail various market mechanisms used in different basins, including the exchanges that took place through Water Exchange Centres run by the basin agencies. All market inter-basin exchanges involved transfer of water from the Southern Castilian Plateau and from the headwaters of the Guadalquivir basin to the most arid areas in the Southeast of Spain (Murcia and Almería). The chapter summarizes the findings of two workshops devoted to discuss the market experiences, with water officers, market participants and scientists. A list of recommendations to reform water markets regulation and monitoring is offered in the concluding section.

Keywords: water markets, regulation, inter-basin transfers, market price, water banks, water exchange centres

1 INTRODUCTION

The Mediterranean region will probably be among the areas most affected by climate change in terms of reduced precipitation and increased frequency of extreme events (see Chapter 15). Market mechanisms that value water resources and provide compensation through voluntary transfers of water rights or water use rights can become an essential instrument for coping with water scarcity challenges. Establishing water markets (WMs) is an alternative means for improving water economic use and efficiency. Nevertheless, in Europe, most Member States do not envisage establishing WMs to fight water scarcity, except UK, Belgium and Spain (EC, 2011).

In 1999 the Spanish Water Law was reformed to allow holders of water rights to exchange their water by leasing-out temporally or till maturity their concessions. But exchanges were very rare before 2004, when the onset of the 2004–2008 drought created conditions for private gains-from-trade and the Government gave clear support to the proposed exchanges. This chapter reviews the Spanish WM experience, including its weaknesses and opportunities. It concludes with a number of proposals to upgrade WMs and make them more efficient, liquid and sustainable.

2 WATER MARKETS IN SPAIN

2.1 Institutional set-up

Water rights issued by the Water Authorities are made available through publicly built infrastructures or privately built with permission of the State (hydroelectricity). According to the 1985 Water Act, rights can also be granted to pump groundwater or divert resources directly from surface water bodies. A competitive process (public tender for licenses) for potentially interested agents is used only for hydropower applicants. Irrigators and urban users must go through a technical and administrative process, which aims at establishing the socio-economic interest of the request and its technical and environmental feasibility.

Water use rights are defined by the abstraction point, type of use, calendar, plots and crops to be irrigated and irrigation technologies, usable volume or flow and return flows. The type of use, location, abstraction or return points cannot be changed without an explicit approval by the River Basin Agency (RBA). Rights differ in the priority of their access to water depending on the type of use (domestic, environmental, agricultural, hydropower or industrial).

With the approval of the 1985 Water Act, water became a good under the sphere of the public domain. Nonetheless, holders of private rights over groundwater were given the choice of keeping them as a private right or else converting them into temporal water concessions. A vast majority (more than 80% of right holders according to Llamas *et al.*, 2001) opted for the first option. Any new rights over groundwater granted after 1985 would exclusively be a concession of use of a public good – water. The 1999 reform of the Water Act introduced the legal possibility of voluntary exchanges of public water concessions, but with many restrictions. It only allows the temporary exchanges of public water use rights. Before the 1999 reform only private rights could be formally traded; water flows pumped from private wells could be leased, auctioned or sold.

The 1999 Water Law Reform identified only two ways to exchange public water use rights: i) Right-holders that voluntarily agree on specific terms of trade and jointly file a request in the Agency to lease-out for a number of years the water to which right-holders are entitled; ii) Water banks operations (or *water exchange centres*, as they are called in the 1999 Water Law). Initiated and administered by the RBAs, water banks are set up as public tenders for interested right-holders who would be willing to relinquish their water rights temporally or for the remaining maturity period. The bank's water supply operation involves procuring volumes from voluntary sellers, and making them available for other users, including environmental restoration purposes. Bank's operations may also acquire permanent water rights and operate in exceptional situations of drought or over-exploitation of aquifers (WWF, 2005). In practice, these *water exchange centres* have only functioned as buyers of water or water rights. Water has not been sold to other users. Instead, purchased water has been made available to other users free of charges in the form of new water concessions or devoted to maintaining environmental flows.

2.2 Barriers and limits to trade

The Spanish regulatory framework can be best defined by reviewing the barriers that limit the type of exchanges. Three different barriers can be distinguished: legal barriers, institutional barriers and environmental barriers.

There are two kinds of legal barriers: i) market barriers, that may be erected based on evidence of monopoly power; market barriers result from public agencies' responsibilities and service, and without them, the market would be environmentally harmful and poorly enforceable; ii) barriers related to water rights' definition. In Spain, legal specialists differ in interpreting whether the rights definition necessarily hampers the market (Ariño & Sastre, 2009) or simply enforces the Water Act tenets (Embid Irujo, 2010). Water rights in Spain were originally not designed to be tradable (Garrido & Calatrava, 2009); they were made tradable under certain circumstances in the 1999 Water Law Reform.

Among the different regulatory elements identified by Ariño & Sastre (2009: 100–101) there are some that can restrict the functioning of WMs. First, rights to consumptive uses cannot be sold to holders for non-consumptive uses (hydropower) and vice versa. Second, there are restrictions on potential water buyers, as rights can only be leased out to other rights holders of an equivalent or higher category in the order of preference established by river basin planning or in accordance with the Water Act. Third, there are limits to the spatial extent of trading: licenses for the use of public infrastructure connecting different river basin areas may only be authorized if they come under the National Hydrological Plan or other specific laws. Fourth, there are limits on price setting; regulations may determine maximum price limits for water licenses. Competitive pricing can be substituted by administrative intervention.

The following institutional barriers can be identified: i) regional or area-of-origin barriers. These barriers result from the restrictions, or even upfront opposition, to trading by area-of-origin representatives. For instance, the Regional Government of Aragón blocked water transfer to the Barcelona area during the 2008 drought. It also has in its Autonomy Statute (more or less a regional Constitution) an explicit obligation on Aragón's President to prevent any water transfer out of the region's borders; ii) inter-sectoral barriers occur when representatives of one sector collectively fights exchanges that go against its political standing within the hierarchy of water rights and political priorities. This is generally the case of irrigators. A huge literature (see Easter *et al.*, 1998) exists that shows farmers being initially reluctant to sell water out of the sector. For example, irrigators in the Ebro basin made their water rights available to the Barcelona city during the severe water supply crisis of the 2005–2008 drought, but they would not accept any monetary payment for transferring their resources. There are strategic reasons for combating out-of-sector water sales, chief among them the fear that eventually their tradable rights will be questioned and perhaps irrigators will be deprived of them.

Environmental barriers are those enforced by public agencies responsible for the stewardship of the ecological quality of rivers and water bodies. In general, these barriers, such as minimum environmental river flows, are based on modelling evidence, and are hardly contested. Occasionally, an *environmental tax* is imposed as a proportion of the volume/flow to which the traded right is entitled and which should be left in the natural source.

2.3   A review of Spanish water market experiences

Since 2005, WMs have become more frequent in Spain, although traded volumes represented less than 1% of all annual consumptive uses. During the 2005–2008 drought, WM exchanges alleviated the conditions of those basins where water scarcity was most severe. Water trading takes place in many different ways. First, there are informal exchanges at the local level in many southern and eastern regions, taking place even before the 1999 Reformed Water Law was enacted. Second, there is trading of private groundwater pumping rights. Third, there are formal lease contracts and trading of public concessions under the 1999 Reformed Water Law. Some of these exchanges take the form of purchases of land upstream to transfer the water to other downstream areas of the same basin. Fourth, during the 2005–2008 drought the government allowed, under Royal Decree 15/2005 and subsequent Decrees, inter-basin temporary trading. Last, *water exchange centres* have been used to make purchases by Water Agencies (Offers of Public Purchase of Rights) for environmental or urban uses. These different types of water exchanges are not necessarily exclusive but complementary, as they satisfy different users' supply needs (Garrido & Calatrava, 2009).

Table 1 summarizes some of the existing market experiences and schematically lists both the currently existing exchanges typology and their characteristics.

2.4 Exchanged volumes and prices: economic interpretation

In this section, we report the exchanged volumes and prices in the most important typologies of WMs in Spain. These are the most recent and relevant market experiences in terms of exchanged volumes.[1]

Case I Operations of the Guadiana exchange centre

In the Guadiana basin (central Spain) the *Special Plan for the Upper Guadiana* was approved in order to solve the environmental problems affecting groundwater bodies due to aquifer overexploitation (see Chapter 20). A public water bank was established to acquire rights to reduce pumping rates by 250 hm^3 by 2027 [hm^3 = cubic hectometre = million m^3 = 10^6 m^3]. There were three public offers (October 2006, March 2007 and September 2007) targeted to irrigators, but required the means to acquire land rights with appurtenant water rights, to prevent further irrigation consumption in these lands. Right-holders located in areas closer to river banks or protected areas were prioritized among the lowest bidders. Maximum prices were set at 10,000 €/ha for land without permanent crops, 6,000 €/ha with permanent crops; the minimum price was 3,000 €/ha. In 2010, six operations were completed. With a total budget of 84.5 M€, only 66 M€ were spent to acquire 6,900 ha, with 29 hm^3 of registered groundwater rights, of which 13.6 hm^3 were transferred to the Regional Government of Castilla-La Mancha, which then allocated them in the form of public concessions to farmers that complied with certain requirements. The remaining 15.4 hm^3 correspond to the difference between the nominal water allotment of the purchased water rights (4,500 m^3/ha) and the effective amount of water available to farmers because of existing pumping restrictions (about 2,200 m^3/ha).

---

1 Some types of water markets, such as informal trading or trading of groundwater rights, are very difficult to document. The reader will find more information in Calatrava & Gómez-Ramos (2009) and Garrido & Calatrava (2009).

Table I Current Spanish water trading experiences.

Features										
Type of exchanges	Geography	Hydro-logical settings	Agents	Exchanges	Market structure	Potential environmental impacts	Regulatory framework	Role of water agency	Current existing trading	Observations
Informal trading of surface resources	Intra-basin (same irrigation district)	More frequent in dry years	Farmers within the same irrigation district (same public water concession)	Temporary	Bilateral agreements	Unlikely	Limited to the same district. Require permission of the irrigation district	None	Limited. Very difficult to document	Irrigation districts allow these exchanges only if no monetary compensation exists
Trading of private ground-water rights	Intra-basin (same area)	Drought and normal water availability periods. More frequent in dry periods	Sellers: holders of private groundwater rights. Buyers: other farmers, private societies, etc. Temporary buyers of water: mostly farmers	Permanent and temporary	Bilateral agreements	Excessive pumping and aquifer overdraft	Trading restricted to the same hydrological area. Prices in permanent trading must be above a minimum legal price established by regional governments. Restricted by the existence of transportation infrastructures	None	Very active in dry periods in the South-East basins	Low market transparency. Difficult to assess the extent of trading. Speculative nature of the market: in many cases right-holders are not users but only sellers. Large gains-from-trade, especially for sellers
Formal lease contracts	Intra-basin	Drought and normal water availability periods	All type of users	Temporary	Bilateral agreements	Only if environmental and/or return flows from upstream uses are affected	Water agency must permit the change in the place of use of the water. Restricted by the existence of transportation infrastructures	Gives permit to the water transfer. Owns main Transportation infrastructures. Establishes transportation fees	Very little activity documented	Potential was expected to be significant. In practice it has been quite limited

(Continued)

Table 1 (Continued).

Features Type exchanges	Geography	Hydro-logical settings	Agents	Exchanges	Market structure	Potential environmental impacts	Regulatory framework	Role of water agency	Current existing trading	Observations
Purchases of land	Intra-basin	Drought and normal water availability periods	Sellers of land: individual landowners. Buyers of land: irrigation districts. Buyer and seller of water is the same agent (irrigation district)	Permanent	Bilateral agreements in agricultural land markets	Only if environmental and/or return flows from upstream uses are affected	No limits to irrigated land trading. Water agency must permit the change in the place of use of the water. If the purchased land belongs to an irrigation district, this must permit the transfer. Restricted by the existence of infrastructures	Gives permit to the transfer of water. Owner of transportation infrastructures. Establishes fees for transportation	Some cases (e.g. Totana and Aguilas irrigation districts, Segura basin)	Potential for this type of trading is quite reduced. In terms of regulation they are similar to the formal lease contracts
Inter-basin trading	Between users in Tajo and Segura basins, and between users in the Guadalquivir and the Andalusian Mediterra-nean basins	Dry periods	Sellers: irrigation districts. Buyers: irrigation districts and the adminis-tration for domestic and environ-mental uses	Temporary	Bilateral agreements. Water administra-tion has played an active role as intermediary	Only if environmental and/or return flows from area-of-origin uses are affected. No effects have been documen-ted. Potential environmental benefits in the area of destination	Water agency (MAGRAMA*) must permit the change in the place of use of the water. Restricted by the existence of transportation infrastructures	Gives permit to the transfer. Owns main transportation infrastructures. Establishes fees for transportation	Some experiences with large exchanged volumes (Estremera -SCRATS, Canal de las Aves-MCT, Guadalquivir irrigation districts-Aguas del Almanzora)	Large potential. Strong opposition from regional governments in some areas-of-origin of water

Public Exchange Centres/ Water Banks	Intra-basin	Drought and normal availability periods	All. Potential sellers: agricultural users. Buyers would be other agricultural users and the administration for domestic and environmental uses	Temporary in the Segura; permanent in other basins	Via a central public agent	Only if environmental and/or return flows from upstream uses are affected	Restricted by the existence of transportation infrastructures. Prices set by water agency	Buyer and intermediary between sellers and buyers	Experiences in different Spanish basins (Júcar, Segura, Guadiana…)	Strongly supported by public budgets
Option Contracts	Intra-basin and inter-basin	Drought periods	All. Potential sellers: agricultural users. Buyers would be other agricultural users and the administration for domestic and environmental uses	Temporary	Bilateral agreements	Only if environmental and/or return flows from area-of-origin uses are affected. Potential environmental benefits in the area of destination.	Water agency must permit the change in the place of use of the water. Restricted by the existence of transportation infrastructures	Gives permit to the transfer of water. Owner of transportation infrastructures. Establishes fees for transportation	One recent experience between users in the Tajo and the Segura basin	

Source: Own elaboration.

* MAGRAMA: Ministerio de Agricultura, Alimentación y Medio Ambiente.

One subtlety of the Guadiana scheme is the fact that, while farmers entering the program must surrender their private rights, those that gain access to them will be granted 30-year *concession* rights (which is a more attenuated property than the others). So the Guadiana basin will have more users with *concessions* than with private rights (Garrido & Calatrava, 2009).

The Guadiana public offerings were planned to continue in 2008 and the following years but the effects of the global economic crisis brought the Special Plan for the Upper Guadiana to a sudden stop.

Case 2   Operations of the Júcar basin exchange centre

Its objective was to increase the water table levels to ensure that the Júcar River did not dry out during the dry spell of 2005–2008, as had occurred during the previous drought in the 1990s. The purchases were for just the 2006/07 and 2007/08 irrigation seasons. The aim was to reduce extractions by 100 hm³ in the Upper Júcar aquifer to enhance flows for the lower part of the basin. Farmers were given the option to lease-out their rights for one year in return for a compensation that varied between 0.13 to 0.19 €/m³, depending on the distance of the seller's location to the associated wetlands or to the river alluvial plain. The 2006/07 program had a budget of 12 M€ and purchased 27.3 hm³ for 5.5 M€. The second program in 2007/08 had a similar budget and required three rounds of acquisition offers (December 2007, February 2008 and March 2008) to acquire 50.6 hm³ for 12.7 M€ (CHJ, 2010). This exchange centre did not meet its purchase objectives, as there were not enough bidders to cover the entire budget and target volume.

Case 3   Operations of the Segura basin exchange centre

The Segura basin, in the southeast of Spain, is the most water-scarce basin in the country. There is a wide gap between water supply and demand, mainly because of increasing consumption, caused by the huge increase of irrigation schemes developed over the last few decades. The Segura exchange centre issued two public offers targeted to rice farmers in the upper part of the basin who were willing to temporarily lease their surface water. Two public offers were established in 2007 and 2008 with a budget of 700,000 € each, and a maximum price of 0.18 €/m³. In 2007, 2.93 hm³ were purchased at an average price of 0.168 €/m³ and with a total budgetary cost of 495,000 € (Calatrava & Gómez-Ramos, 2009). 41 lease contracts were signed with small farmers accounting for 371.5 ha. The result of the 2008 offer was similar to the 2007 one. Purchased volumes were intended for maintaining environmental flows in the Segura and Mundo River in the Albacete province (Castilla-La Mancha) but only once the domestic demands were satisfied. In practice, all the purchased volumes were for maintaining environmental flows.

Case 4   Formal lease contracts under the 1999 Reformed Water Law provisions

There are only a few documented experiences of formal lease contracts since the 1999 Reformed Water Act. Contrary to what was initially expected, many users have been reluctant to formally exchange their water or concessions. Maybe the most important experience in terms of volume was in the Tagus River in 2002, between the *Mancomunidad de*

Canales del Sorbe (Guadalajara), a large urban retailer (buyer), and the irrigation district of *Canal de Henares* (seller). 20 hm³ were transferred, at a fixed price of 38,000 €/year, plus a variable quantity of 0.04 €/m³ for the first 4 hm³, and 0.02 €/m³ for the rest of the total volume. In the Segura basin, 35 formal lease contracts were authorized between 2000 and 2005, for a total volume of 10.1 hm³, less than 1% of total annual water consumption in the basin (Calatrava & Gómez-Ramos, 2009). In the Guadalquivir, some exchanges represented just one right-holder permuting his own rights from the lower basin (with more salinity concentration) with his rights in the upper basin.

Case 5   Inter-basin exchanges under Royal Decree 15/2005

According to the 1999 Reformed Water Law, exchanges involving different river basins (jurisdictions) require the explicit approval of the Ministry of Environment. In 2005–2008, Spain suffered a drought that prompted the Spanish Government to permit inter-basin water exchanges (Royal Decree 15/2005). There are two important inter-basin aqueducts that would enable exchanges across basins (the Tagus-Segura Transfer and the Negratín-Almanzora Transfer, the latter between the Upper Guadalquivir basin and the Almanzora basin, in Almería). There are others operating in the country, but no exchange request has yet been filled.

Across-basin exchanges were contracted in 2006 (six in number, totalling 75.5 hm³), 2007 (17, representing 102 hm³), and 2008 (two, with 68 hm³). Farmers in the area-of-origin (Tagus and Upper Guadalquivir basins) leased out their water rights to farmers and urban users in the recipient basins of Segura (*Sindicato Central de Regantes del Acueducto Tajo-Segura* and *Mancomunidad de los Canales del Taibilla*) and the Andalusian Mediterranean basins (*Aguas del Almanzora*, which mainly services irrigators). In the Tagus basin, the sellers were the over-supplied irrigation districts of *Canal de Estremera* and *Canal de las Aves*. Farmers received a payment of 2,400 €/ha for fallowing their irrigated land, which in those years was more than the value of the crops (maize) they would have grown under normal conditions.

The amounts bought by users in the Segura basin from the Tagus basin only in 2006 largely surpassed those of all the exchanges approved among users in the Segura basin between 1999 and 2005. The *Mancomunidad de Canales del Taibilla*, the major urban water supplier in the Segura basin, signed an agreement in 2006 with farmers in the Upper Tajo basin (*Canal de las Aves* irrigation district) to buy up to 40 hm³ at a price of 0.28 €/m³. In 2007, 36.9 hm³ were bought at a price of 0.23 €/m³. The price in 2006 was greater because when the agreement was reached the selling farmers had already incurred in some cultivation costs (Calatrava & Gómez-Ramos, 2009).

The contract between the *Canal de Estremera* Irrigation District and the *Sindicato Central de Regantes del Acueducto Tajo-Segura* (SCRATS) has been active during 4 years. SCRATS paid 6 M€/year for 31 hm³/year. The price was 0.19 €/m³ in 2006 and increased up to 0.22 €/m³ in 2008 (Calatrava & Gómez-Ramos, 2009).

In 2007 and 2008, when almost no water could be transferred to the Almanzora Valley through the Tagus-Segura aqueduct due to the prolonged drought, farmers in the Almanzora looked for alternative resources (25 hm³/year) and established two type of agreements: i) They acquired 1,400 ha of irrigated land in the Marshes of Guadalquivir; and ii) established formal lease contracts with different irrigation districts in the Middle Guadalquivir (*Bembézar* and *Guadalmellato* irrigation districts) and the

Genil catchment (Corominas, 2011). This author calculates the profit obtained by the sellers in the Guadalquivir entering the latter-mentioned lease contracts as the difference between the income losses due to lower use of water and the received compensation. This profit was 220 €/ha (Guadalquivir) and 280 €/ha (Genil). Corominas (2011) stated that for prices of 0.15 €/m³, both buyers and sellers could obtain gains from the exchanges in the Guadalquivir River basin (in practice, the price was 0.18 €/m³).

The exchanging system in the former case involves three geographical sites in the arrangement: water rights linked to land in the lower Guadalquivir basin (i), were transferred to the Andalusian Mediterranean basins (ii), using the Aqueduct Negratín-Almanzora (iii), whose abstraction point is in the Upper Guadalquivir. However, there was only one agent, i.e. the company *Aguas del Almanzora*, which acted as buyer and seller at the same time. To reduce the environmental and third-party impacts a volumetric tax of 50% was enacted, which implied that the contractor was given permission to transfer only 50% of the water rights attached to the land purchased.

Aguas del Almanzora also established five-year water lease agreements with farmers in the Middle Almanzora Valley (*Pago de la Vega del Serón* irrigation district) with concessions from the Negratín reservoir (Guadalquivir basin) at prices in the range of 0.15–0.18 €/m³.

A common element in both across-basins exchanges is the fact that the MARM (Spanish Ministry of Environment) decided to exempt the exchanging parties from paying the fees applicable to all regular aqueduct beneficiaries, on the grounds that there was an extreme drought situation in which these exchanges took place. In the case of the inter-basin Tagus-Segura Aqueduct the fees ranged from 0.15 €/m³ for irrigators to 0.21 €/m³ for water agencies supplying municipalities in the recipient region.

3   CONCLUSIONS

More than twenty different experts and stakeholders were consulted in the course of two meetings during 2011. All consulted experts were knowledgeable of the market experiences reviewed earlier in depth, directly or indirectly. A wide consensus exists among them about water markets (WMs) being considered an interesting tool to help water allocation in Spain. They agreed on considering WMs as having a great potential in solving critical situations related to water scarcity and drought. However, most consulted stakeholders found several weaknesses or problems in the current Spanish WMs system. The lack of transparency has been identified as one of the main challenges of water management in Spain (see Chapter 17). There is hardly any public information about who uses the water and for what, or what are the potential benefits and externalities. The lack of information is exacerbated in a context of liberalized water reforms. In the absence of robust water governance and effective surveillance, it is very unlikely that WMs will be efficient and socially accepted.

Other important issues that were identified as critical were:

- The need for more flexibility in the priorities criterions used to allocate water as established in the Law or in the Water planning documents.
- The need for national legislation to clarify the conditions under which those exchanges that involve more than one region could be made. The existing legislation should

clarify aspects such as the spatial and temporal restrictions to trading or the criteria for the approval or rejection of water exchanges by the Water Authorities. Also, the integration of water trading in the process of hydrological planning would be desirable.

– Water prices were too high because sellers had in most cases a dominant position. So there should be more transparency in price-setting.
– Public Exchange Centres (Water Banks) have been mainly used in Spain to solve environmental problems related to the overexploitation of water bodies in different basins. They should have a more active role in pursuing other formats of market exchanges.
– There are only a few documented experiences of formal lease contracts between right holders. Trading has been concentrated in the southeastern part of Spain with limited numbers of participants. In general, the participation of individual right holders has been limited while the participation of governmental bodies or public water agencies as buyers has been the rule rather than the exception. Moreover, in general terms, there were not enough bidders to cover the entire budget when a public water exchange centre was established.
– During the 2005–2008 drought period, the Spanish Government permitted interbasin market exchanges using the pre-existing water infrastructures. The role of the central government was instrumental in facilitating exchanges across basins, but new regulations are needed to review and process them in a more transparent and expedient way.

One example of water market reform came in 2010 with Andalucía's new Water Law, after this region assumed almost all competencies in terms of water management in its territory in 2007. Although constrained by the National Water Law, the Andalusian Water Law established some changes related to the water management with the WFD criteria. One of the main differences with the National Water Law is the change in the priority system. Now, irrigation is on the same level as other users such as industries (for example, thermo-solar plants), so exchanges between these two uses are allowed. When allocating water, economic and environmental efficiency, third party effects and other aspects will be taken into account. As many stakeholders believe that the priorities system should be more flexible, the change in the Andalusian Water Law could serve as a precedent.

REFERENCES

Ariño, G. & Sastre, M. (2009). Water sector regulation and liberalization. In: Garrido, A. & Llamas, M.R. (eds.), *Water Policy in Spain*. CRC Press, Taylor & Francis, Leiden: 95–106.

Calatrava, J. & Gómez-Ramos, A. (2009). El papel de los mercados de agua como instrumento de asignación de recursos hídricos en el regadío español [The role of water markets as an allocative instrument of water resources in Spanish irrigation sector]. In: Gómez-Limón, J.A.; Calatrava, J.; Garrido, A.; Sáez, F.J. & Xabadia, A. (eds.), *La economía del agua de riego en España* [The economics of irrigation water in Spain]. Fundación Cajamar, Almería, Spain: 295–319.

CHJ (Confederación Hidrográfica del Júcar) (2010). *Memoria 2004/2009* [Report 2004/2009]. Confederación Hidrográfica del Júcar [Jucar River Basin Authority].

Corominas, J. (2011). *Aplicación a los regadíos del Guadalquivir del intercambio de derechos del agua en situaciones de dotación restringida.* [Lecture]. Botín Foundation Seminar on Water Markets in Spain. June 2011, Madrid, Spain.

Easter, K.W.; Rosegrant, M.W. & Dinar, A. (1998). *Markets for Water: Potential and Performance.* Kluwer Academic Publishers.

EC (European Commission) (2011). Third Follow up Report to the Communication on water scarcity and droughts in the European Union. COM (2004) 414 final. Available from: http://eurlex.europa.eu/LexUriServ/LexUriServ.do?uri=COM:2011:0133:FIN:EN:PDF [Accessed 28th May 2011].

Embid Irujo, A. (2010). The foundations and principles of modern water law. In: Garrido A. & Llamas, M.R. (eds.), *Water Policy in Spain.* CRC Press, Taylor & Francis, Leiden: 109–116.

Garrido, A. & Calatrava, J. (2009). Trends in water pricing and markets. In: Garrido, A. & Llamas, M.R. (eds.), *Water Policy in Spain.* CRC Press, Taylor & Francis, Leiden: 129–142.

Llamas, M.R.; Fornés, J.M.; Hernández-Mora, N. & Martínez Cortina, L. (2001). Aguas subterráneas: retos y oportunidades [Groundwater: challenges and opportunities]. Botín Foundation and Mundi-Prensa. Madrid, Spain.

WWF (2005). *Los mercados de aguas y la conservación del medio ambiente: Oportunidades y retos para su implantación en España* [Water markets and environmental conservation: Implementation opportunities and challenges in Spain]. WWF. Available from: http://assets.wwfspain.panda.org/downloads/posicion_wwf_sobre_mercados_de_aguas.pdf [Accessed 14th May 2011].

Chapter 17

Public participation and transparency in water management

Lucia De Stefano[1], Nuria Hernández-Mora[2],
Elena López-Gunn[1], Bárbara Willaarts[3] &
Pedro Zorrilla-Miras[4,5]
[1] Water Observatory of the Botín Foundation; Department of
  Geodynamics, Complutense University of Madrid, Madrid, Spain
[2] Founding Member, New Water Culture Foundation, Madrid, Spain
[3] Water Observatory of the Botín Foundation;
  CEIGRAM, Technical University of Madrid, Madrid, Spain
[4] Autonomous University of Madrid, Madrid, Spain
[5] Terrativa Sociedad Cooperativa, Madrid, Spain

ABSTRACT: Public participation is broadly considered to have a positive impact on the quality of governance. Transparency is the first step in the public participation *ladder* since it implies that people have access to the necessary information to make informed contributions to decision-making. This chapter gives an overview of the main challenges for the Spanish water sector in terms of public participation in the water planning process of the Water Framework Directive (WFD), and presents results of an assessment of information transparency of the Spanish water authorities. Although the WFD has contributed to improving the situation, in Spain the tradition of public accessibility to data and public participation in water management decisions is still rather poor. In addition to making all relevant information publicly available, the most compelling challenge is possibly ensuring its reliability and consistency. Another key issue is making the information accessible to different target audiences by adapting it to their level of interest and technical capacity.

Keywords: public participation, transparency, accountability, Water Framework Directive, water planning process

1 INTRODUCTION

Public participation is a process where people -individuals, groups and organizations- are allowed to influence the outcome of plans and working processes that affect them. Thus, public participation is linked to public decision-making and, although there is still little conclusive evidence on its direct impacts on the policy process and political decision-making (Abelson & Gauvin, 2006), it is broadly considered to have a positive impact on the quality of governance (López-Gunn, 2002). Transparency is the first step in the public participation *ladder*, since it implies that people have access to the necessary information to make informed contributions. The following steps, with an increasingly higher level of involvement of stakeholders, are consultation and active involvement.

This chapter gives an overview of the main challenges for the Spanish water sector in terms of public participation in water-related decisions. To do so, it first gives a brief overview of the opportunities (and related pitfalls) for regulated public participation in the water planning process, as provided by the Water Framework Directive (WFD). This is distinct from participation practices in decisions related to, for example, new water infrastructure design and approval, which are not analyzed in this chapter and are usually framed within standard environmental impact assessment procedures. Second, it focuses on the level of access to information in the Spanish water sector. It concludes with some considerations of the opportunities and threats for an improved public participation in water management planning.

2   PUBLIC PARTICIPATION IN THE WFD PLANNING PROCESS

The first international declaration that explicitly addresses the importance of public participation in water management dates back to January 1992, at the Dublin International Conference on Water and the Environment. In the same year, both the Rio Declaration and the Helsinki Convention on transboundary waters echoed the Dublin principles in relation to participation. In 1998, the Aarhus Convention[1] transposed the Rio principles relating to access to information into a legally binding document. By then, the European Union (EU) had already established minimum standards for public access to information and public participation in some environmental issues, but the Aarhus Convention extended the EU requirements, giving broader definitions of environmental information and public authority, and recognizing the right of citizens to turn to courts of justice when environmental rights are infringed.

Since 2000 a number of EU Directives have promoted several participatory elements, namely a Directive on public access to environmental information (Directive 2003/4/EC); the partial transposition of the Aarhus Convention; a Directive for public participation when developing certain plans and programs relating to the environment (2003/35/EC); and a Directive on the assessment of certain plans and programs on the environment (2001/42/EC). All these Directives complement the public participation provisions defined specifically for water resources planning and management under the WFD.

The WFD states that "to ensure the participation of the general public[2] ..., it is necessary to provide proper information of planned measures and to report on progress with their implementation, with a view to the involvement of the general public before final decisions on the necessary measures are adopted" (WFD, Preambles 46). Specific legal provisions are then described in Article 14 of the WFD, which distinguishes three forms of public participation in the planning process (with an increasing level

1 The Convention on Access to Information, Public Participation in Decision-making and Access to Justice in Environmental Matters, or Aarhus Convention, was signed on June 25, 1998 and entered into force in 2001.
2 In this chapter in some cases we distinguish between the general public and stakeholders. By the first we mean society as a whole, by the second we mean individuals or organized groups of individuals that have a direct stake in decisions taken.

of involvement): provision of information; consultation on draft planning documents; and active involvement. According to the Directive, the first two are to be ensured, while the latter should be encouraged.

Since the approval of the WFD several assessments on the quality of public participation have been undertaken both at EU and Spanish levels (for EU-wide evaluation see De Stefano & Schmidt, 2012; for Spain see Espluga *et al.*, 2011; Hernández-Mora & Ballester, 2010; Espluga & Subirats, 2008; FED, 2007). Drawing from the results of these assessments, it is possible to make a strengths-weaknesses-opportunities-threats analysis (SWOT) of participation in the framework of the WFD (Table 1).

Table 1 shows that, while the main strengths and opportunities are related to legal obligations to ensure participation and the long time frame of the planning process, human and political factors may hinder public participation efforts. In particular, the inertia to change current practices, as well as the complexity of the planning process, may create frustration and fatigue among stakeholders. Indeed, participation entails a difficult balance between representative democracy -where elected or appointed representatives have the task of making decisions- and the expectation by stakeholders to have an impact on policy making. The actual or perceived lack of impact of stakeholders' engagement may detract from the legitimacy of the participatory process or may eventually lead participants to withdraw from it. Moreover, the role of lobby groups -with explicit or vested interests in the decisions to be made- is ambivalent and can be difficult to manage, within and outside of regulated participatory processes. On one side, interest groups can help in widening the scope of the discussion and conveying widespread concerns to decision-makers. On the other side, however, and

Table 1 SWOT analysis of the WFD and its implementation process regarding public participation.

Strengths	Weaknesses
Legal obligation to encourage/ensure participation.	Poor participatory tradition in some countries.
Increased financial resources associated with the water planning process.	Limited human and financial resources.
3-step process, stimulating public and stakeholders and managing their expectations.	Tight implementation schedule.
Three planning cycles to gradually adjust approaches and tools.	Technical complexity of the planning process.
	Inadequate national transposition of EU legal requirements in some countries.
Opportunities	**Threats**
Aarhus Convention requirements.	Stakeholder fatigue and disillusionment.
EU Directive on access to environmental information.	Inertia of well-established planning processes to adapt to more participatory approaches.
Support and input from research programs and projects.	Resistance of traditionally powerful lobbying groups to the increase in transparency and participation scope.
Increased credibility of public authorities.	Lack of political commitment to participation processes and outcomes.
	Budget cuts due to the EU economic crisis.

Source: Modified from De Stefano & Schmidt (2012).

depending on their lobbying effectiveness, they can shift the focus of decisions from the achievement of the *common good* to the support of a specific sector of society.

Participatory processes are also the fora where conflicts or tensions among uses and interests are expressed. This entails an opportunity to conciliate different interests and at times it can lead to a deadlock situation in the decision-making process.

3 INFORMATION TRANSPARENCY

Transparency is a core component of the so called second generation institutional reform, and it is increasingly associated with better socio-economic development, as well as with higher competitiveness and lower corruption, which ultimately can improve policy outcomes (Bellver & Kaufmann, 2005). Transparency for example, can facilitate participation and collective action by stakeholders and is at the heart of water governance, fair allocation to users and sound incentives for efficient water use. Transparency in the work of public administration is considered to be the key in the fight against corruption in the public sector. It is also essential for an effective public participation, as only a well-informed public can effectively take part and actively contribute to public decision-making. Some authors distinguish between *transparency* OF *governance* and *transparency* FOR *governance*. The first relates to empowering society in observing "the actions either of *regulators* to whom they have delegated power or other powerful actors in society" (Mitchell, 2011), while the second refers to the disclosure of information by government as a means to influence the behaviour of corporations or other organizations. This section focuses on the first concept in the context of water management in Spain, and considers only access to information by the general public and interested parties. Nonetheless it is important to highlight that, in order to ensure real and full transparency, access to information must be accompanied by: 1) a guarantee about the quality and reliability of the information provided; 2) the opportunity of participating in the associated processes of decision-making; and 3) access to justice when the right to information or participation is not granted.

In Spain the concern about corruption in the public sector and the perceived lack of transparency has triggered several initiatives to improve transparency and accountability in public decision-making and the management of public resources. In 2008 Transparency International-Spain (TI Spain) developed an index of transparency for local authorities, which measures the degree of transparency in the functioning and operation of public administration at the municipal level, and is calculated on an annual basis, with public presentations to the press and wide distribution of the results. This was followed in 2010 by an index of transparency for Autonomous Communities (regional governments), also carried out by TI Spain[3]. The most recent and significant initiative is the proposed Law of Transparency, introduced by the newly elected conservative government in March 2012 as a key element in its effort to curb corruption and increase government legitimacy.

In the case of the water sector, in 2003 the New Culture of Water Foundation[4] undertook a preliminary diagnosis of the nature of corruption and mismanagement in

---

3 For the methodology and results of these annual indexes see: [http://www.transparencia.org.es/].
4 Private not-for profit organization dedicated to advocacy, education and outreach on water policy and management (www.fnca.eu).

the water sector in Spain (Martínez & Brufao, 2006). The report included a discussion on limitations to the right to participation and access to environmental information. In 2010 TI Spain -with the support of a technical team formed by the authors of this chapter- developed the Index of Transparency in Water Management (from now on referred to as INTRAG, its Spanish acronym), an aggregate of 80 indicators aimed at assessing transparency in River Basin Organizations (RBOs) in Spain. INTRAG has been calculated for two years running (2010 and 2011) and the results are discussed in this chapter. It is worth noting that this is not an isolated initiative. In effect, the Spanish Association of Public Water and Sanitation Providers (AEOPAS) and the Institute of Fiscal Studies of the Spanish Ministry of Economy, developed in 2011 an assessment index for urban water providers that includes, among other issues, indicators for transparency and public participation in urban water resources management inspired in part by the INTRAG methodology.

3.1 Assessing transparency in the water sector

When assessing the ease of access to information, at least two levels of transparency can be clearly distinguished: proactive information and access to information upon request. INTRAG focuses primarily on the first level of transparency. The second level is also very relevant for good governance, and in Spain it is still a pending issue. Despite the existence of a clear regulatory framework for access to environmental information, processes to actually obtain the requested information can be long and cumbersome.

INTRAG comprises a set of 80 indicators organized into six areas: Information about the RBO; Relationships with stakeholders and the public; Transparency in the planning process; Transparency on water use and management; Economic and financial transparency; and Transparency in contracts and tenders.

The calculation of the INTRAG index is based solely on information available in the websites of the evaluated RBOs. It is important to underline that INTRAG evaluates the presence or absence of information important for transparency in the management of water, and not the quality of that information (De Stefano *et al.*, 2011). Too much information, or information presented in a way that is inaccessible or inappropriate to the needs and capabilities of the target audience, can also result in opacity. INTRAG focuses on the information available through the web pages of RBOs because it is accessible to the wider public and does not require direct access to managers and policy makers, which has traditionally limited access to information to organized stakeholders and pressure groups.

3.2 INTRAG 2010 and 2011

The results obtained in the application of the INTRAG in 2010 and 2011 show that there is a clear need to improve transparency in water management in Spain[5] (Figure 1). Indeed, in 2010 only seven out of fourteen RBOs evaluated obtained a

---

5 Each indicator may have two values: 0 (information not available) and 1 (information is available). Therefore the maximum achievable score for each RBO is 80. In the final calculation of INTRAG this score is expressed in a scale of 100 for better communicability of the results.

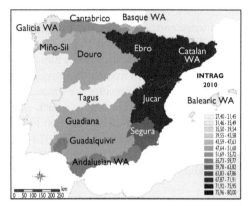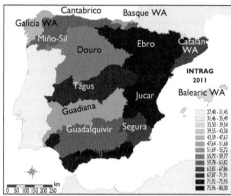

Figure 1 Overall scores of INTRAG 2010 and 2011. (Source: Own elaboration with data from TI Spain [http://www.transparencia.org.es]).

total score higher than 50% of the maximum score attainable, with no RBO scoring more than 71%. In 2011 the results improved slightly, with 11 RBOs obtaining a score higher than 50% and three scoring higher than 71%.

Figure 1 shows the scores for the RBOs evaluated. The figure shows that, in both 2010 and 2011, the most transparent basin organizations were the Ebro and Júcar RBOs. The Catalan Water Agency achieved top marks in 2010 but received a lower score in 2011, after a new management changed policy priorities, among them budgetary cuts and a decreased emphasis on public participation and transparency. At the other end of the spectrum, the entities that have achieved the lowest overall scores both years were the Galician Water Agency and the Balearic Water Agency. The Tagus RBO, on the other hand, evolved from a failing grade in 2010 (27%) to a fourth position in 2011, also as a result of a change in the Presidency of the Agency and of an increased emphasis on governance and transparency. These results show that the ranking of a specific water authority can rapidly change if there is a restructuring of its webpage, which can respond to changes in top-down defined policy priorities. Below we focus on the overall picture to identify the main strengths and weaknesses in information transparency as detected by INTRAG 2010 and 2011.

The breakdown of results by thematic areas can be seen in Table 2 and shows that, for both years, there is a significant amount of information available about the organizational structure and legal context of the RBOs (*Information about the RBO*) and about the planning process underway in the context of the WFD (*Transparency in the planning process*).

The thematic areas relative to *Relationships with the public and stakeholders* and *Transparency in contracts and tenders* have a certain degree of transparency, but need to be enhanced. In the first area, it is important to improve transparency on the composition as well as on the content of the debates and deliberations of the councils, boards and committees that are a part of the RBO's decision-making bodies (Governing Boards, Water Councils, Competent Authorities Committees, etc.). Another area that needs improvement is the publication of annual statistics referring

Table 2 Average scores of INTRAG 2010 and 2011 by thematic area.

Thematic area	Average score 2010 (%)	Average score 2011 (%)
Information about the River Basin Organisation	72.0	85.7
Relationships with the public and stakeholders	58.0	56.1
Transparency in the planning process	88.0	79.9
Transparency in water use and management	32.0	47.0
Economic and financial transparency	35.0	43.8
Transparency in contracts and tenders	48.0	51.9
Overall average	51.2	59.6

Source: TI Spain [http://www.transparencia.org.es].

to public requests for information from RBOs and their responses to those requests (an integral part of the legal right to information on environmental matters), or the publication of the RBOs' annual activity reports. For the area relative to contracts and tenders, there has been some improvement in 2011, but more information is still needed about modifications of projects, end-of-construction settlements and on the major contractors and suppliers of each Agency.

The largest deficiency in transparency is found in the areas of *Transparency on water use and management* and of *Economic and financial transparency*, although there has been some improvement between 2010 and 2011 (average scores increased from 32% to 47% and from 35% to 43.8%, respectively). In the first area there was a general deficiency of information on statistics of water use (updated information on existing water use permits, new permit requests and annual volumes extracted); information on compliance with existing regulations on water quality and environmental flows; and the legally required feasibility reports for new hydraulic infrastructures regulated under the *public interest* regime[6]. With regard to *Economic and financial transparency* there is generally little information available on: cost recovery for investments in new public hydraulic works; water tariffs and dues; financial relationships with water user communities and irrigator associations; and budget execution by the Agency.

4 CHALLENGES FOR THE FUTURE

The assessment of public participation in the Spanish water sector shows that, despite the presence of a favourable legal framework, it is still necessary to move from formal participation to real participation of the public and all interested parties in water-related decision-making processes (for more information see Hernández-Mora & De Stefano, 2011). Information transparency is key to that. In general, INTRAG 2010 was positively received by the RBOs and there is evidence that their transparency is slowly improving. Nonetheless, there is still a long path ahead for having all the relevant information available online. For example, it is necessary to improve

---

6 The public interest regime or *régimen de interés general* is a legal concept that allows for preferential funding and fast tracking of the permitting process of public infrastructures.

information on indicators relating to the economics and finance of RBOs because this kind of information is crucial to improving institutional credibility and public trust.

In addition to facilitating access to information, the Spanish water sector faces several challenges in terms of transparency. The most compelling challenge is possibly ensuring the reliability and the consistency of the information made available by public administrations. Another key issue is making the information accessible to different target audiences by adapting it to their level of interest and technical capacity. It is also important to create participatory processes that motivate and facilitate not only stakeholders, but also the general public to becoming involved in public decision-making regarding water management.

The WFD has contributed to increasing the quantity and quality of public participation and transparency in the Spanish water sector. The learning process of the first planning cycle (2009–2015) as well as the progressive activation of some sectors of civil society in relation to water are likely to set a fertile ground for an increased and improved participation in the future. A key consideration in this sense is that water managers should be aware that processes and practices applied in the first planning cycle of the WFD are a step forward but do not represent a catalogue of *best practices*. Therefore the results obtained from these processes are not an indicator of the potential of public participation as an integral part of public decision-making, but rather a starting point from which to draw lessons and make improvements in the future.

REFERENCES

Abelson, J. & Gauvin, F.P. (2006). *Assessing the Impacts of Public Participation: Concepts, Evidence and Policy Implications*. Research Report P06. Public Involvement Network. Canadian Policy Research Networks.

Bellver, A. & Kaufmann, D. (2005). *Transparenting Transparency. Initial Empirics and Policy Applications*. The World Bank.

De Stefano, L. & Schmidt, G. (2012). Public participation and water management in the European Union: experiences and lessons learned. In: Cosens, B. (ed.), *The Columbia River Treaty Revisited: Transboundary River Governance in the Face of Uncertainty*. A Project of the Universities Consortium on Columbia River Governance. University Press, Oregon State, USA.

De Stefano, L.; Hernández-Mora, N.; López-Gunn, E.; Willaarts, B.; Zorrilla Miras, P. & Llamas, M.R. (2011). Transparencia en la gestión del agua en España: fortalezas y debilidades [Transparency in water management in Spain: strengths and weaknesses]. In: Hernández-Mora, N. & De Stefano, L. (coords.), *Transparencia en la Gestión del Agua en España* [Transparency in water management in Spain], SHAN Series no. 4, Botín Foundation. Available from: http://www.fundacionbotin.org/case-studies_publications_water-observatory_trend-observatory.htm [Accessed 31st May 2012].

Espluga, J. & Subirats, J. (2008). Reflexiones sobre los procesos participativos en marcha [Considerations on the ongoing participatory processes]. In: *Actas del VI Congreso Ibérico sobre Gestión y Planificación del Agua*. New Water Culture, Bilbao, Spain. Available online from: http://www.fnca.eu/congresoiberico/documentos/p0403.pdf [Accessed 31st May 2012].

Espluga, J.; Ballester, A.; Hernández-Mora, N. & Subirats, J. (2011). Participación pública e inercia institucional en la gestión del agua en España [Public participation and institutional intertia in water management in Spain]. Reis, 134, April–June 2011: 3–26. doi:10.5477/cis/reis.134.3.

FED (Fundación Ecología y Desarrollo) (2007). *Percepciones y opiniones de 40 de los principales agentes de la sociedad civil sobre la implementación de la Directiva Marco del Agua y el proceso de participación en la Demarcación del Ebro* [Perceptions and opinions of 40 of the main civil society actors about the implementation of the Water Framework Directive and the participation process in the Ebro River Basin District]. Ebro River Basin Authority. Available from: http://oph.chebro.es/DOCUMENTACION/DirectivaMarco/ParticipacionC iudadana/40agentes.pdf [Accessed 31st May 2012].

Hernandez-Mora, N. & Ballester, A. (2010). Public participation and the role of social networks in the implementation of the Water Framework Directive in Spain. *Ambientalia*, Special Issue: Ten years of the Water Framework Directive: An Overview from Multiple Disciplines. Available online from: [http://www.ambientalia.org].

Hernández-Mora, N. & De Stefano, L. (cords.) (2011). Transparencia en la Gestión del Agua en España [Transparency in water management in Spain]. SHAN Series no. 4, Botín Foundation. Available from: http://www.fundacionbotin.org/case-studies_publications_ water-observatory_trend-observatory.htm [Accessed 31st May 2012].

López-Gunn, E. (2002). La participación de los usuarios y de los ciudadanos en la gestión de las aguas subterráneas: el caso de Castilla-La Mancha [Users and citizens participation in groundwater management: the Castilla-La Mancha case]. *Jornadas sobre presente y futuro del agua subterránea en España y la Directiva Marco Europea*. Ponencia 7.2. Zaragoza, Spain. AIH-GE. Available from: http://www.aeuas.org/resources/pdf/gunn.pdf [Accessed 31st May 2012].

Martínez, J. & Brufao, P. (2006). *Aguas limpias, manos limpias: Corrupción e irregularidades en la gestión del agua en España* [Clean waters, clean hands: corruption and misbehaviours in water management in Spain]. Ed. Bakeaz, Colección Nueva Cultura del Agua.

Mitchell, R.B. (2011). Transparency for governance: The mechanisms and effectiveness of disclosure-based and education-based transparency policies. *Ecological Economics*, 70(11): 1882–1890. 15 September 2011, Special Section – Earth System Governance: Accountability and Legitimacy.

Chapter 18

Taming the groundwater chaos[1]

Elena López-Gunn[1], Marta Rica[1] &
Nora van Cauwenbergh[2]
[1] *Water Observatory of the Botín Foundation; Department of*
  Geodynamics, Complutense University of Madrid, Madrid, Spain
[2] *UNESCO-IHE, Delft, The Netherlands*

ABSTRACT: This chapter deals with how to tame the Spanish groundwater chaos by identifying examples- defined by the absence of actual control or order in the governance and management of the resource combined with physical deterioration. Aquifer use has intensified in the last 50 years, in many cases over the stipulated recharge rate. The Spanish law articulated since 1985 developed measures to regulate and control abstractions by declaring an aquifer overexploited, yet these measures have failed in most cases to make users comply and ultimately improve the quantitative and qualitative status of the resource as required under the Water Framework Directive (WFD). Yet there have been spontaneous user led initiatives, framing collective action institutions. These young groundwater collective institutions developed along a spectrum of available organizational formats, both in the public and in the private domain, reflect the diversity of groundwater rights. An evolution of collective action is emerging with the focus on reducing risk through the development of a portfolio of water resources: surface, groundwater, desalinated, recharged or recycled. The most important development has been the introduction of flexibility of access to multiple types of water resources. An example on this latter aspect is presented, with the case of three groundwater bodies in Almería: Campo de Dalías, Medio-Bajo Andarax, and Campo de Níjar, where a set of diverse institutional settings has been established in order to tame the *chaos*, but which leaves some questions unanswered on the overall resilience of the overall system to intensive groundwater use.

Keywords: chaos, overexploitation declaration, collective action, adaptation, Almería

1 INTRODUCTION

Groundwater has been largely *out of sight out of mind*, yet in the last decades a global silent revolution has come to the fore in many emergent and populous countries in the world. The case of Spain, which underwent this silent revolution four decades ago (Llamas & Martínez-Santos, 2005) offers useful opportunities for lesson drawing and learning in relation to groundwater management and collective institutions.

---

1 The title of this chapter is in honour of the classic book by W. Blomquist looking at the case of California (USA) (see Blomquist, William. 1992. Dividing the Waters. San Francisco: ICS Press).

Groundwater is currently one of the most extracted natural resources as well as globally the largest stock of freshwater resources, with increased accessibility due to technological advances. It is a new frontier in resource use when e.g. surface water is fully allocated and it is often (politically) easier to look for additional resources than to re-allocate between competing uses. Yet while groundwater intensive use is on the rise globally, groundwater governance is often lagging behind. Groundwater is a classic common pool resource (Ostrom, 1990), and this nature offers both management opportunities and inherent problems. In this chapter groundwater chaos refers to the often absent or ineffective control of groundwater use which often leads to physical deterioration and reengineering by farmers (Shah, 2009). At the heart of taming chaos lies the twin dilemma of ease of access and difficulty in excluding users (or closing the resource). In addition, important inherent resource qualities (like low upstart costs, on site availability, resilience to droughts, etc.), combined with increased uncertainty due to climate variability and change, make groundwater an increasingly attractive resource, and therefore even more pertinent to try and devise workable solutions for taming groundwater chaos. This chapter deals with the cause and consequences of the Spanish groundwater chaos, highlighting the strategies that have been taken by zooming into one specific yet notable example in the region of Almería to adapt to the consequences of intensive groundwater use.

2   THE ORIGIN AND MAGNITUDE OF CHAOS

Groundwater use in Spain has increased dramatically over the last few decades (see Chapter 7) with the total volume pumped growing from 2,000 hm^3/year[2] in 1960 to more than 6,500 hm^3/year in 2006 (Hernández-Mora et al., 2007; Dumont et al., 2011a). This is higher than the 4,000 hm^3/year estimated by MIMAM (1998) this latter estimate does not include groundwater abstracted informally (Dumont et al., 2011b; De Stefano & López-Gunn, 2012), but recent estimates put this figure at 7,000 hm^3 (see Chapter 7). Aquifer intensive use has been the subject of long debates (Custodio, 2002; Llamas & Martínez-Santos, 2005), especially to find a solid definition which links to parallel debates on sustainable yield or the groundwater balance. A growing number of scholars have highlighted the simplistic nature of utilizing a decrease in aquifer reserves and annual discharge or water table level decline as indicators for aquifer overexploitation, since it may mean that the aquifer is evolving to a different equilibrium (Martínez Cortina, 2011). The preferred term of intensive use does not carry normative judgments, while taking into account the modification of the hydrogeological functioning of the aquifer regarding water abstractions (Llamas & Custodio, 2002). Beyond scientific discussions, the Spanish water law includes the concept and procedures to follow for overexploited aquifers, with criteria based on a negative balance between water abstraction and recharge, and where basin boards could declare an aquifer overexploited. Once this declaration is made final, it carries substantive changes in management (see Table 1). In the late 1990s a total of 77 aquifers or hydrogeological units, out of 467 were identified as having problems (ITGE, 1997). However,

---

2 hm^3 = cubic hectometre = million m^3 = 10^6 m^3.

Table 1 Comparing chaos: groundwater intensive use regulation in Spain.

	1985 Spanish water law	*2000 WFD adaptation*
Management units	467 hydrogeological units, of which 77 are classified as overexploited, and 17 officially declared overexploited.	Approx. 700 groundwater bodies of which 343 are declared as in poor quantitative and qualitative status.
Management implications	Aquifers declared over-used: 1 Close aquifers to new use. 2 Restrict existing water rights. 3 Creation (top down) of a groundwater user group. 4 Establishment of an Annual Abstraction Plan with a cap on annual abstraction.	Implementation of programme of measures to reach the objective of good status at the next planning stage, including the creation of groundwater body communities or *Comunidad de Usuarios de Masas de Agua Subterránea* (CUMAS), Andalusian water law since 2010
User Groups	User communities (wide diversity of both public and private institutions)	Will have to be re-arranged as CUMAS

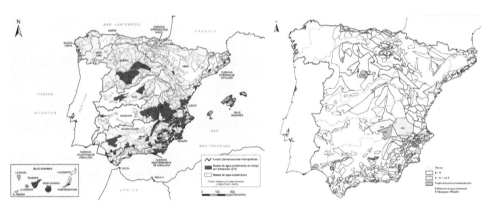

Figure 1 Map of groundwater bodies at risk (left) and map of hydrogeological units (right). (Source: Varela (2009) and MAPA (2001)).

only 17 have been declared legally overexploited and only two with a definite declaration of overexploitation. Accumulated experience has demonstrated that measures and designation of overexploitation have been heavily influenced by political reasons, since a restriction in water use has significant socioeconomic impacts in the area of application.

The European Water Framework Directive (WFD) uses a different terminology to refer to intensive groundwater abstractions. The units for management are designated as groundwater bodies, and the aim is to ensure good qualitative and quantitative status, understood as "the levels of groundwater in the groundwater body such that the available groundwater resource is not exceeded by the long-term annual average rate of abstraction". There are approximately 730 groundwater bodies in Spain and 297 were identified as either in chemical or quantitative risk or both and 41 are still under study (see Chapter 7). We see that in terms of number, the severity of groundwater at risk has increased with the WFD criteria (Figure 1). One of the pending challenges to tame the groundwater chaos will be to align the Spanish law with the European mandate where, for example, the declaration of overexploitation could be contemplated under the programme of measures (Rodríguez Cabellos, pers. comm). Nevertheless, taking into account the socioeconomic impact of this declaration, different criteria used, and the difficulties for the administration to make users comply with restrictions, it seems that the WFD provides an opportunity to review this declaration of overexploitation and to find workable measures to make this effective.

3 TAMING THE CHAOS: COLLECTIVE ACTION

It has been shown that users, with a joint objective, can cooperate for the conservation and management for resources used in common, not leading necessarily to a Common Pool Resource dilemma (López-Gunn, 2007). The emergence of collective institutions for groundwater management is part of the solution to chaos as heralded by Ostrom (1990), as a way to address the problem of information asymmetry and transaction costs. This is because the lack of administrative resources is a primary burden towards making a reliable inventory of water rights and extractions, as well as to control individual users' behaviour towards water use (Sahuquillo et al., 2009), even when information technology can help. We see a comparative advantage that local users can have over government in controlling and monitoring groundwater use. In fact partnerships or collaborative approaches can act as complements to the *command and control approach*, particularly when these have not been successful (Rica et al., 2012). Thus it becomes vital to somehow grasp the information held by users as a precondition or necessary step, to tame an inherent problem of uncoordinated – formal or informal – actions by thousands of users.

Spain, together with countries like Mexico or India, has accumulated valuable experience on a range of self-regulation initiatives led by users. Well known globally for its millenary tradition of surface water irrigation communities (see the case of *Heredades* in the Canary islands – Chapter 22), in Spain almost 60% of irrigated land is in hands of irrigation collectives (Valero de Palma, 2011). In the case of groundwater, there are 40 years of accumulated experience by the oldest groundwater user associations (GWUAs) and a new trend in the emergence of new organizations. By exploring the Spanish case, this chapter provides information on lesser known aspects in collective institutions, like the increasing number and diverse range of user communities, focused on groundwater and more recently on other resources like desalinated, recycled or recharged water. The emergence and evolution of collective institutions experienced three waves. The first wave refers to the long and well documented history of surface water irrigation communities. The second wave refers to younger groundwater collective institutions. The third wave marks the appearance of user collectives linked to the use and exploitation of a new range of water resources like desalinated or recycled water, made possible due to technological advances and knowledge. This chapter will only reflect on the last two waves.

Analysing the second wave of collective action, groundwater collective institutions have developed mainly through user initiative along a spectrum of available organizational formats both in the public and in the private domain, reflecting the diversity of groundwater rights (Figure 2). Until recently the nature of the water right (public or private) in many ways marked a path dependence in the nature of the organization. The first groundwater collective institution was established in 1976 in Delta del Llobregat, and since then another 19 collective institutions have emerged. Of these, only a minority (3) were created because of state *dictat* following the Water Act. The majority emerged spontaneously due to user self-interest (Rica *et al.*, 2012), as a reaction to droughts or, in most cases, farmers organizing to defend their private water rights when confronted by a potential declaration of aquifer overexploitation. Another important factor to define the nature of the organization is tradition and culture of the region. Many GWUAs along the Mediterranean coast opted for civic associations such as Societies for Agrarian Transformation or Societies of Goods, due to their historical prevalence around Valencia, whereas in Catalonia the choice was to seek the protection of the administration by opting for state water concessions and the constitution of a public Water User Community. Apart from this division according to law, there are other typologies based on their constitutional structure and the categorization by the Water Act as third, second and first order, third order being the most complex (López-Gunn & Martínez Cortina, 2006).

However, there is increased evidence that the nature of the institution, whether public or private has not been a determining factor in effectiveness and performance. More than the juridical nature of the water rights and collective institutions, a number of factors have been important for the effectiveness of institutions: first, issues like secure and agreed resource entitlements, joint infrastructure ownership, or collaborative

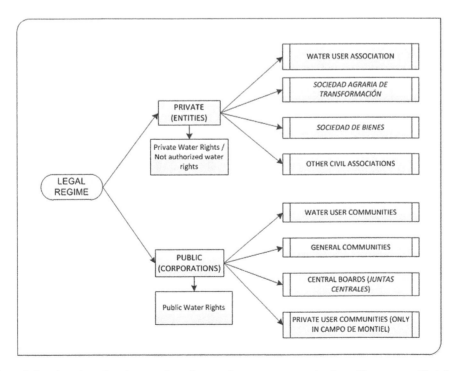

Figure 2 Legal options for the creation of groundwater user organizations. (Source: modified from Rica *et al.* (2012)).

Figure 3 Newspaper reference to aquifer recharge management model in the Douro basin. (Source: Rico (2011)).

development between users and the administration of abstraction plans, strong internal monitoring and sanctioning mechanisms, and finally, legitimacy of water rights and recognition and support of water user collectives by higher level authorities such as the water boards (López-Gunn & Martínez Cortina, 2006). A third wave of collective action is emerging (as will be analysed below in the case of Almería), with the focus on reducing risk through the development of a portfolio of water resources, which in effect *enlarge* available resources: surface, groundwater, desalinated, recharged or recycled. The most important development has been to introduce flexibility of access to multiple types of water resources. Therefore it no longer makes sense to sectorialize collective action on origin of water or type of water right. For example, in the Júcar basin, *Juntas Centrales* are by norm responsible for groundwater and surface water, acknowledging the complementarity in the use of both resources. Meanwhile in the Douro, there have been interesting experiments with the creation of communities for recharged aquifers (Huertas, 2011) (see Figure 3). Two GWUAs in El Carracillo and Cubeta de Santiuste, established after the recharge projects led by the administration, are facing the complexity of managing a conjunctive use of surface and groundwater, where there is still juridical uncertainty regarding recharged water use (Huertas, 2011).

4 CASE STUDY: GROUNDWATER IN ALMERÍA, INSTITUTIONAL AND RESOURCE DIVERSITY AS ADAPTATION TO CHAOS

Surface water irrigation communities were created almost exclusively by state initiative, whereas in the case of groundwater this was mainly user led. In this context the most striking example of private entrepreneurship is the case of *plasticulture* in Almería, an extensive area of greenhouse agriculture, with up to 27,000 ha the largest in the world, in the Southeast of Spain. This offers a microcosm on the emergence and evolution of collective action in groundwater, while it showcases many of the remaining challenges and available opportunities for taming groundwater chaos. Almería is the most productive agrarian province of Spain and the best example of the silent revolution. In less than half a century the region has catapulted itself from one of the poorest regions in Spain to become a European leader in agri-business. To understand this transformation, it is necessary to take a look at the past, and the growth of collectives managing water from wells. The agrarian policy after the civil war from 1939 was a determining factor for the establishment of agriculture in Almería. Irrigation districts were designed by the *Instituto Nacional de Colonización* (INC), whose objective was to support rural development through irrigation projects. In 1971 the INC became the *Instituto de Reforma y Desarrollo Agrario* (IRYDA), and in 1984 in Andalucía it was named the *Instituto Andaluz de Reforma Agraria* (IARA). In Almería, settlers were established in certain areas, particularly in Campo de Dalías, Campo de Níjar and Huércal-Overa. Technicians from these agrarian reform institutes researched on greenhouse technology on artificial soil to improve land productivity, and the transformation began (Rivera, 2000). Irrigator communities emerged in different ways. The pioneering case was the irrigation districts designed by INC, whose management was then transferred to its users. For example, in Campo de Dalías a number of wells provided groundwater to six irrigation sectors, with land plots linked to a certain well.

This initiative was adopted by the rest of the population, who either contributed with financial resources or labour efforts to build a common well to irrigate their land (Cuadrado, pers.comm.), or bought a small plot from big land owners who sold their hours from the well to land tenants (Jiménez, pers.comm.). These initiatives became *Sociedades de Bienes* or *Sociedades Agrarias de Transformación*, a formal associative figure to regularize the situation of water sharing and land under private law.

Three contiguous aquifers all located in the province of Almería offer insights into existing and future challenges and opportunities for groundwater management (see Table 2). The three aquifers share similarities like climatic conditions and for all, groundwater is a key factor for economic development, based on the export of highly profitable greenhouse crops (estimated at 60,000 €/ha/year) (Dumont *et al.*, 2011b). Water demand has grown at a higher rate than the available water resources, leading to a situation of groundwater level decline, made worse by deteriorating groundwater quality and marine intrusion. The three aquifers hold, totally or partially in certain areas, overexploitation declarations from the late 1980s. However, the irrigated surface experienced an increase of more than double, from 15,000 to 30,000 ha approx., that only seems to have stabilized due to the current economic crisis. In terms of actions to address problems with groundwater quality and quantity, a number of initiatives are taking place: the first initiative is based on drilling deeper wells and/or intensifying greenhouse activity with the introduction of more and better water saving devices (see Chapter 19). The second initiative has been collective action by users around organizations within a nested GWUA, making collective efforts to get better energy prices after sector liberalization (Poveda, 2011), together with a better knowledge on the aquifer. And finally, a third initiative is based on water recycling and desalination techniques, allowing both diversification of risk and access to supplementary water resources.

Table 2 Groundwater resources in case study areas.

Name	Campo de Dalías-Sierra de Gádor	Medio-Bajo Andarax	Campo de Níjar
Area	797,09 km²	341,9 km²	466,15 km²
Declaration of over-use (1985 Spanish Water Law)	Declared over-exploited in 1986 and 1995. No management plan.	Declared over-exploited in the lower area in 1986. Management plan in construction. Inventory and process of regulation for private water rights.	Declared overexploited in 1987. Initiative to develop management plan, never implemented.
Groundwater body in new Basin Plan	Gw. body No. 060.013. Bad global state, 2010.	Gw body No. 060.012. Bad global state, 2010.	Gw. body No. 060.011. Bad global state, 2010.
Aquifer	Mixed: carbonitic in mountains and lower aquifer; detritic in coast.	Mixed: carbonitic in mountains; detritic in plains.	Detritic nature.
Water Distribution control	Each GWUA manages own well water rights over the excess water from Beninar reservoir.	Groundwater and surface water in wet years, GWUA manage own well, marketing water to most demanding.	Desalinated water through Water User Associations, farmers manage own well in addition to GWUAs.
Agricultural model	20,940 ha of greenhouses (95% of irrigation in the area)	2,400 ha of greenhouses and 4,200 ha of fruit trees, olives and citric produce.	4,500 ha of greenhouses, 700 ha of horticulture, olive and fruit trees.

Water User groups	Individuals, 60 GWUA (3 delivering surface water partially). *Comunidad General Usuarios/ Junta Central Usuarios.*	Individuals, around 53 WUAS; 14 delivering groundwater totally or partially; 1 using recycled water from city of Almería. *Junta Central de Usuarios* within groundwater body limits.	Individuals, 37 WUAS (2 delivering surface water totally or partially; 1 delivering totally desalinated water to mix with ground-water)
Associative milestones	*Junta Central de Usuarios del Poniente Almeriense,* public corporation created on users' initiative since 1991 incorporating: 7 municipalities, 3 industries, 38 GWUAs from public and civil regimes and around 118 individual users. In parallel there is *Comunidad de Usuarios del acuífero Sierra de Gádor,* which is also a public corporation entity composed by different GWUAs, who refuse to take part in the *JCUAPA* as a General Community.	*Junta Central de Usuarios del Medio-Bajo Andarax* is responsible as a public corporation, including the main WUAs, municipalities, industries and individual users. 16 private entities, and integrating 4,000 members. Besides these small communities that use groundwater, a larger one, *Las Cuatro Vegas de Almería* is responsible for the tertiary treatment by ozone and more recently chlorine of the secondary treated waste-water from Almería city, and for the delivery of the recycled water among its associates.	*Comunidad de Usuarios del Campo de Níjar (CUCN)* was created to manage desalinated water for irrigation from the desalination plant constructed. This is a public corporation in charge of delivering desalinated water, regulated by users.

Source: Own elaboration, based on Junta de Andalucía (2010), and Van Cauwenbergh&Francés (2010).

These three case studies highlight a dual strategy to *augment* water supplies which are intensively used, while the *institutional or formal arrangements* to redistribute good quality resources and stop abstracting in sensitive spots are being worked out (Domínguez Prats & Franqueza Montes, 2009). In the interim, user communities' representatives admitted to informal arrangements to share water with those that had serious quality problems as well as selling *water turns* at the same price or at a higher price, sometimes reinvesting the money on the community and sometimes representing extra income for the water turn owner. This issue requires more attention, since it is

---

Box I   Comparative data on the Campo de Dalias, Medio-Bajo Andarax and Campo de Nijar aquifers

Source: Junta de Andalucía (2009).

Campo de Dalías: Regional studies from the Spanish Geological Institute are currently assessing the relocation of wells from areas that are highly sensitive to salinization from abstraction (Domínguez Prats & Franqueza Montes, 2009). Water overexploitation is currently located in the lower aquifer which has better quality and better storage, while upper layers are no longer being used, causing waterlogging as these layers are recovering their recharge balance. There is a plan to diversify the origin of the water, giving particular attention to the nearby reservoir-Benínar, about 3–6 hm³/year, the de-brakishing of the upper aquifer in an emerged wetland for 2 hm³/year, the reuse of wastewater from the main cities up to about 10 hm³/year and desalination for up to 30 hm³/year. Groundwater wells reach depths of 300 m with pumping costs estimated at 0.13 €/m³ to 0.19 €/m³ (Martínez, 2011).

Source: Junta de Andalucía (2009).

Medio-BajoAndarax: The main problems are related to groundwater overexploitation and salinization, contamination of surface and groundwater with badly treated wastewater and diffuse contamination of agricultural origin (Van Cauwenbergh *et al.*, 2008, and Van Cauwenbergh & Francés, 2010). The WUA has a temporary license to use wastewater from Almería city, with a maximum of 12 hm³/year for agricultural purposes. In 2009, 6.6 hm³ of recycled water were delivered, where WUA members are not obliged to buy recycled water. However, the use of recycled water does not mean that less water is being abstracted from the aquifer. In wet years, this amount decreases as the WUA then uses available surface water. The price of recycled water is 0.25–0.30 €/m³, in comparison to the estimated pumping price of 0.13–0.20 €/m³). Both are significantly more expensive than surface water (estimated at 0.01–0.02 €/m³) (Pérez Sánchez, pers. comm.). Quality problems with wastewater effluent call for a renewal of reuse infrastructure that has to go hand in hand with establishment of new organizational structure.

Campo de Níjar: The alternative resource to increasingly saline groundwater is desalinated water from a desalination plant in Carboneras, the largest one in Europe, with a capacity to generate 42 hm³/year, not yet used to its full potential. The plant and the secondary distribution network were built mostly through public investment, with users responsible for water delivery in the tertiary network and management. Its construction purpose was to reduce pressure on the aquifer, while securing the relevant economic activity in the area. However, it is not clear whether the use of desalinated water has meant a parallel reduction in groundwater use. Users blend desalinated water with saline groundwater from the aquifer, with price and final water quality being the determining factors on the amount of water used from each source. Depending of the crop produced more (e.g. raff tomato) or less (e.g. watermelon) saline water mix will be used. The user price of desalinated water is 0.48 €/m³, compared to 0.10–0.30 €/m³ of cost for groundwater pumping (López, pers. comm.).

an informal way of water redistribution, an informal water market (see Chapter 16 on Water trading in Spain).

In summary, taking into account the tendency to substitute or supplement the use of groundwater resources under stress with other supplies in order to meet demand, it could be argued that groundwater mismanagement acts like a magnet for additional resources. Groundwater chaos makes users vulnerable since there is no insurance against mining the resource in terms of quantity or deteriorating water quality, to the point that it could threat users' livelihoods. Groundwater users to minimize risk – if no management measures are introduced – start to look for additional, non-conventional, resources to secure water availability. These additional resources tend to be more expensive. Yet users pay subsidized prices, which do not internalize the large infrastructure investment, and – when combined with cheaper groundwater resources since environmental externalities are not included (like e.g. reduction in groundwater quality) – means that there is no incentive, signal or internalisation of environmental externalities and no reduction in intensive groundwater resource use. These new water resources are not a substitute, but rather become both additional resources and insurance for risk to a potential lack of water. These non-conventional (better quality) water resources are then blended with existing (poorer) quality groundwater. However, when claiming for additional water, groundwater users are *de facto* reacting to a deteriorating aquifer of which they are both victims and executioners. Yet the necessary signals for social learning from mismanagement are masked, and both the emergence and user engagement in new types of user organizations go hand in hand with a process of legitimizing claims to water and securing access to water. It also gives groundwater users of overexploited aquifers the grounds to apply for subsidies in order to construct expensive infrastructure.

5   CONCLUSION: THRIVING IN CHAOS: THE PROBLEM OF INCENTIVES AND MOTIVATION

In this chapter we have analysed the inherent problems and opportunities of groundwater as a common pool resource, its intensification in use and a range of emergent institutions and strategies that have been adopted. Taming groundwater *chaos* is partly a problem of lack of information by the Water Boards which impinges on the capacity to control and manage. This is partly the reason why the declaration of overexploitation was a not a successful measure and should be revisited in the on-going WFD adaptation. We have demonstrated how the declaration was ignored in the province of Almería, where groundwater use for irrigation intensified after being declared overexploited. This chapter has highlighted that collective action is a spontaneous emerging property of chaotic systems. Through collective action, current lack of information regarding water use, e.g. in terms of inventories of groundwater use rights, could be overcome. The chapter has also identified that whereas debates before were centred on the nature and characteristics of water rights, on the lack of definite water rights inventories and the problem of over-allocation of water rights, now it should also consider the search for additional resources. This is triggered by the inherent difficulties in establishing clear resource boundaries. Regulatory frameworks like the European Water Framework Directive however, raise questions on what kind of incentives can help to keep resource use within its natural (fluctuating) resource boundaries. The question is whether, in order to maintain the resilience of the system, there should be additional water sources as a palliative measure to satisfy current demands or, instead whether before new resources are brought into play, this is made conditional on a previous necessary step like e.g. the development of a groundwater management plan, which maps how the economy can adjust gradually to existing resource limits. That is in many ways the opposite concept to *managed depletion* used in some Western USA states. The examples of recent collective action in Almería show how the introduction of new non-conventional (desalinated, recycled water resources) is an effective way to minimise the risk that lack of water means for a high value agricultural activity, where water is an essential production factor with no substitute. However, using new water resources does not necessarily mean that the quantitative and qualitative status of the aquifer is improving. Therefore, it seems that technological improvements have allowed the socioeconomic system to function. This might however go against the motivation and collective action potential of users to preserve the resource, catalyse social learning and trigger adaptation when faced by resource limits, thus halting necessary innovation. Hiding signals from the system on its vulnerability due to intensive use, if no alternative sources were available, prevents learning from reaching resource limits. In these areas, strategies have emerged to maintain system resilience in a collective way: first, actions towards a decrease in energy price and water efficient infrastructure; second, the acceptance on the need to devise and agree on a groundwater abstraction management plan; and finally, use of alternative sources such as recycled water or desalination. All these strategies point towards a more efficient use of resources. However, only the first two would increase the resilience of both the socioeconomic system and resource base, thus *taming the chaos*. In the final option, users have opted to draw upon external resources rather than self-regulate, breaking the dependency from the groundwater resource. Furthermore, legitimizing GWUAs gives users the possibility to reclaim the financial support needed to develop new infrastructure for alternative sources to be available. Groundwater chaos in the case studies discussed, known as *the orchard of Europe* highlights that taming groundwater chaos is at a cross roads between solutions which emerge and are locally driven and contained, and solutions which require of (subsidised) external resources to make up the water deficit.

REFERENCES

Custodio, E. (2002). Aquifer overexploitation: what does it mean? *Hydrogeology Journal*, 10(2): 254–277. doi:10.1007/s10040-002-0188-6.

De Stefano, L. & López-Gunn, E. (2012). Unauthorized groundwater use: institutional, social and ethical considerations. *Water Policy*, 14(2012): 147–160.

Domínguez Prats, P. & Franqueza Montes, P. (2009). La investigación de los acuíferos en apoyo a la mejor gestión de los mismos: la experiencia del Instituto Geológico y Minero de España con los acuíferos del Poniente almeriense. [The investigation of aquifers for their improved management: the experience of the Geological and Mining Institute of Spain in the western Almeria aquifers] In *AEUAS conference, "Use, protection and restoration of groundwater bodies"*, 29–30 October, Almería, Spain.

Dumont, A.; De Stefano, L. & López-Gunn, E. (2011a). El Agua Subterránea en España según la Directiva Marco del Agua: una visión de conjunto. [Groundwater in Spain under the Water Framework Directive: an overview] *VII Iberian Congress on Water Management and Planning "Iberian Rivers + 10. Looking ahead after 10 years of DMA"* 16/19 February 2011, Talavera de la Reina, Spain.

Dumont, A.; López-Gunn, E. & Llamas, M.R. (2011b). *La Huella hídrica extendida de las aguas subterráneas en el Campo de Dalías (Almería, Spain)*. [The extended water footprint of groundwater in the Campo de Dalias (Almería, Spain)]. Iberian Congress on Groundwater, Zaragoza, Spain, 14–17 September 2011.

Hernández-Mora, N.; Martínez Cortina, L.; Llamas, M.R. & Custodio, E. (2007). *Groundwater issues in southwestern EU member states: Spain country report*. European Academies of Sciences Advisory Council (EASAC). Fundación Areces, Madrid, Spain.

Huertas, R. (2011). Retos y oportunades en la gestión colectiva de las aguas subterráneas. [Challenges and opportunities in the collective management of groundwater] *Seminar on collective action for groundwater management in Spain*. Botín Foundation Water Observatory, Madrid, Spain.

ITGE (Instituto Tecnológico GeoMinero de España) (1997). *Catálogo de acuíferos con problemas de sobreexplotación o salinización* [Catalogue of aquifers overexploitation or salinisation problems]. Madrid, Spain.

Junta de Andalucía, (2009). *Proyecto de Plan Hidrológico de la Demarcación Hidrográfica de las Cuencas Mediterráneas Andaluzas*, [Draft Water Plan of the River Basin Watershed Andalusian Mediterranean] Seville, Spain. Available from: http://www.juntadeandalucia.es/medioambiente/site/portalweb/menuitem.7e1cf46ddf59bb227a9ebe205510e1ca/?vgnextoid=3bba6ff4a9743310VgnVCM2000000624e50aRCRD&vgnextchannel=75b3e6f6301f4310VgnVCM2000000624e50aRCRD [Accessed 21st June 2012].

Junta de Andalucía (2010). *Inventario de regadío de Andalucía de 2008* [Inventory of irrigation in Andalusia, 2008]. CD-ROM. Consejería de Agricultura y Pesca. Seville, Spain.

Llamas, M.R. & Custodio, E. (2002). Acuíferos explotados intensivamente: conceptos principales, hechos relevantes y algunas sugerencias. *Boletín Geológico y Minero*, 113(3): 223–228.

Llamas, M.R. & Martínez-Santos, P. (2005). Intensive Groundwater Use: Silent Revolution and Potential Source of Social Conflicts. *Journal of Water Resources Planning and Management*, 131(5): 337–341.

López-Gunn, E. (2007). Self-regulation as an alternative for the future? Groundwater management in Spain. In: NGWA Press (ed.), *The Global Importance of Groundwater in the 21st century*: 345–350. Westerville, USA.

López-Gunn, E. & Martínez Cortina, L. (2006). Is self-regulation a myth? Case study on Spanish groundwater user associations and the role of higher-level authorities. *Hydrogeology Journal*, 14(3): 361–379. doi:10.1007/s10040-005-0014-z.

MAPA, (2001). *Unidades Hidrogeológicas con problemas de sobreexplotación. Plan Nacional de Regadíos Horizonte 2008* [Hydrogeologic units with problems of over exploitation. National Irrigation Plan Horizon 2008] Madrid, Spain.

Martínez, F. (2011). *Estudio de la huella energética del abastecimiento urbano de agua de la provincia de Almería*, [Study of the energy footprint of urban water supply in the province of Almería. Research Work of the Master Water and Environment in semiarid areas (AQUARID)]. Department of Hydrogeology and Analytical Chemistry. University of Almería, Spain.

Martínez Cortina, L. (2011). *Balances hídricos y estimación de la posible evolución futura del sistema hidrológico de la cuenca alta del Guadiana*. [Water balance and estimation of the possible future evolution of the hydrological system of the Upper Guadiana basin]. Conference for the 10th anniversary of AEUAS The future of groundwater management", Ruidera, Spain.

MIMAM. (1998). *Libro Blanco del Agua en España* [White Paper on Water in Spain] Madrid, Spain.

Ostrom, E. (1990). *Governing the Commons: The Evolution of Institutions for Collective Action*. Cambridge University Press.

Poveda, J.A. (2011). Uso conjunto a través de la gestión colectiva e incentivos para un uso eficiente (Agua-Energía) [Conjunctive use through incentives and collective management for efficient use (Water-Energy)]. *Seminar on collective action for groundwater management in Spain*, Botín Foundation Water Observatory. Madrid, Spain. Available from www.fundacionbotin.org/agua.htm [Accessed 14th July 2012].

Rica, M., Lopez-Gunn, E., & Llamas, R. (2012). Analysis of the emergence and evolution of collective action: an empirical case of Spanish groundwater user associations. *Journal of Irrigation & Drainage*, 61(Supplement S-1), 115–125. doi:10.1002/ird.1663.

Rico, M. (2011). La comarca del Carracillo exporta su modelo de recarga del acuífero. *El Norte de Castilla*, [The region of Carracillo export its model of aquifer recharge]. Segovia, Spain.

Rivera, J. (2000). *Agrarian Colonization Policy in Campo de Dalías (1940–1990)*. Instituto de Estudios Almerienses, Cajamar. Almería, Spain.

Sahuquillo, A.; Custodio, E. & Llamas, M.R. (2009). La gestión de las aguas subterráneas. *Tecnología del Agua*, first part: February, pp. 60–67; second part: March, pp. 54–67.

Shah, T. (2009). *Taming the Anarchy. Groundwater Governance in South Asia*, Washington DC, USA: Resources for the Future, IWMI.

Valero de Palma, J. (2011). Retos Institucionales de la gestión colectiva del agua en España. [Institutional Challenges of collective management of water in Spain *Seminar on collective action for groundwater management in Spain*. Botín Foundation Water Observatory. Madrid. Spain. Available from www.fundacionbotin.org/agua.htm [Accessed 14th July 2012].

Van Cauwenbergh, N. & Francés, I. (2010). *Proyecto Altaguax: Los retos de la gestión del agua en la cuenca del río Andarax*. [Altaguax Project: The challenges of water management in river basin Andarax]. Reports of participation workshops during 2009–2010. Available from: www.altaguax.org [Accessed 20 May 2012].

Van Cauwenbergh, N.; Pinte, D.; Tilmant, A.; Francés, I.; Pulido-Bosch, A. & Vanclooster, M. (2008). Multi-objective, multiple participant decision support for water management in the Andarax catchment, Almeria. *Journal of Environmental Geology*, 54: 479–489.

Varela, M. (2009). Conocimiento de las aguas subterráneas en España. [Knowledge of groundwater in Spain] *Conference Groundwater in Spain: situation and policy*. 24 November 2009, Madrid, Spain.

Chapter 19

Implications of the modernization of irrigation systems

Elena López-Gunn[1], Beatriz Mayor[1,2] & Aurélien Dumont[1]
[1] *Water Observatory of the Botín Foundation; Department of Goedynamics, Complutense University of Madrid, Madrid, Spain*
[2] *University of Wageningen, Wageningen, The Netherlands*

ABSTRACT: This chapter looks at the relevance of an *ex post* analysis of the National Irrigation modernization programme implemented in Spain, possibly the largest in terms of surface area and investment in the whole of Europe and one of the largest programmes in the world. This plan was a state led effort to increase water efficiency in irrigation and generate water savings at plot and basin level, particularly to reduce water stress during drought periods. This was within a paradigm where irrigated agriculture had traditionally primed over other economic activities and water uses, and which promised the achievement of substantial water savings. There is now some evidence, after the completion of the programme, that planned water savings have not been met, and in some cases the increase in water efficiency application seems to have entailed an expansion in irrigated land and/or some crop changes, leading to potentially a higher overall local irrigation water consumption. However there are other unintended consequences and in some cases co-benefits in terms of reduced use of fertilisers due to fertirrigation, and better traceability and control on water use due to technological improvements. Yet the lack of reliable and consistent information on the actual aggregated consequences of this large public investment programme highlights the need for a detailed assessment on the consequences and logic of the modernization process. A set of indicators and a range of ways to implement these programmes are suggested to help identify and make a balance of the results *vs* the investment required, with views to a future modernization projects.

Keywords: modernization process, unintended consequences, water savings, water efficiency, energy efficiency

1 INTRODUCTION

This chapter offers an *ex post* analysis of one the largest irrigation modernization programs undertaken at national, European and global scale, which is conceived as part of a new turn in water policy towards demand management measures. Complementing previous emphasis on increasing supply through e.g. reservoirs and transfers, irrigation modernization has become a major thrust of Spanish water policy in the last 15 years. The chapter looks at the central role irrigation plays in Spanish water policy for two reasons: first, to trace its historical and cultural importance dating back to the 19th century; and second, because of the volume that irrigation represents for the Spanish water budget, consuming around 70% of water, and thus central to any debates or discussions on saving water. Examples of such debates are whether to leave more water for the environment or how to re-allocate water to other productive sectors, like energy and solar-thermal plants (see Chapter 14), tourism or public water supply. In order to analyse the intended and un-intended consequences of a major public investment programme for water efficiency and water savings, the chapter is based on a number of methods: literature review, in depth expert interviews, the analysis of data on the implementation of these plans, some specific case studies and methods to evaluate *water efficiency*, *water savings* and *water accounting*. This chapter complements other previous studies on this topic (Berbel & Gutiérrez, 2004; Camacho Poyato, 2005; Cánovas, 2008; Cots, 2011; Rodríguez-Díaz *et al.*, 2004; Varela-Ortega, 2006) and is a shorter version of separate paper (López-Gunn *et al.*, 2012).

2   THE LOGIC FOR IRRIGATION AND ITS MODERNIZATION IN SPAIN

In Spain the hydraulic paradigm permeates water-related decision making and policy frames. An expert interviewee, the Ex-Deputy Director for irrigation, pointed out that within this paradigm, agriculture (together with hydropower) has been – historically for valid reasons – a privileged user (Garrido & Llamas, 2009). A hard economic crisis in the end of the 19th century created tension between a growing bourgeois society amidst a largely rural and illiterate country. Water thus became a symbol of prosperity and modernity, and irrigated agriculture was seen as pivotal for change in a rural Spain (López-Gunn, 2009). Meanwhile in the Spain of the 21st century irrigation is now center stage because water has many productive uses, of which irrigation is one. In an urban society, the protection of natural or ecological values increasingly raises questions on the traditional dominant use by irrigated agriculture. As Allan (2010) identifies, the challenge in many semi-arid countries are decisions on the allocation of the *big* water rather than *small* gains to be made on efficiencies in the public supply sector (see Chapter 13).

In Spain, as a semi-arid country, irrigation is important, where Spain accounts for almost a third of the total irrigated area in the whole of the European Union and its potential irrigated area is almost fully utilised. Irrigation is considered a strategically crucial sector since it consumes around 70% of total water resources and uses 50% of the water kept in Spanish reservoirs in a regular year. Irrigation is key to the agricultural sector since it accounts for 60% of the total agricultural produce (i.e. 13,000 M€ out of an estimated 20,500 M€ and 80% of total farmer exports). Yet it only represents 14% of the agricultural area, although productivity is six times higher than in rain fed agriculture and an income four times higher than in dryland farming.

Increased production, higher incomes, direct and indirect employment and contribution to agricultural Gross Domestic Product (GDP) explain both the inertia and drive for irrigation in Spain. Most of the country is naturally water scarce due to its geographical location and this water scarcity- framed within a hydraulic paradigm- was a problem solved through state intervention, to augment or control water resources, mainly via technological or infrastructural measures. Supply management measures were taken in the 20th century to increase available water through water reservoirs and transfer construction. This supply approach started to be challenged in the 1990s, with the push for so called demand management measures seeking *water efficiency* and *water savings*, where the modernization of agricultural irrigation was seen as the main strategic measure.

The farming sector has also experienced a dramatic transformation, with the co-existence of traditional farming alongside a thriving and dynamic competitive agri-food sector. Thus the sector has become sensitive to a public image as an old fashioned, wasteful and inefficient user of water. This *public image* became particularly poignant in the mid 1990s, when during a prolonged drought, many cities in Southern and Mediterranean Spain and a total of 12 million Spaniards experienced water service interruptions, whilst fields continued to be irrigated. At this point, in the midst of the discussion on the 1993 National Hydrological Plan, the Spanish Parliament asked for a review of irrigation and a National Irrigation Plan. The focus centred on the almost half of the irrigated land, which was still irrigated through traditional gravity fed surface systems, and where technically *water savings* could be achieved through technology-change. Traditional irrigation systems in Spain were portrayed as having low water efficiency (60% on average) (Barbero, 2006) due to substantial water losses in old conveyance networks in extensive flood irrigation systems. Before 2002, 700,000 ha were irrigated by ditches often through a network of concrete channels more than 60 years old, and where large water losses were reported on 400,000 ha. In 2002, 60% of the irrigated areas was still irrigated by flood irrigation, with less than a third of irrigated land having a guaranteed water supply. Sprinkler irrigation was used in only 24% of the irrigated area and only 17% had drip irrigation. After more than seven years of data collection, research and analysis, Royal Decree 329/2002 was enacted in 2002, the starting point for the *National Irrigation Plan – Horizon 2008* (or NIP 2008) to modernise the sector (MAPA, 2001; Barbero, 2006).

3   DEMAND MANAGEMENT AND IRRIGATION MODERNIZATION

The NIP 2008 took a staged approach, with a stated policy objective to modernize 1,134,891 ha by 2008 (i.e. 1/3 of the irrigated area). It covered the period 2002 to 2008, using the framework of river basins and specifying areas to be modernised in terms of hectares per region (Díaz Eimil, 2001). Modernization was based on the lining of old canals and improving the irrigation system and storage facilities, farmer training on good irrigation practices, and on improving water quality and drainage (canals were substituted by tubes in most cases). This Plan represented a shift away from big water infrastructure like reservoirs or water transfers, opting instead for modernization, thought to be cheaper per m³, while it had the added value of taking into account other social and environmental aspects. The aim was to ensure "that each m³ had a name and surname" (i.e. the traceability and control of water systems), as stated by the secretary of the General National Irrigators Association in one of the interviews. Farmer organizations and the Federation of Irrigators Communities became active campaigners for modernization, a strong lobby able to glue all the different (often) competing organizations into a common objective: a major investment programme for irrigation modernization. The programme was implemented through the coordination of the Ministry of Agriculture and the Ministry of Environment, which after the 2008 national election were merged into a single Ministry. It was largely executed by State companies who acted as catalysts for State investment, to encapsulate the advantages of the private sector and speed up investment (Díaz Eimil, 2001). The modernization programme relied on three parallel modernization tracks to succeed: i) modernization *upstream* or so called *en alta* or *wholescale*, in e.g. regulation of the main irrigation networks and reservoirs managed by water authorities; ii) *downstream*, *retail distribution* or *en baja* by agricultural agencies (i.e. the lining or substitution of canals); and finally, iii) by *farmers* modernizing their farm at plot level. All these pieces in the puzzle had to fit if the overall targets on efficient water use were to materialize. The NIP 2008 had ambitious targets in terms of projected water savings estimated at 2,100 hm³ (or 1,850 m³/ha) [hm³ = cubic hectometre = million m³ = 10^6 m³]. In terms of irrigation technology there has been an evolution from flood irrigation towards drip irrigation. Whereas in 2002 1.3 million hectares were irrigated by gravity, greater than the 1.1 million hectares with drip irrigation, by 2009 the accounted 1.6 million hectares of drip irrigation exceeded those with flood irrigation, which diminished to 1.05 (MARM, 2010). This also indicates that there was not only a conversion on part of the flooding in irrigated fields to drip irrigation, but also an extension of the irrigated land.

A severe drought in 2006 triggered a second Plan, the *Shock Plan for Irrigation Modernization* (or SP 2006), to achieve additional water savings of 1,420 hm³/year on top of the planned water savings foreseen by the NIP 2008 (Ariza, 2006). Passed as an urgent measure, 2,680 M€ were invested to modernise an initially intended irrigated area of 0.87 million hectares, which rose up to 1.32 million. Therefore the total investment considering both NIP 2008 and SP 2006 was more than 7,000 M€ to generate planned water savings in the order of 3,100 hm³. According to a study by the Public Policy Evaluation Agency (AEVAL, 2010) on the Segura and the Guadiana basins, the water savings have been of 94 hm³ in the Guadiana and 65 hm³ in the Segura. However this has not necessarily translated into reduced withdrawals which would reduce the pressure on the basin or free up water for other uses, as already observed by Molle & Turral (2004) for other cases in different parts of the world. This is due to a number of reasons: first, numbers included in the NIP 2008 and the SP 2006 were savings estimated at plot level (or classical efficiency as described by Lankford, 2012); second, assumptions were made on the type of crop remaining constant; and third and most important, that there would be no increase in the irrigated area.

Zooming into the implementation of the SP 2006, the regions with the largest investment were Andalusia (corresponding largely to the Guadalquivir basin and internal regional basins), Valencia (Júcar basin), Castille-León (Duero basin) and Aragón (Ebro basin) (see Figure 1 and Table 1). Investment however was not necessarily proportional to a corresponding amount in terms of water savings since the type of action under irrigation modernization was varied. However what is also relevant is that an analysis of water savings does not shed light or facilitate debate on the *logic of modernization*. It does not demonstrate a clear explicit *ex ante* decision criteria on where or why money was invested. The case of Andalusia shows that the water savings obtained increase with the intensity of the reforms, which in turn depend on the level of investment per hectare (Corominas, pers. comm.). Thus it would be possible to estimate which areas are more suitable for investment based on the initial efficiency of the irrigation system to identify those that have a higher potential for the most suitable

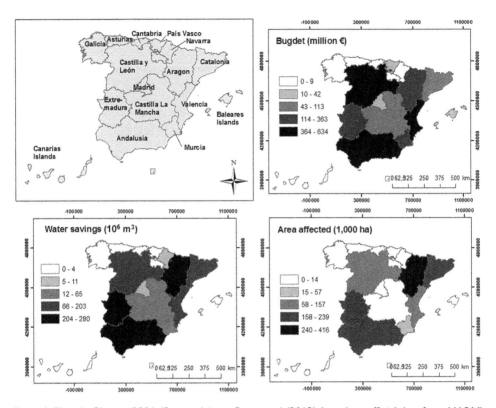

Figure 1 Plan de Choque 2006. (Source: López-Gunn *et al.* (2012), based on official data from MARM).

Table 1 Regional costs per m³, ha and farmer.

Region	M€/hm³	€/ha	€/farmer
Andalusia	2	3,377	53,618
Valencia	3	2,931	8,606
Castille-León	2	3,496	27,739
Aragón	1	871	28,921
Extremadura	1	958	25,265
Murcia	3	3,883	12,788
Catalonia	1	613	6,328
Castille-Mancha	2	6,051	32,177
Basque Ctry.	11	27,775	63,899
Balearic Isl.	16	23,142	53,169
Madrid	4	49,233	–
Cantabria	18	19,770	45,503
Canary Isl.	2	24,725	11,749
Galicia	3	5,132	55,714
Navarre	0	1,518	–
Asturias	20	17,143	40,000
AVERAGE	**5.5**	**11,913**	**33,248**

Source: López-Gunn *et al.* (2012), based on official data from MARM.

modifications to the irrigation infrastructure in order to achieve an increase in water use efficiency. However, the debate still remains on what happens with this *saved water*: whether it remains in the rivers and aquifers or it is put to some additional use (e.g. the irrigation of more land that could not be irrigated before due to technological or water availability limitations) or a change to water intensive crops.

4   THE WATER/FOOD/ENERGY NEXUS IN SPAIN: THE LAW OF UNANTICIPATED CONSEQUENCES

In public policy analysis, often as important as the evaluation of the stated objectives are also the unanticipated consequences. In the case of irrigation modernization, Gleick *et al.* (2011) has raised the importance to look beyond basin efficiency and new water towards basin productivity, and so called *co-benefits*, and a more comprehensive evaluation of what sustainable water management in agriculture means (Fereres & Connor, 2004). One positive unintended consequence (in this case co-benefit) of the modernization programme identified by several interviewees is the reduction in pollution due to the onset of collective fertirrigation, whereby farmers irrigate and apply nutrients at the same time in a more efficient manner. This was stated by the secretary of the Irrigators General National Association, the ex-water director for regional water agency and the technical director of a water authority, which coincided in identifying in having had a clear positive externality in the reduced pollution to streams and aquifers since lower doses of agro-chemicals are applied, and especially because the costs are also reduced by half (e.g. from 300–400 €/ha to 200 €/ha).

A second consequence has been a large investment into the sector (irrigation technology and alternative water supplies like desalination and re-use) which has good prospects in the coming years as an export industry, in competition at the global level for the adjudication of contracts to modernise or strengthen the agricultural sector.

A third unintended consequence of the modernization programme is related to the water-energy nexus (Rodríguez-Díaz *et al.*, 2011). In the year 2008, after a process of energy liberalization, and just when the NIP 2008 was coming to an end, the preferential binomial tariffs, the so called tariff R were removed (see Chapter 14), with an increase in prices between 50% to 80%. Whereas the cost of water was estimated at 80–100 €/ha, the costs of energy was around 200 to 300 €/ha. That is, the irrigation communities are now paying three times the cost of water in energy costs. This, considering the increase of around 20% on the energy tariff registered in the last four years due to market liberalisation, together with the high loans they had to acquire to face the cost of modernization, has led many of these communities to quite difficult economic situations. According to Head of the Agricultural Public Investment company there has been an average drop in 30% water use (abstraction) at plot level due to modernization, and this lower water use may compensate for the increase in energy prices in those areas with low application efficiencies. The result has been a stronger incentive for farmers to be more efficient in water due to the associated energy costs of using water resources (Corominas, 2009). For example, in the draft Plan that could represent the continuation, or third stage, (MARM, 2010) of Irrigation Modernization energy efficiency is considered as a key priority (IDAE, 2008; Hardy & Garrido, 2010).

A fourth unintended consequence relies in the already observed fact (Ward & Pulido, 2008) that sometimes specific targeted policies to promote water use and irrigation efficiency in agriculture can lead to the opposite effect, the so called Jevons paradox or rebound effect. For instance, the re-use of water savings generated at plot level for the intensification of irrigation or the expansion in irrigated area, ultimately leads to an overall rise in consumptive use (Ward & Pulido, 2008).

Two independent in-depth study is carried out by Cots (2011) and Lecina et al. (2010) in the Ebro basin seem to be indicative of these different issues. At least four relevant effects of modernization were observed. First, there may be an overall rise of water consumptions, since the resulting water savings do not compensate for the increased demand brought by the expansion in the irrigated area and the shift to higher value and more water consumptive crops enabled by higher irrigation application efficiencies. Second, there was a reduction in irrigation return flows, which before would have been reused or gone back into the system, thus reducing water supply downstream. Third, there was an increase on energy consumption and energy dependence brought by the generalized mechanization of irrigation. And last, due to modernization there was an increase in water and economic productivities in agriculture, even when constrained by market fluctuations and increasing energy prices. However, when trying to assess whether these patterns are mirrored at the macro-scale for the Spanish NIP 2008 and SP 2006 there is limited aggregated data available to judge on, and the existing data is imprecise and often based on normative judgements. Therefore it reflects a pressing need to make an overall *accounting* and rigorous analysis of the results and impacts of the modernization process, which in line with Gleick et al. (2011) also incorporates the concepts of *productivity* and *co-benefits*, particularly if additional large modernization investment is planned for the future.

5 A METHOD TO EVALUATE THE EFFECTS OF MODERNIZATION

It has already been explained how the main justification for modernization plans was the expected water savings, among other associated benefits. These water savings would result in the reduction of water diversion from rivers and aquifers thanks to a better conveyance and application efficiency. However, this approach is overlooking the fact that, often surplus water applied in less efficient system, returns to the basin for a downstream use and is not *lost*. In order to assess the results of modernization, this section proposes a set of indicators which includes all the main aspects (see Table 2).

The first indicator, water quantity, allows for the identification and quantification of the different flows going in and out from the river or the aquifer into the irrigation system through the use of a combination of water accounting methodologies proposed by Molden & Sakthivadivel (1999) and Perry (2007), and the water footprint (WF) as defined by Hoekstra et al. (2011). Calculating the WF of the irrigation area *before* and *after* modernization will provide a better estimation of water savings, since this methodology computes only the water that, after use, is not available for further use in the same basin. In the case of agrarian production, the WF includes: i) the evapotranspiration of the crops; ii) the water that flows to *sinks* (the sea or water bodies where it becomes unusable); and iii) the water incorporated in the production (minimal fraction compared to the others). Additionally, beneficial and unproductive consumption can be distinguished, which allows to consider independently evaporation from the soil and plants transpiration.

Table 2 Set of Indicators to evaluate modernization projects.

Indicator	Aspects to be considered	Analytic method or tool
Water quantity	Types of uses Consumption estimate Water available for other uses	Water accounting method Blue, green water footprint tool Time and location factors
Water quality	Diffuse pollution	Concentration levels of different pollutants in the sinks: nitrates, phosphates, pesticides, etc.
Economic welfare	Productivity. Technological innovation and expertise. Potential technological exports.	Crop (tons/ha). Water productivity (tons/m³). Side opportunities: Development of technologies and knowledge, business agreements and international cooperation.
Social welfare	Quality of life for farmers. Economic solvency (depending on how funds have been distributed).	Labour hours, RPC. Amortization costs/Economic yield. Intangible personal gain (expertise, technical knowledge, social cooperation and participation, collective power).
Environmental	State of the environment components (aquifers, river flows, soil quality, natural ecosystems maintenance: terrestrial and wetlands).	Evolution of aquifers reserves. Evolution of river flows. Soil quality. State of natural ecosystems.
Energy demands	Evolution of energy demand and dependence. Development of onsite energy sources for self-provision.	Energy demand (kWh/ha). Local energy supply (kWh/ha). Source of energy supply: renewable, thermoelectric, etc. Price of energy (€/ha).

Source: Own elaboration.

Many international research institutions aim to define a general methodology that can be applied worldwide in any basin. The downside, like any generic model made to fit many possible and different cases, is that this might not fit completely for all of them. As an alternative approach Snellen (pers. comm.) sustains that: "many indicators are used by researchers, never by irrigation agencies themselves. An alternative approach is to use *a service approach* [see Dolfing & Snellen, 1999], which also uses performance indicators, but only for checking whether services are delivered as agreed upon. Therefore the main characteristic of these indicators is that these indicators are convenient for the people directly involved: the service providers and the water users". Under this option, the best set of indicators for a certain basin, to fit a specific basin context, is achieved when selected, negotiated and adapted jointly by the water agencies and the water users themselves. This is possibly a more practical approach, difficult to generalise but which might be conducive to achieve the intended (negotiated) overall outcomes[1].

1 For more information on this approach see Dolfing & Snellen (1999).

6   ANALYSIS AND REFLECTIONS ON THREE CASE STUDIES

A sample of three diverse case studies are presented below to highlight the real life complexity from the implementation of modernization projects on the ground, and how these fit within the framework of river basin planning required under both Spanish water law and the EU Water Framework Directive. According to a Deputy Water Comissariat, it should be linked to the planned evaluation on the cost effectiveness of measures.

Case 1   A structural modernization project of groundwater irrigation for a variety of crops in Alicante

A modernization plan of 70 M€ investment affecting 22,200 ha has been implemented. The plan was aimed at reducing irrigation consumption to stop the intensive use of local aquifers, in combination with a hypothetical water transfer from the Júcar River. This transfer which represented an investment of 370 M€ and water reallocation of 6 hm³ has been delayed for 8 years. The modernization works included: a pressurized pipe system to plot level, 16 storage ponds interconnected through a pipe network and the installation of a remote control system to allow farmers to control the exact timing and amount of irrigation via the internet which allows an accurate water abstractions register. In terms of water savings, it has resulted on a lower reduction of irrigation consumptions than planned (5 hm³ compared to 20 hm³ initially planned), however it has ensured the continuity of local irrigated agriculture. This is because modernization has given farmers the flexibility to be able to pump during cheap energy tariffs, cushioning farmers from the rise in energy prices. On the other hand, it has entailed a high level of debt for an average (>60) years old hobby farmer, faced with current low market prices for crops cultivated in the area.

Case 2   A modernization for greenhouse agriculture with high value export crops in Almería

The coast of Almería province (Andalusia) is famous for the process of agricultural intensification based on greenhouses to grow high value vegetables. This however has resulted into a drastic fall on aquifer levels and subsequent marine intrusion. In this area the major part of irrigation water comes from groundwater in a deep confined aquifer. The upper shallow aquifer contains poor water. Because of the deep water table, the pumping costs are high, which has encouraged farmers to adopt drip irrigation to ensure an efficient use of groundwater. Thus, potential new investments is targeted to an already very efficient system, and where investment is geared for system maintenance, but where the potential water savings are low since the application of water is already highly efficient. Investment is justified by the high added value of the crop. The modernization scheme has two main strategies. The first consists on the construction of a new pressurized network to reduce leakage in the water distribution system. It would benefit around 9.000 ha, i.e. almost half the irrigated area. Moreover, this network would interconnect all the different wells, and link them to regulation reservoirs which would allow for a complete monitoring of water characteristics from pumping to distribution. This latter most expensive part of the project is not directly linked with leakage reduction (which is the basis for water savings), it is centred instead on increasing water security through the availability of water (interconnection) and a possibility to pump during the cheapest hours of electricity, which results in financial savings for farmers. The estimated water savings would be around 2.6 hm^3 (i.e. which represents 4.6% of the Water Footprint in the area estimated at 58 hm^3, Dumont et al., 2011), for a total budget of more than 27 M€. The second strategy was targeted at generating alternative water resources through two desalination plants and a water treatment plant which provides high quality recycled water (a large investment paid 70% by the government with 20% from European funds and 10% by Irrigation Communities). The plan is therefore focused on the reduction of risk, by increasing reliability for a high value agriculture with a substantial amount of public investment (by the Spanish State and EU) subsidies.

Case 3　The modernization of irrigation canals for rice and cereals in Extremadura

In Extremadura region, the riversides of the Guadiana River basin have traditionally been occupied by irrigated fields where surface irrigation of cereals and rice was and remains the main practice (44.6%). A slight trend has been observed towards drip irrigation. However, the most important modernization works were oriented towards improving the state and efficiency of the irrigation network, and especially of irrigation canals. The first project consisted on the substitution of the old pipe system for a new impermeable concrete one and the installation of flow meters reaching to plot level. The second project entailed the installation of an automated telecontrol system which allows for the programming and monitoring of precise irrigation programs via internet or mobile phone, while flow and water quality measurements are periodically taken. The investment required by these two projects accounted for 34 M€ with estimated 129.6 hm^3 water savings. These estimations though, have been made by the calculations on the theoretical improved efficiencies. These however have not yet been contrasted with figures based on real measurements. It does not mean it is not possible since the projects include the installation of flow meters which allow for a close tracking of water consumption.

Source: Own elaboration on the basis of fieldwork and expert interviews.

Table 3 Main characteristics of the three study cases.

Case	General characteristics Main crops	Farmer characteristics % active pop.	Age	Type	Economic yield (€/ha)	Modernization plan Overall cost (M€)	Theoretical savings (hm^3)	Real savings (hm^3)
Alicante	Vineyard, olive, almonds, vegetables	6–10	>60	High % hobby farmer	1,500	70	20	5
Almería	Paprika, watermelon, melon, aubergine	50	40	High % commercial farmer	26,470	27	2.6	Lost due to higher ETP
Guadiana	Cereals, rice, forage crops, olives	13.8	30–40	Commercial farmer	1,290	34	129.6	Not measured

Source: Own elaboration from SEIASA (2010).

At the micro or operational level these specific cases highlight the complexity of the modernization process, where the implementation of similar patterns of investment in different areas gives way to very different realities, especially in the social and economic spheres e.g. the capacity to cope with the investment and pay back costs is not the same for the average 60 year old farmer with a middle rent than for a high income 40 year old business farmers. These differences were probably not considered *a priori* to evaluate the extent to which in some areas these large public and private investments were justified considering a large number of potential consequences or indeed to set some criteria beforehand to decide on the destination and distribution of funds. Applying the theory of the policy cycle to the modernization process, the issue on how to invest scarce resources of *time* and *money*, and how effective results are according to the original aims (achievement of objectives or unexpected consequences), is one of the uncertainties a government has to cope with. However, the only way to tackle this uncertainty relies on the elaboration of a systematic planning process which includes three main steps: a characterization and pre-evaluation of the initial situation, a close monitoring programme to keep track of the evolution, and an *ex post* evaluation of the final results (Figure 2). In this sense, it is important to distinguish the *Agenda de Regadíos* 2015 of Andalusia (Junta de Andalucía, 2010a; 2010b), which includes a thorough analysis of modernization projects and presents a coherent and well conceived plan for tracking the effectiveness of the plan.

The evaluation step is designed to identify possible deviations or changes on the initial conditions to then readapt policies, or learn from them. The lack of this step is actually the most important gap in the NIP 2008, which had three important consequences:

i   First, well justified criticisms on the realization of water savings can eclipse some other important positive aspects and successes achieved, such as the implantation of an accurate measurement system which keeps detailed track of the

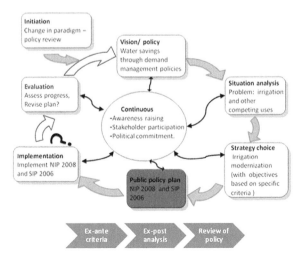

Figure 2 The incomplete policy cycle of irrigation modernization. (Source: Own elaboration).

water consumptions in important basins. This represents a huge advance in GIS characterization and in irrigation technologies and important increases on agricultural productivity and farmer's quality of life. It is a very powerful tool for water planning giving access to better or more accurate data on actual use.

ii Second, well intentioned initiatives can fail if the appropriate means and strategies are not implemented, and there are (expensive?) missed lessons to learn from the implementation of these two major public investment programmes: there were good intentions to make plans together with the Irrigation Communities. However, the eagerness for speed and execution of works by the State building companies, in common with other infrastructure projects, has led in some cases to the delivery of rushed plans and the need to repeat some works and consequently higher investments than initially estimated.

iii And third, important failures or counterproductive actions if not detected could be repeated later on, or with alternative policy paths not being considered. This is especially relevant considering that a third phase of the Irrigation Modernization Plan Horizon 2015 is planned (MARM, 2010) with important public investments. In this context lesson drawing on the approach pursued focused on technical (irrigation) water use efficiency, by eliminating any unproductive loss and generating or importing new water resources. However, to adjust supply to irrigation demands it is important to also consider whether the new plan could instead look for other strategies like more sustainable and appropriate crop patterns or activities, such as solar energy, and adapt to the real existing and available resources.

7 CONCLUSIONS

In the 1990s wasting water in irrigation was bad for the public image of agriculture and for irrigation, when in the middle of a drought cities experienced restrictions, while fields continued to be irrigated. Ten years later in the mid-2000, during another prolonged and severe drought there was no need for the introduction of restrictions. According to the representative of the irrigation associations of Spain, this is both tribute and evidence on the success of the irrigation modernization programme. However, other voices are still engaging in the debate on the future and logic of irrigation and a large public investment in modernization. Some sectors demand that water flows again in rivers and is preserved for other sectors, as required in a modern society, to reduce the pressure on aquatic systems. In a modern country where only 4% of the population is directly employed in agriculture (see Chapter 6) and a great portion of it is aged, there is a new dichotomy since this co-exists with a highly dynamic, entrepreneurial agri-business sector thriving despite or because of the current economic crisis. This dichotomy forces a re-think on the logic for irrigation in Spain as a mature water economy, and where the Spain of the 21st century bears little resemblance to the Spain at the end of the 19th century. It also brings out the uncertainty on the maintenance of a stable primary sector which ensures food security in the future, not only in Spain but in the whole of Europe, given the ageing process of rural population and the crucial role of the younger generations to take over.

One of the upcoming questions in Spain after a major public investment programme of around 7,000 M€ over the last ten years into irrigation modernization (like other major programmes undertaken on roads, or high speed trains), is to establish the logic and parameters for returns on this large investment in irrigation modernization. This is particularly pertinent in the present situation, when public investment should be thoughtfully re-assessed and with a third phase of modernization plans on the way. As analysed before in this chapter, irrigation technical efficiency is neither necessarily equivalent to water savings, nor the only means to achieve them. This chapter argues that a focus on water conservation would provide a clearer end goal for policy, and open the door to a wider range of measures on water management to be combined in the ideal policy mix to fit different circumstances, i.e.: water tariffs (OECD, 2010), water quotas, water markets, self-regulation, conditional water licenses and crop switching to less water consumptive crops (as already implemented in the *Agenda de Regadíos de Andalucía, Horizonte 2015*). In this context, the use of a complete variety of indicators and water accounting methods will also have to consider other aspects such as farmer welfare and water productivity to shed light, at national and regional level, on the social, economic, hydrological and environmental viability of future irrigation modernization processes. Rural development is no longer synonymous with agriculture, but rather seek a more diversified and integrated economic balance between sectors (e.g. renewable energy and solar farms, rural tourism, etc.).

After all, it is ironic that with irrigated agriculture caught at a crossroads between two external policies (the reform of the Common Agricultural Policy and the compliance with the EU Water Framework Directive) and despite the lack of firm conclusions about the magnitude of real water savings, it is the unexpected impact of the parallel policy of energy liberalisation that has opened a new space for decision making in Spain on the overall efficiency gains to be made in resource use (water and energy).

ACKNOWLEDGEMENTS

We are grateful to Dr. B. Snellen (University of Wageningen) for his kind revision and valuable contribution to build an *out of the box* vision. We are also grateful to all the experts interviewed during the preparation of this chapter and for all people interviewed in the case study areas.

REFERENCES

AEVAL (2010). *Evaluación de la gestión y funcionamiento de las Confederaciones Hidrográficas. Agencia de Evaluación y Calidad, Ministerio de la Presidencia* [Evaluation of the management and operation of Spanish Water Boards. Evaluation and Quality Agency, Ministry of the Presidency], Madrid, Spain.

Allan, T. (2010). Prioritising the processes beyond the water sector that will secure water for society – farmers in political, economic, and social contexts and fair international trade. In: Martínez Cortina, L.; Garrido, A. & López-Gunn, E. (eds.), *Re-thinking water and food security*. Chapter 6. Taylor and Francis, London, UK. 96–106.

Ariza, M. (2006). Obras del Plan de Choque en la zona de Actuación de SEIASA del Sur y Este S.A. [Shock Plan Works in SEIASA DEL Sur y Este S.A.'s Area of Action] *Agricultura: Revista agropecuaria*[Agriculture: Agropecuary magazine], 890: 880–883.

Barbero, A. (2006). The Spanish National Irrigation Plan. In: *OECD. Workshop on Agriculture and Water: Sustainability, Markets and Policies*, Session 5 – Institutions and policies for agricultural water governance, 14–18 November, 2005, Australia.

Berbel, J. & Gutiérrez, C. (2004). *I Informe sobre la Sostenibilidad del Regadío*. [I Report on the Sustainability of Irrigation]. Ed. Feragua.

Camacho Poyato, E. (2005). Análisis de la eficiencia y el ahorro del agua en el regadío de la cuenca del Guadalquivir [Analysis on the efficiency and water savings of irrigation in the Guadalquivir river basin]. *Agricultura: Revista agropecuaria [Agriculture: Agropecuary magazine]*, 880: 880–887.

Cánovas, J. (2008). *Modernización de los regadíos. Ahorro de* Jornada sobre ahorro, eficiencia en el uso del agua y gestión de la demanda, Foro de ciencia y tecnología [Irrigation modernization, Water savings] In: Seminar on savings, efficiency in water use and demand management; Science and Technology Forum, March 26, 2008, Madrid, Spain. Madrid, University of Alcala de Henares.

Corominas, J. (2009). Agua y energía en el riego en la época de la sostenibilidad [Water and energy in irrigation in the sustainability age]. In: *Jornadas de Ingeniería del Agua* [Water Engineering Workshop], 27–28 October, Madrid, Spain.

Cots, L. (2011). *Desarrollo y calibración de un modelo de simulación de recursos hídricos aplicado a la cuenca del río Corb dentro de la zona regable de los canales de Urgell (Lleida)* [Development and calibration of a water resources simulation model applied to the Corb river basin, within the irrigable area of Urgell Canals (Lleida)]. PhD. Agricultural Engineering School. University of Lleida.

Díaz Eimil, C. (2001) La modernización del regadío a través de las SEIASAS [Irrigation modernization through the SEIASAS]. *Agricultura: Revista agropecuaria [Agriculture: Agropecuary magazine]*, 850: 244–247.

Dolfing, B. & Snellen, W.B. (1999). *Sustainability of Dutch water boards: appropriate design characteristics for self-governing water management organizations*. ILRI (International Institute for Land Reclamation and Improvement) Special Report, 45 p.

Fereres, E. & Connor, D.J. (2004). Sustainable water management in agriculture. In: Cabrera, E. & Cobacho, R. (eds.), *Challenges of the new water policies for the 21st century*. A.A. Balkema, Lisse, the Netherlands: 157–170.

Garrido, A. & Llamas, M.R. (eds.) (2009). *Water policy in Spain*. A.A. Balkema, CRC Press, London, UK: 234 pp.

Gleick, P.H.; Christian-Smith, J. & Cooley, H. (2011). Water-use efficiency and productivity: rethinking the basin approach. *Water International*, 36(7): 784–798.

Hardy, L. & Garrido, A. (2010). *Análisis y evaluación de las relaciones entre el agua y la energía* Papeles de Agua Virtual [Analysis and evaluation of the relationship between water and energy, Virtual Water Papers], no. 6. Botín Foundation, Madrid, Spain. Available from: www.fundacionmbotin.org/agua.htm [Accessed 30th October 2011].

Hoekstra, A.Y.; Chapagain, A.K.; Aldaya, M.M. & Mekonnen, M.M. (2011). *The Water Footprint Assessment Manual. Setting the Global Standard*. Earthscan Ltd., London, UK.

IDAE (Instituto para la Diversificación y Ahorro de la Energía (2008)). *Ahorro y eficiencia energética en la agricultura* [Institute for Energy Diversification and Savings, Energy savings and efficiency in agriculture]. General Secretariat of Energy, Ministry of Industry, Tourism and Trade. Madrid, Spain.

Junta de Andalucía (2010a). *Agenda del regadío andaluz. Horizonte 2015* (Agenda Andalusian Irrigation. Horizon 2015). Seville, Spain.

Junta de Andalucía (2010b). *Inventario de regadíos y su evolución en la última década.* [Inventory of irrigation and its evolution over the last decade]. Council of Agriculture and Fisheries. Seville, Spain.

Lankford, B. (2012). Fictions, fractions, factorials and fractures; on the framing of irrigation efficiency. *Journal of Agricultural Water Management.* 108: 27–38.

Lecina, S.; Isidoro, D.; Playán, E. & Aragüés, R. (2010). Irrigation modernization and water conservation in Spain: The case of Riegos del Alto Aragón. *Agricultural Water Management,* 97: 1663–1675.

López-Gunn, E. (2009). Agua para todos: A new regionalist hydraulic paradigm in Spain. *Water Alternatives,* 2(3): 370–394.

López-Gunn, E.; Zorrilla, P.; Prieto, F. & Llamas, M.R. (2012). Lost in translation? Water efficiency in Spanish agriculture. *Journal of Agricultural Water Management.* 108: 83–95.

MAPA (Ministerio de Agricultura, Pesca y Alimentación) (2001). *Plan Nacional de Regadíos Horizonte 2008.* [National Irrigation Plan Horizon] Ministry of Agriculture, Fisheries and Food, 2008.Madrid, Spain.

MARM (Ministerio de Medio Ambiente, Rural y Marino) (2010). *Estrategia nacional para la modernización sostenible de los regadíos, Horizonte 2015.* [National strategy for sustainable modernization of irrigation, Horizon 2015]. Madrid, Spain.

Molden, D. & Sakthivadivel, R. (1999). Water accounting to assess use and productivity of water. *Water Resources Development,* 15: 55–71.

Molle, F. & Turral, H. (2004). Demand management in a basin perspective: is the potential for water saving overestimated? Paper presented at the *International Water Demand Management Conference,* Dead Sea, Jordan. Available from: http://www.iwmi.cgiar.org/assessment/files/pdf/publications/conferencepapers/demand%20 management%20 in%20 a%20basin%20perspective(1).pdf [Accessed 4th October 2011].

OECD (Organization for Economic Co-operation and Development) (2010). *Economic Surveys: Spain.* December 2010, OECD, Paris, France.

Perry, C. (2007). Efficient irrigation; inefficient communication; flawed recommendations. *Irrigation and Drainage,* 56: 367–378.

Rodríguez-Díaz, J.A.; Camacho Poyato, E. & López-Luque, R. (2004). Application of data envelopment analysis to studies of irrigation efficiency in Andalusia. *Journal of Irrigation and drainage engineering,* May–June: 175–183.

Rodríguez-Díaz, J.A.; Pérez-Urrestarazu, L.; Camacho-Poyato, E. & Montesinos, P. (2011). The paradox of irrigation scheme modernization: more efficient water use linked to higher energy demand. *Spanish Journal of Agricultural Research,* 9(4): 1000–1008.

SEIASA (2010). *Memoria informe 2010* SEIASA (Sociedad Estatal de Infraestructuras Agrarias) or (State Society of Agricultural Infrastructure), [Annual Report 2010]. Ministry of Rural Affairs and Marine Environment].

Varela-Ortega, C. (2006). Socio economic, institutional and policy considerations for irrigation modernization. In: FAO-IPTRID-NOSSTIA, *International Symposium on irrigation modernization: constraints and solutions,* 28–31 March 2006, University of Damascus, Syria.

Ward, F. & Pulido, M. (2008). Water conservation in irrigation can increase water use. *PNAS: Proceedings of the National Academy of Sciences,* Nov. 25, 105(47): 18215–18220.

Part 5

Case studies

Chapter 20

Tablas de Daimiel National Park and groundwater conflicts[1]

Elena López-Gunn[1], Aurélien Dumont[1] & Fermín Villarroya[2]

[1] *Water Observatory of the Botín Foundation; Department of Geodynamics, Complutense University of Madrid, Madrid Spain*
[2] *Department of Geodynamics, Complutense University of Madrid, Madrid Spain*

ABSTRACT: Located in the Upper Guadiana basin, the Tablas de Daimiel National Park represents a paradigmatic illustration of the effects of abstractions on a groundwater dependant wetland. The *groundwater silent revolution* of the second part of the 20th century led to the dramatic reduction of flooded area and numerous associated ecological damages, converting the area as a laboratory of policies to both regulate the extractions (for instance, the declaration of over-exploitation in 1994) and to limit the economic and social impacts on the local economy, which relies on irrigated agriculture. This situation leads also to a generalized informal use of groundwater. We focus on the last attempt of regulation, the Special Plan for Upper Guadiana, as an attempt to reorganize the water rights structure and reduce extractions to obtain water for ecological flows to the Tablas de Daimiel. However, its cost has limited its full implementation and only the results of the first phase (up to June 2011) are assessed using the water footprint accounting methodology. A substantial possible reduction of 10% of the extractions and opportunities for a new basis for the management beyond a purely quantitative view are identified. There are new opportunities open like: greater emphasis on quality products, sharpen up monitoring system and sanctioning campaigns, a more diversified economy which also includes other sectors like agritourism, ecotourism and renewable energy.

Keywords: intensive groundwater use, wetlands, water footprint, Tablas de Daimiel, socio-ecological system

I INTRODUCTION

Tablas de Daimiel is one of the most iconic wetlands in Europe and Spain due to the dramatic changes experienced over the last 40 years, coinciding with the intensive use of the Western Mancha aquifer (WMA), on which the wetland is largely dependent (Llamas & Custodio, 2003; Martínez Cortina *et al.*, 2011). Located in the Upper Guadiana basin, Tablas de Daimiel is the best known wetland part of a natural eco-region known as *Mancha Húmeda*, a UNESCO Man and the Biosphere Reserve,

---

1 Part of this chapter is an abridged version of an article by López-Gunn, Zorrilla & Llamas (2011), published by Stockholm International Water Institute (On the Water Front): [http://www.siwi.org/documents/Resources/Best/2010/2011_OTWF_Elena_Lopez_Gunn.pdf].

which encompasses a series of wetlands of different types, and where it is now estimated that only 20% of the original area remains, with very few of the wetlands functioning naturally (De la Hera, 1998). Although this example reflects a specific case study in Spain, many of the issues raised have echoes in other areas in the world, facing similar development dilemmas over the economic incentives that drive the use of natural resources and how to find opportunities for change in the behaviour of key *water managers* like farmers, who globally are responsible for more than two thirds of global water use. The intensive groundwater use in the 5,000 km^2 WMA has over the last 40 years represented an abstraction of around 20,000^2 hm^3 of which 3,000 hm^3 came from groundwater storage. This fuelled a spectacular socio-economic development of what used to be a poor and backward region, with strong migration to urban areas.

2 THE GROUNDWATER SILENT REVOLUTION AND TABLAS DE DAIMIEL

Tablas de Daimiel National Park (see Figure 1) has a strong symbolic value for a number of reasons. The name of *Daimiel* in Spanish is translated as the land that gives honey. It represented a landmark in a largely arid and poor region, which thanks to natural springs and so called *tablas fluviales*, provided sustenance for the local population from fisheries, crabbing and associated land uses like small orchards. It is also a landmark in Spanish conservation history as a symbol to reverse existing policies. On the same year that the Park was designated (1973), some watercourses were channelled and modified to increase agricultural land. The *Ley Cambó* (Cambó Law) dating back to 1918 was re-vitalized with a new Act in 1956 to facilitate the reclamation of marshland into agricultural land and drain wetlands because of their perception as insalubrious areas full of malaria, wastelands of little value. In parallel, the rationale for incentivizing irrigated agriculture was the depopulation in the area, which in 1981 had regressed to numbers similar to the 1930s. The regional government gave soft

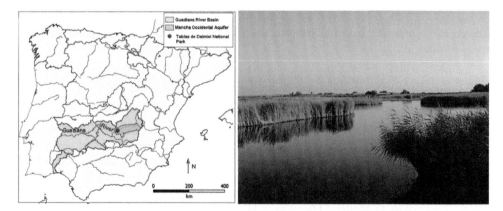

Figure 1 Location Guadiana basin and Tablas de Daimiel National Park.
(Source: López-Gunn et al. (2011)).

2 hm^3 = cubic hectometre = million m^3 = 10^6 m^3.

loans to farmers to encourage the irrigation of a dry land farming system of vines and cereals. This process of providing funding was based on the belief by many analysts (both academic and political leaders) that saw the role of groundwater as the engine driving regional development, encouraging farmers to tap the *sea beneath their feet*.

Initially a few and eventually thousands of farmers drilled wells to tap groundwater resources which had been inaccessible due to lack of technology and knowledge. Wells were authorized to irrigate maize and barley, replacing a traditional, extensive dry land Mediterranean agriculture of olives, vines and wheat (see Figure 2).

Over a period of a relatively short time the area experienced a deep process of socio-ecological transformation with large environmental externalities, as groundwater stopped overflowing from the Western Mancha Aquifer (WMA) to Tablas de Daimiel. Over a short period between 1974 to 1984 groundwater use grew from a mere 200 hm³/year to 500 hm³/year, when the estimated renewable resources were around 260 to 300 hm³/year. The area under irrigation over the same period almost tripled, from 30,000 ha to 85,000 ha (see Figure 3). The level of abstractions were influenced by three main factors: the rainfall regime, the price of the crops and the associated EU Common Agricultural Policy (CAP) subsidies (direct, e.g. to cereals; or indirect, like alcohol distillation) and more recently, to the costs of pumping, because of a sharp increase in energy costs due to the recent de-regulation of the energy sector (see Chapter 14).

The consequences of this intensive groundwater use were felt particularly from the mid to late 1980s when dry years coincided with the expansion of irrigated land. The drop in aquifer levels reached 40 and 50 meters in some areas, with many farmers deepening their wells and some drying up (the so called *war of the well*).

The Guadiana River Basin Authority declared the WMA legally over-exploited in 1994. The implications of this declaration were a series of tough restrictions: i) the forbidding of drilling of new wells; ii) the compulsory top down formation of Water User Associations; iii) the delimitation of the aquifer perimeter; iv) the ruling

Figure 2 Old and new agriculture. (Photo credits: Zorrilla (2009)).

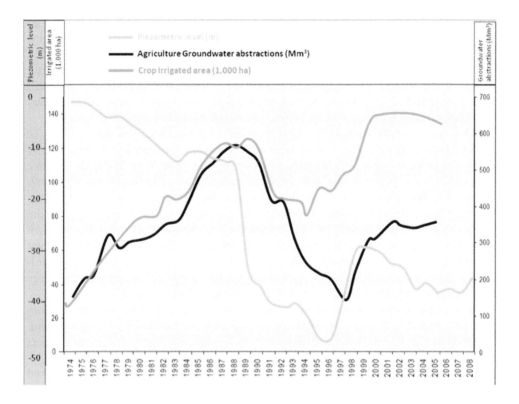

Figure 3 Aquifer levels, groundwater abstractions and irrigated area. (Source: Zorrilla (2009)).

that existing wells could not be deepened; and v) a strict reduction on the water use allowed per hectare (water quotas). These restrictions applied under an annual abstraction regime. However, the Guadiana River Basin Authority did not follow through to ensure compliance due to a lack of capacity and large transaction costs of monitoring an estimated 17,000 farmers. On the other hand, the regional government – responsible for agricultural policy but with no regulatory powers over water use – had calculated that implementing the annual abstraction regime would cost the region 7,700 million pesetas (at 1990 prices) that are equivalent to 46 M€ in income lost (Viladomiu & Rosell, 1996).

3   THE MOST RECENT PUBLIC POLICY: THE SPECIAL PLAN FOR THE UPPER GUADIANA

There have been a number of public policy measures that have been implemented in the area to address the conflict between agriculture and wetland protection (Zorrilla *et al.*, 2010). This chapter analyses the most recent public policy initiative: the Special Plan for the Upper Guadiana (or SPUG) (CHG, 2006). The origin of the SPUG was to address the situation in the Upper Guadiana basin and its structural problem of an over-allocation of water rights by 50%. The SPUG was a legal requirement under the Spanish Water Plan of 2001, which after an estimated 22 versions, was finally passed in 2008. This final Plan reflects a consensual agreement between the main stakeholders in the area, namely farmers, and farming unions, the water user groups, the municipalities, the regional government, the environmental non-governmental organisations and the central administration. The plan was to run from 2008 until 2027 with a budget of 5,000 M€. It was considered the main measure within the Programme of Measures of the future Guadiana Hydrological Plan in order to recover the two groundwater bodies that cover the WMA to good status in accordance with the Water Framework Directive (WFD) (both quantitative and qualitatively) (Barcos *et al.*, 2010). It includes a series of agricultural and regulatory measures to reduce groundwater abstractions to 200 hm³/year. The key action foreseen is the acquisition of groundwater rights, through a Rights Exchange Centre allocating 810 M€ for the purchase of water rights and land in strategically important areas for the National Park.

The Plan can be analysed as a large scale effort for both ecological restoration and distributive justice, by incorporating social aspects since it aims to re-distribute access to water between farmers. The main cornerstone therefore is the purchase of water rights to be re-allocated to small, professional farmers (30% of bought rights) and also for environmental purposes (70% of purchased rights). The water rights would be bought from farmers irrigating cereals and redistributed for the irrigation of economically more productive crops like vine and vegetables. A series of criteria has been defined on social grounds to establish the priority for water rights bought that will be granted to farmers, which are currently using groundwater without formal water rights. The criteria defined have been based on plot size, age and professional status, to grant formal rights to small farms (10 ha on average) belonging to young farmers (up to 40 years old), and which have agriculture as a main source of income. The price of purchase has been fixed to a maximum of 10,000 €/ha. Another important measure of the SPUG is the emphasis on the control of extractions with the use of remote sensing control and metering devices.

By June 2011 only the regularization relative to vine had been implemented (Requena, 2011). Figures 4 and 5 represent the effect of the implementation of the SPUG on the water footprint (WF) of the WMA if fully implemented or as implemented in June 2011, respectively. To understand the effect more clearly, and because there are now strong uncertainties on the implementation of the measure for vegetables due to lack of financing, we only consider the effect of the regularization of the vine WF. It is essential to note that vine regularization only allows a volume of extraction of 700 m³/ha (i.e. half the right of legal users and up to three times less than estimated illegal water use). This volume would ensure the basic water necessities of the plant rather than boosting yield.

In the case of vine, the regularization almost reached the planned area. However, the allocation of 70% to the environment as planned in the SPUG was not respected, as the big majority of the purchased rights (81%) have been allocated for the regularization of non-authorized users (Requena, 2011). As a consequence, only 2.6 hm³ have been dedicated directly (i.e. coming directly from bought rights) to the environment and this represents less than 1% of the total WF of the WMA, which could lead to question the efficacy of the SPUG. However, an important point to consider

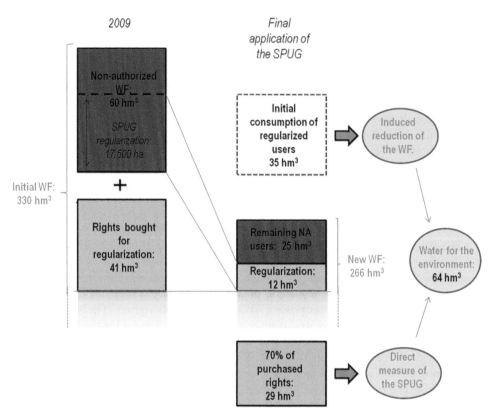

Figure 4 Effect of the SPUG implementation (for vine) on the WF of the WMA.

NA: Non-authorized / If action is taken against remaining non-authorized users after the application of the SPUG, the total WF could reach 240 hm³.

is that regularized users -if monitored effectively- will no longer consume their initial use (since their new WF will correspond to the part of the purchased rights that they received) and therefore this initial WF also goes to the environment. This indirect water saving represents more water than the direct application of the measure of the SPUG and amounts to 31.4 hm³ in June 2011. The total lowering of the WF at this date is therefore of 34 hm³ (10% of the WF of 2009), giving a completely different perspective on the effectiveness of the SPUG. The purchase of the rights to respect the initial distribution between regularization and environment would imply a final reduction of the WF of 64 hm³ (20% of the initial WF).

A basic assessment of the cost-effectiveness of the SPUG as applied in June 2011 shows that the water rights were bought for around 5 €/m³. However, on the basis of the WF accounting and the drop of the WF of 34 hm³, the recuperation of water for the environment was obtained for less than 2 €/m³. Moreover, if we consider that the final objective of the SPUG is not only quantifiable in terms of extraction reduction, but is also to build the basis for a new local governance of the groundwater bodies,

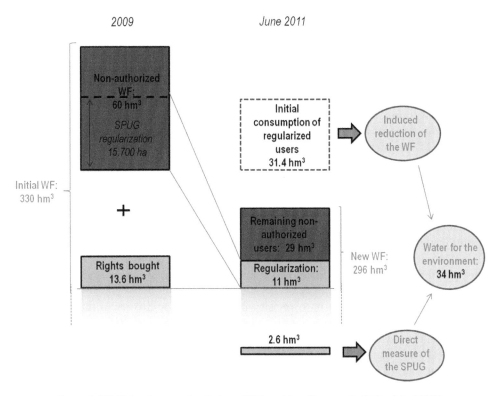

Figure 5 SPUG implementation in June 2011 and its effect on the WF of the WMA.

thanks to the incorporation of the informal irrigators and a possible new cooperation between the stakeholders, the cost of the SPUG is relative including some of the
advances in the social consensus achieved.

4   CONCLUSION

In the year 2011 Tablas de Daimiel became a *mirage*, when thanks to two wet years,
the Park increased its flooded area from virtually zero ha to 2,000 ha (maximum
possible) and aquifer levels recovered by an estimated 17 meters on average. Farmers
have abstracted less water, using green water from rain and soil moisture to produce
their crops (although rising electricity costs must be kept in mind) and this has created a window of opportunity. At the same time, as explained in the previous section,
the water rights purchased by the SPUG have changed a number of important basic
conditions and introduced a number of opportunities:

−   First, it has in many ways shifted water use, cutting down consumption for cere-
    als, whilst allowing the use of water rights for more productive and labour inten-
    sive crops like vineyards (Dumont *et al.*, 2011).

- Second, it has regularized the use of water for vine: most of the informal irrigators (with the exception of informal use for vegetables which is much smaller in number and quantity of water) have now been incorporated into the water rights systems thanks to the purchase and re-allocation of water rights.
- Third, there has been a parallel campaign to sharpen up the monitoring and sanctioning, regime which becomes a ground zero, a clean slate which from this point can ensure the strong collaboration between the regulator (the water authority) and the farming community. The control and monitoring of water use goes hand in hand with a negotiated and agreed abstraction targets to meet the WFD requirements, whilst guaranteeing a viable and thriving agricultural economy and most important avoiding the continued spiralling in informal water use.

The downside is that the cost of the Plan – in the current economic climate with an estimated budget of 5,000 M€ over 2008–2027 for 2% of the land in Spain – is simply unaffordable to the public budget, even when the current serious Spanish economic crisis is overcome. Moreover, the SPUG together with the window of opportunity offered by a series of generous wet years, are only one part of the equation.

A much needed measure is to revisit one of the major drivers in the region for groundwater abstractions, the economic incentives for irrigation. Reform has to go to the heart of the existing payment structure and deep inertias in the incentive structure for farmers to irrigate. The reform of the CAP in 2013 offers an ideal opportunity to decouple the local economy from intensive water use, to help achieve the WFD objectives over time, limiting the impacts on the local economy. Under the concept of Payment for Environmental Services and the Rural development pillar of the CAP the incentive structure could help to change from irrigated agriculture by adding an environmental premium for dry land agriculture in the form of payments for the services derived from soil conservation, and multifunctional agriculture (Wilson, 2010), with a more holistic vision on agriculture to produce not just food but also clean water, the co-production of environmental and agricultural goods, and the re-direction towards a green economy model and, taking into account that the amount of rain-fed land doubles the irrigated area, the training of rain-fed farmers on ecological rain-fed agriculture and green marketing for instance.

In addition, a more diversified economy which also includes other sectors like ecotourism, renewable energy (like thermo-solar, i.e. *growing* kWh *rather than maize*, as farmers in the region acknowledged, even though solar energy is subsidized) and high quality agricultural products would make the local economy more diversified, resilient and less dependent on intensive aquifer exploitation.

The challenge is to change the current agricultural model, where the social system is thriving at the expense of the ecological system, towards a more balanced model which aims to allocate water more equitably than at the moment between all users to include the Biosphere Reserve and the Park itself. In the case of water in arid environments, which face strong competition between users, moving away from the current trade-offs (and stand offs between sectors) requires identifying win-win solutions.

REFERENCES

Barcos, L.; López-Gunn, E.; Villarroya, F.; Fernández-Lop, A. & Llamas, M.R. (2010). *The Special Upper Guadiana Plan: assessment of early effectiveness as main policy measure to implement the EU Water Framework Directive.* [Poster]. European Groundwater Conference, Madrid, Spain, May 20–21.

CHG (*Confederación Hidrográfica del Guadiana*) (2006). *Plan Especial del Alto Guadiana* Confederación Hidrográfica del Guadiana [Special Plan for Upper Guadiana]. Guadiana Water Authority, Badajoz, Spain.

De la Hera, Á. (1998). *Análisis hidrológico de los humedales de la Mancha Húmeda y propuesta de restauración de un humedal ribereño: El Vadancho (Toledo).* [Hydrological analysis of wetlands in the Wet Spot and proposed coastal wetland restoration: The Vadancho (Toledo)]. PhD, Dpt. of Geodynamics, Geology Faculty, Universidad Complutense, Madrid, Spain.

Dumont, A.; Willaarts, B.A.; Novo, P. & López-Gunn, E. (2011). *The Extended Water Footprint and non-authorized groundwater use in the Upper Guadiana Basin (Spain): can increased productivity explain non-compliance?* Groundwater 2011 – Gestion des resources en eau souterraine, 24th European Regional Conference of ICID, March 14–16, Orléans, France.

López-Gunn, E.; Zorrilla, P. & Llamas, M.R. (2011). The Impossible Dream? The Upper Guadiana system: Aligning Changes in Ecological Systems with Changes in Social Systems. *On the Water Front*, SIWI Special Publication. Available from: http://www.siwi.org/documents/Resources/Best/2010/2011_OTWF_Elena_Lopez_Gunn.pdf [Accessed 30th May 2012]

Llamas, M.R. & Custodio, E. (eds.) (2003). *Intensive Use of Groundwater: Challenges and Opportunities.* Balkema Publishers, Dordrecht.

Martínez Cortina, L.; Mejías, M.; Díaz Muñoz, J.A.; Morales, R. & Ruiz Hernández, J.M. (2011). Cuantificación de recursos hídricos subterráneos en la cuenca alta del Guadiana. Consideraciones respecto a las definiciones de recursos renovables y disponibles. [Quantification of groundwater resources in the Upper Guadiana. Considerations regarding the definition of renewable and available resources] *Boletín Geológico y Minero*, 122(1): 17–36.

Requena, R. (2011). *El Centro de intercambio de derechos de uso del agua en el Alto Guadiana.* [The trading centre for water use in the Upper Guadiana]. 6th National Seminar on Formal Water Markets in Spain. Water Observatory, Botín Foundation, June 27 2011.

Viladomiu, L. & Rosell, J. (1996). Medio Ambiente y PAC: una primera aproximación a los programas agro-ambientales españoles [Environment and CAP: a first approach to the Spanish agri-environmental programs]. *Boletín Económico ICE*, December–January: 49–58.

Wilson, G. (2010). Multifunctional *quality* and rural community resilience. *Transactions. Institute of. British. Geographers*, 35(3): 364–381.

Zorrilla, P. (2009). *Análisis de la gestión del agua en el acuífero de la Mancha Occidental: construcción de una red bayesiana mediante procesos de participación pública.* [Analysis of water management in the aquifer of the Western Channel: a Bayesian network construction through public participation processes.] PhD, Departament of Ecology, Autónoma University of Madrid, Spain

Zorrilla, P., Carmona, G., De la Hera, Á., Varela-Ortega, C., Martínez-Santos, P., Bromley, J. & Jorgen Henriksen, H. (2010) Evaluation of Bayesian Networks in Participatory Water Resources Management, Upper Guadiana Basin, Spain. *Ecology and Society*. [Online] 15(3), 12. Available from: http://www.ecologyandsociety.org/vol15/iss3/art12/ [Accessed 17th August 2011].

Chapter 21

Intensively irrigated agriculture in the north-west of Doñana

Jerónimo Rodríguez[1] & Lucia De Stefano[2]
[1] *Humboldt University, Berlin, Germany*
[2] *Water Observatory of the Botín Foundation; Department of Geodynamics, Complutense University of Madrid, Madrid, Spain*

ABSTRACT: The Doñana region in southwestern Spain is one of the largest and most complex natural systems in western Europe. Groundwater resources constitute a key component of its natural processes and a fundamental input for the intensive irrigated agriculture that began in the 1970s and it is now one of the most important economic drivers in the region. This agricultural activity and its associated groundwater use affects the hydrological cycle of the Doñana both in terms of quality and quantity and puts pressure on its sensitive ecosystems. Since the 1990s, several initiatives were undertaken to find a balance between socio-economic development and nature conservation and avoid reaching a deadlock situation. The complex nature of the resource system, the presence of unauthorized extractions, unsolved ownership issues and a fragmented institutional structure hinder the establishment of a sustainable resource use regime.

Keywords: groundwater, institutional structure, irrigation district, strawberries, Doñana

1 INTRODUCTION

With a total area of 2,700 km², one third of it under some level of protection, the Doñana region constitutes one of the richest natural systems in western Europe (Oñate *et al.*, 2003). Located on the Atlantic coast of Andalusia between the estuary of the Guadalquivir River on the east, the Odiel and Tinto Rivers on the west (Figure 1), the Doñana region boasts the largest wetland system in the European Union.

The term Doñana is usually associated with marshes, sand dunes, beaches, and protected natural areas, as part of the Doñana region is a National Park created in 1969, surrounded by a Natural Park. However, from an ecological perspective, Doñana goes much further, including the territory of 14 municipalities in the provinces of Cádiz, Huelva and Seville, and can be classified into 4 eco-districts (Montes *et al.*, 1998). The system known as Aquifer 27 or Doñana aquifer is the main source of water for human activities and for the maintenance of a rich system of wetlands and streams.

The region offers a key point on the route of migratory birds, a nesting site for waterfowl, and the last stronghold for heavily endangered fauna, including the emblematic Iberian lynx and the Spanish imperial eagle. The wetlands system provides important environmental services like the regulation of the local hydrologic

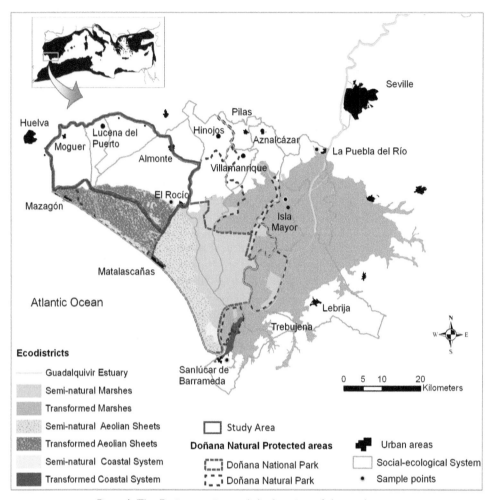

Figure 1 The Doñana region and the location of the study area.
(Source: Adapted from Palomo *et al.* (2011)).

cycle, the production of biomass and provision of landscape and other aestetic goods (Martín-López *et al.*, 2011). Today, the Doñana National Park belongs to the RAM-SAR Convention, the Man and Biosphere Programme of the UNESCO, the EU Natura 2000 Network, and is catalogued as a UNESCO World Heritage Site.

Historically, the region remained underdeveloped and sparsely inhabited for long time and land uses were limited to extensive cattle breeding, traditional lumbering activities, rain-fed crops, and hunting campaigns of the nobility. By the end of the 19th century, projects to transform areas into farmland, and promote land colonization began to be implemented to foster local development (Ojeda & del Moral, 2004; Swyngedouw, 1999). In the 1960s, supported by the developmentalist vision of the time, Doñana is perceived as an important economic opportunity, where commercial forestry, irrigated agriculture, beach tourism and natural beauty provide the means to help improve the local hard living conditions (Ojeda & del Moral, 2004). In this context, one of the most important changes was the increasing use of groundwater to irrigate the zone located in the western part of the Doñana region. Two main categories of crops developed: a capital-intensive rice sector around the town of Villamanrique de la Condesa on the east, and a labour-intensive fruit and horticultural sector in the west.

Water became the driver of a spectacular socio-economic development and, at the same time, a factor of conflict. The development of intensive irrigated agriculture brought about drops in the groundwater levels, contamination by leaked agrochemicals, and territorial fragmentation due to the creation of new arable land (Corominas, 1999; Manzano & Custodio, 2005; Custodio et al., 2008). All these activities have affected the hydrological dynamics of the aquifer system (Rodríguez et al., 2008) and the long-term evolution of the Doñana aquatic and terrestrial ecosystems (WWF/Adena, 2009). Since the 1990s the debate has focused on how to achieve harmony between the agricultural sector and the natural values of the area (Ojeda & del Moral, 2004), but no long-term solution has been found so far.

This chapter aims at shedding light on the factors that determine the complex interaction between human activity and nature conservation in areas devoted to intensive irrigated berries[1], citrus and other fruits, located north-west of the Doñana area. The interest of this area lies in the significant economic value of the agricultural production – mainly strawberries for export to the European markets – and in the complexity of the factors that determine the success (or the failure) of any policy action in the area.

The chapter is based on fieldwork carried out between January and May 2011. It included a survey among horticulture producers; semi-structured interviews with local stakeholders; and a review of the local socio-economic conditions, the regulatory framework and the biophysical characteristics of the system. The study looks at the area considered in the Farmland Reorganization Plan published in 2010 (Junta de Andalucía, 2010b) as a social-ecological system (Ostrom, 2009), where interactions of the resource system, the users and the governance structure build a complex structure of nested subsystems and crossed interdependences.

2   WATER USE IN THE NORTH-WEST OF DOÑANA

The region under study lies on north-western part of Doñana, on the semi-transformed aeolian sheets of the deltaic unit of the Doñana aquifer (Figure 1, location of the study area) and includes the territories of the municipalities of Moguer, Rociana del Condado, Lucena del Puerto, Bonares, Almonte (including the village of El Rocio) and a part of Palos de la Frontera of the Huelva Province. Some of these areas were part of the Almonte-Marismas Agricultural Transformation Plan (PTAAM) launched during the 1970s by the Spanish government and intended to transform 38,000 ha into irrigated land using up to 150 hm^3/year [hm^3 = cubic hectometre = million m^3 = 10^6 m^3] of groundwater. Finally the area affected by the Plan was reduced, because it could severely affect water-related ecosystems inside and outside of the National Park.

---

1 The area produces many red berries, strawberry being the dominant one.

The study area is located on the Doñana aquifer, which extends over 3,400 km² and is composed of a series of connected unconfined or semi-confined hydrogeological units (Manzano & Custodio, 2007). The diffuse boundaries of the resource system, the temporal gap between abstraction affections and the moment they can be measured, and the inter-annual oscillations of the precipitation regime (Custodio et al., 2008), make it difficult to forecast the behaviour of the system and predict its hydrogeological evolution.

Agriculture is the main water user: groundwater abstraction from the whole aquifer is estimated to be 45–60 hm³/year, equalling almost one third of the total aquifer recharge (through precipitation, an average of 200 hm³/year, with large inter-annual variability) (Manzano & Custodio, 2007; Rodríguez et al., 2008; CIED, 1992). In the area, the main irrigated crops are 5,800 ha of red berries and 1,700 ha of citrus (Junta de Andalucía, 2010b; Fuentelsaz et al., 2011). Other uses of groundwater are urban supply for the local towns (5.3–7 hm³/year), some of which are important tourist destinations. A small amount of surface water is imported from the Chanza-Piedras dam and is used for strawberry irrigation and urban supply.

Intensive use of land and groundwater is affecting the Doñana ecosystem and its biodiversity. In the strawberry area visible impacts are: the fragmentation of the territory with negative impacts on some mammals like the Iberian lynx, the alteration of the river-aquifer dynamics of La Rocina, Cañada, del Partido and other creeks (with the associated drops in the flow, the feeding of the wetlands, the transformation of permanent ponds and lagoons into temporary ones and the disappearance of others), localized groundwater pollution by agrochemicals and salinization, increased soil erosion rates, and the retreat of the phreatophytic vegetation, replaced by vegetation characteristic of more arid environments (Custodio et al., 2008; Manzano & Custodio, 2007; Rodríguez et al., 2008).

3 THE SOCIO-ECONOMIC VALUE OF GROUNDWATER USE

With an annual production value of 250 M€, 1,700 farmers, 12,000 permanent and 50,000 temporary employees in the province of Huelva, from which the largest part is located in the area of Doñana (Junta de Andalucía, 2010b), horticulture has transformed a formerly economically deprived region into one of the largest strawberry producers in the world. The sector produces around 25% of the strawberries in the EU15, and employs between 24% and 48% of the active population in the towns of the study area (Junta de Andalucía, 2009; 2010b).

In terms of the economic value of water use, the apparent productivity of strawberries in this area has been estimated to be 8.50 €/m³ for blue water (irrigation) and 5.90 €/m³ for both blue and green water (total water consumption) (Aldaya et al., 2010).

In the area, horticultural production is a highly mechanized and competitive activity, deploying modern irrigation systems and high productive early varieties. Specialized in supplying northern Europe during the first months of the year, the sector is very sensitive to the decisions of distributors and consumers in those markets. Hence, most of the producers have good production practices certifications awarded by third party organizations. These certifications focus on traceability, agrochemical

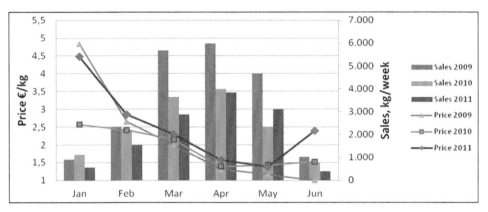

Figure 2 Price and volume traded of Huelva strawberries in the Madrid wholesaler market 2009–2011.
(Source: Adapted from Mercamadrid Price Information System).[2]

management and labour issues. Main drivers of innovation in water use are profitability and risk hedging. Production and prices have an important seasonal component and are negatively correlated (Figure 2), with prices starting at high levels in January and falling constantly until May.

Between June and the beginning of the following season, production is marginal and, given the high specificity of the assets (knowledge, infrastructure), there are few chances to recover eventual losses during the rest of the year. Given this short window, the priority is to start harvesting and reach the production peak as soon as possible. To achieve this, a sufficient and timely water supply is fundamental. Thus water saving considerations become secondary, reinforced by the relatively small share of irrigation in the total production cost (only 4% to 8%). Producers say they use between 5,300 and 7,500 m³/ha and consider that further reductions in water consumption are difficult to achieve.

4   THE STRAWBERRY PRODUCING AREAS: A MIRROR OF A COMPLEX SOCIO-ECOLOGICAL SYSTEM

The red berry-producing area can be divided in three main sectors, depending on the relation with the resource (land and water) and the level of institutional development of water users: the most western one (here called *Sector 1*) includes the cultivated areas of Moguer and a part of Palos de la Frontera municipality; the second one (*Sector 2*) includes Rociana del Condado, Bonares, Lucena del Puerto and the northern part of Almonte; and a third sector (*Sector 3*) surrounds El Rocío (see Figure 1).

Sector 1 is where greenhouses were first established and is the most advanced in institutional terms. Producers are of medium scale (7–12 ha), and mostly managing their own private farms and associated in large cooperatives. Producers in this area started their activity using unlicensed wells and in a second stage some of them accepted

2 [http://www.mercamadrid.es/index.php?option=com_estadisticas&task=mensuales&Itemid=159].

Table 1 Main characteristics of the three sectors of the study area.

	Sector 1: Palos, Moguer	Sector 2: Lucena, Bonares, Rociana, North of Almonte	Sector 3: El Rocío (Almonte)
Sector size (ha)	2,017	2,288	4,647
Main water source	Surface water	Groundwater. Pumping from local creeks (Arroyos del Horcajo, del Fresno, del Avispero Hondo)	Groundwater
Main crops	Redberries	Redberries and citrus	Redberries and citrus
Average well depth (m)	15	20–30	90
Land ownership	Private	Public concession	Public concession, Private
Average farm size (ha)	Mid (7.7)	Small (2–5.4)	12–38; >100
Role of ICs and other cooperatives	Surface water supply. Infrastructure provision. Commercialization	Groundwater supply. Commercialization and input procurement	Limited
Legal situation	ICs are registered legal entities. Most of the farms are licensed	Numerous unauthorized land or water uses	Larger users, unlicensed extractions. Some derived from the PTAAM, have officially granted licenses
Main environmental problems	Effects on the landscape, territorial fragmentation, pollution through agrochemicals	Unauthorized occupation of public forests, reductions in the piezometric level	Links to La Rocina Creek. Leaks of agrochemicals

Source: Own elaboration.

to switch to publicly provided surface water. Surface irrigation water conducted from the Chanza-Piedras dam is supplied by Irrigator Communities (ICs), which charge a fixed rate per unit of area plus a volumetric rate. The Communities are organized under a public-private structure with mandatory membership, and have become strong organizations since their creation at the beginning of the present century.

Surface water costs are higher than for groundwater, but the better physico-chemical properties of water resources and other technical advantages[3] allow users to

---

3 Water is already pressurized when delivered, sparing pumping costs, and there is a complete monitoring infrastructure, including water meters in good conditions.

Figure 3 Redberry greenhouses (left); groundwater extraction and storage facilities in public forest area (right). (Photo credits: J. Rodríguez (2011)).

avoid additional costs and annoyances. Despite the availability of surface water, some producers, mainly in Moguer, keep their wells operative, as they consider groundwater access as an acquired right and a risk hedging asset.

In the second strawberry sector (Sector 2), farmers are small-scale producers who work independently or are associated in cooperatives, in some cases gathering up to 200 ha. These cooperatives coordinate the provision of technical assistance services, the procurement of inputs and the commercialization process. An important share of the production is located on public forest land given under concession (*parcelas en canon*) by the municipalities or the province.

Irrigator Communities play a minor role here: in this area water is obtained from private wells or is pumped directly from small streams. The high density of water abstraction points – both authorized and unauthorized – increases the difficulty of the already complicated monitoring task by the competent Water Authority.

For the producer, water-related costs are represented by the drilling of the well, associated infrastructure and pumping. Given the large share of the fixed costs and the relatively low variable costs (energy and maintenance), once the well is open there are few incentives to minimize water use, besides saving on the fertilizers that are applied through irrigation. Moreover, extending the cultivated area – even with unauthorized plantations – translates into lower unit costs. The land property regime also has an effect on farmers' behaviour: as many users do not own the land, and are uncertain about the future administrative and legal viability of their activity, the intertemporal discount rate is affected and present benefits are preferred over uncertain future benefits.

The sector of El Rocío (Sector 3) corresponds to ICs associated with the Almonte-Marismas Agricultural Transformation Plan. These farms, with areas between 12 and 38 ha, rely on groundwater supplied by the ICs and from private wells. In this region the presence of Los Mimbrales and Alamillo farms, which cultivate irrigated citrus and olive trees, also constitute important players. The high expectations associated with the initial goals of the Transformation Plan, the presence of weak ICs, the proximity to the Doñana National Park and the fragile ecosystem of La Rocina creek, make water management in this sector particularly complex. This area displays the largest registered drop of the piezometric levels (Manzano & Custodio, 2007).

5   GOVERNING GROUNDWATER IN AN UNCERTAIN INSTITUTIONAL SETTING

For long time, water management in the study area has been under the jurisdiction of the Spanish central government through the Guadiana River Basin Organization (RBO) in the west and the Guadalquivir RBO in the east. The whole area belongs to the Andalusia region, whose regional government is responsible for land use planning (including agriculture), with some intervention of the provincial and municipal authorities. The lack of coordination between the intervening administrative levels, the limited human and financial resources available to monitor and enforce regulations, and, more importantly, the politicians' reluctance to take unpopular decisions, have negatively affected the institutional credibility of the involved authorities. In 2006, water resources management of the portion of the study area belonging to the Guadiana basin was transferred to a newly created Andalusian Water Agency, with the associated uncertainty that occurs in institutional transitions. Moreover, since 2007 the disputes over the decentralization of water management and planning of the Guadalquivir river basin (see Chapters 3 and 4) have resulted in a situation of increased institutional uncertainty also in the area traditionally managed by the Guadalquivir RBO, making it even more difficult to implement long term water policies.

Since the 1990s, when a severe drought made evident the impact of intensive land use on the Doñana system (Ojeda & del Moral, 2004), several public initiatives

Table 2 Public initiatives related to the organization of the strawberry producing area.

	Main focus	Mapped irrigated land (horticulture)	Water transfer as possible solution
International Commission of Experts (1992)	Build sufficient knowledge about the situation in the area	4,500 ha	No
Plan de Ordenación Territorial de Doñana (2003)	Re-organization of land use	7,963 ha	Yes
Plan de Desarrollo Sostenible de Doñana (2010)	Diagnosis of land use in Doñana and building a vision for the future	Not specified	No
Plan Especial de la Fresa, PECN (2010)	Harmonize environmental protection with irrigated agriculture; Regularize irrigated areas	8,952 ha	Yes

Source: Own elaboration with data from CIED (1992); WWF (2009); Junta de Andalucía (2010a; 2010b).

have tried to conciliate economic development and nature conservation. Even with different focus and nature, some common elements have to be pointed out: the need to map the existing irrigated land and its temporal evolution; and the importance of improving knowledge about the dynamics of the aquifer and the groundwater extractions, as necessary conditions for the design of policy interventions. The transfer of surface water resources from the Guadiana basin as a means to release pressure from the Doñana aquifer has also been considered as an option.

The most recent policy elements are the POTAD (Ordination Plan for the Doñana Area), approved in 2003, and stemming from it, the PECN (Special Plan for the Regulation of Irrigation in the Northern Part of Doñana), which was published at the end of 2010 and whose official approval is still pending. While POTAD has a regulatory value only, PECN also includes a specific budget of 4.5 M€ over a 12-year period and specific actions to reorganize water use in Doñana.

The PECN foresees the creation of ecological corridors to reconnect Doñana with Sierra Morena, in the north, and this requires some strawberry plantations, both authorized and unlicensed, to be relocated or dismantled. For this, decided political will as well as financial means to pay compensations are necessary. Besides the budgetary uncertainty, the PECN implementation faces legal difficulties. As many illegal greenhouses are located on public property, occupiers do not formally own the land. Thus it is not possible to buy the land back, and it will be necessary to design specific legal mechanisms to execute this operation.

In terms of user-based governance, only 32% of the irrigated land in the area belongs to an Irrigation Community (Junta de Andalucía, 2010b). A marked characteristic of the area is the atomization of water users' associations. Agricultural water users cooperation in most of the cases has been limited to specific lobbying activities (e.g. regularization of unlicensed wells, requests for surface water transfer), the provision of technical assistance services, and the coordination of the procurement and commercialization processes.

6 THE WAY AHEAD: IN SEARCH OF A SUSTAINABLE BALANCE

The contribution of intensive irrigated agriculture to the local economy, together with the central role that water issues play in an arid country like Spain, makes control over water resources a major political issue (see Chapter 4). As in the rest of Spain, in Doñana a water policy favourable to local economic interests linked to agriculture brings important electoral revenues, while limiting water use for environmental reasons is often perceived as an economically and politically costly burden that responds to external illegitimate interests.

During the past decade, the enforcement of administrative constraints to the expansion of irrigated land and improvements in the irrigation efficiency have contributed to limit the pace of land use change and to stabilize groundwater consumption. This trend has been uneven, though, with consumptions oscillating between more than 10,000 m^3/ha/year and 5,000 m^3/ha/year. In 2003, the POTAD delimited the areas where irrigated agriculture was permitted, any posterior expansion being illegal. Nevertheless, even if slower, the greenhouse area has kept growing,[4] as a result of the limited enforcement capacity of the competent authorities.

The importance of building up users' associations and the positive effects of their existence can be appreciated in the western sector of the strawberry area, where ICs are active in improving technical standards and formalizing and coordinating efforts and resources. Nonetheless, the heterogeneity of water users, the irregular legal situation of part of them and possibly the lack of a pressing need to cooperate so far have hampered the development of effective user based-resource management structures.

Signs of progress in the collective awareness about the need to reach a compromise between economic welfare and nature can be observed, but the system is still under intense pressure. The study of the strawberry-growing areas as a social-ecological system sheds light on the complexity of the interactions between socio-economic development and nature conservation, thus confirming that there is no *one size fits all* solution not even within the Doñana region.

In Doñana, the inclusion of water and land use-related requirements in the certification systems, the transfer of best practices to improve irrigation efficiency and, more importantly, raising awareness of the benefits – even economic ones – of having well-functioning ecosystems are ways worth exploring in the search for a balance between socio-economic development and nature conservation.

In any case, the development of a higher level of cooperation and trust between users, authorities and other stakeholders, including formal communication and participation mechanisms, seems to be a necessary condition if effective robust and long lasting improvements of water and land governance in the area are to be achieved.

REFERENCES

Aldaya, M.M.; García-Novo, F. & Llamas, M.R. (2010). *Incorporating the water footprint and environmental water requirements into policy: reflections from the Doñana region (Spain).* Papeles de Agua Virtual, 5. Botín Foundation, Santander, Spain.

CIED (Comisión Internacional de Expertos sobre el Desarrollo de Doñana) (1992). *Dictamen sobre Estrategias para el Desarrollo Socioeconómico Sostenible de Doñana.* [Analysis of Strategies for the Sustainable Socioeconomic Development of Doñana] Junta de Andalucía, Seville, Spain.

Corominas, J. (1999). Experiencia sobre el control de las extracciones para uso agrario en el acuífero Almonte-Marismas. [Experiencies concerning the control of the water extractions for agricultural use in the Almonte-Marismas Aquifer] In: Ballester, A.; Fernández-Sánchez, J.A. & López-Geta, J.A. (eds.) *Medida y evaluación de las extracciones de agua subterránea.* Instituto Técnológico GeoMinero de España (ITGE). Madrid, Spain.

Custodio, E.; Manzano, M. & Montes, C. (2008). Perspectiva general del papel y gestión de las aguas subterráneas en el área de Doñana, sudoeste de España. [General Perspective of the role and management of groundwater in the Doñana Area, southwest Spain]. *Boletín Geológico y Minero*, 119(1): 81–92.

Fuentelsaz, F.; Hernández, E., Peiteado, C. & GEOSYS (2011). El uso de la teledetección para analizar incrementos de regadíos y pérdida de zonas forestales. Estudio de WWF sobre

---

4 The POTAD established prohibition or limitations to transform rain-fed land into irrigated plots. In the period 2003–2009, however, the greenhouse area increased by 18.5% relative to the POTAD baseline. The land transformed into greenhouses included forest area (41% of the transformed land) and even highly protected public land (11%) (Fuentelsaz *et al.*, 2011).

cambios en el uso del suelo en el entorno de Doñana entre POTAD y 2009. [The use of remote sensing for the analysis of increases in the irrigated area and the loss of forest area. WWF study of changes in land use between POTAD and 2009] *Conference Proceedings: VII Congreso Ibérico sobre Gestión y Planificación del Agua, Ríos Ibéricos + 10, Mirando al futuro tras 10 años de DMA.* Fundación Nueva Cultura del Agua. [VII Iberian Water Management and Planning Conference. Iberian Rivers +10, Looking towards the future 10 years after the WFD. New Water Culture Foundation] 16–19 February 2011, Talavera de la Reina, Toledo, Spain.

Junta de Andalucía (2009). *Estudio del Mercado Mundial de la Fresa y los Frutos Rojos. Análisis de los Principales Mercados de Destino.* [Study of the World Strawberries and red berries Market. Main markets analysis]. November 2009. Available from: http://www.juntadeandalucia.es/agriculturaypesca/portal/export/sites/default/comun/galerias/galeriaDescargas/cap/servicio-estadisticas/Estudios-e-informes/agricultura/cultivos-horticolas/fresa/Informe_Fresa_y_Frutos_Rojos_mod_Feb-2010Final.pdf [Accessed 31st May 2012]

Junta de Andalucía (2010a). *II Plan de Desarrollo Sostenible de Doñana.* [II Sustainable Development Plan for Doñana] June 2010. Available from: http://www.juntadeandalucia.es/medioambiente/site/web/menuitem.a5664a214f73c3df81d8899661525ea0/?vgnextoid=f12001f21c969210VgnVCM1000001325e50aRCRD&vgnextchannel=3259b19c7acf2010VgnVCM1000001625e50aRCRD&lr=lang_es [Accessed 23rd December 2011].

Junta de Andalucía (2010b). *Plan Especial de Ordenación de las zonas de regadío ubicadas al norte de la Corona Forestal de Doñana.* [Special Plan for the Regulation of Irrigation in the Northern Part of Doñana] December 2010. Available from: http://www.juntadeandalucia.es/obraspublicasyvivienda/obraspublicasyvivienda/portal-web/web/texto/e7c03588-2497-11e0-8da3-6b048da97d96 [Accessed 23rd December 2011].

Manzano, M. & Custodio, E. (2005). El acuífero de Doñana y su relación con el medio natural [Doñana and its relationship with the environment]. In: García Novo, F.; Marín Cabrera, C. *Doñana: agua y biosfera.* [Doñana, water and biosphere] Confederación Hidrográfica del Guadalquivir, Ministerio de Medio Ambiente, Madrid, Spain: 133–142.

Manzano, M. & Custodio, E. (2007). Las aguas subterráneas en Doñana y su valor ecológico [Groundwater in Doñana and its ecological value]. *Enseñanza de las Ciencias de la Tierra,* 15(3): 305–316.

Martín-López, B.; García-Llorente, M.; Palomo, I. & Montes, C. (2011). The conservation against development paradigm in protected areas: Valuation of ecosystem services in the Doñana social-ecological system (southwestern Spain). *Ecological Economics,* 70: 1481–1491.

Montes, C.; Borja, F.; Bravo, M.A. & Moreira, J.M. (1998). *Reconocimiento Biofísico de Espacios Naturales Protegidos. Doñana: una aproximación ecosistémica* [Bio-physical reconaissance of protected areas, Doñana: an ecosystem approach]. Consejería de Medio Ambiente, Junta de Andalucía, Seville, Spain.

Ojeda, J.F. & del Moral, L. (2004). Percepciones del agua y modelos de su gestión en las distintas fases de la configuración de Doñana. [Perception of water and its management models during the configuration process of Doñana] *Investigaciones Geográficas,* 35: 25–44.

Oñate, J.; Pereira, D. & Suárez, F. (2003). Strategic Environmental Assessment of the Effects of European Union's Regional Development Plans in Doñana National Park (Spain). *Environmental Management* 31(5): 642–655.

Ostrom, E. (2009). A general framework for analyzing sustainability of social-ecological systems. *Science,* 325: 419–422.

Palomo, I.; Martín-López, B.; López-Santiago, C. & Montes, C. (2011). Participatory Scenario Planning for Protected Areas Management under the Ecosystem Services Framework: The Doñana Social-Ecological System in southwestern Spain. *Ecology and Society,* 16(1): 23.

Rodríguez, A.; Olías, M. & García, S. (2008). *Manual de Práctica de Campo. Geomorfología e Hidrología del Parque Nacional de Doñana y su Entorno* [Fieldwork Practice Manual,

Geomorphology and Hydrology of the Doñana National Park and its surrounding areas].
University of Huelva – Universidad of Cadiz.

Swyngedouw, E. (1999). Modernity and Hybridity: Nature, *Regeneracionismo*, and the Production of the Spanish Waterscape, 1890–1930. *Annals of the Association of American Geographers*, 89(3): 443–465.

WWF/Adena (2009). *Strawberry farms in Doñana*. World Wide Fund for Nature (WWF). Available from: http://assets.wwf.es/downloads/factsheetstrawberry.pdf [Accessed 23rd December 2011].

Chapter 22

The Canary Islands

Emilio Custodio[1] & María del Carmen Cabrera[2]
[1] *Department of Geotechnical Engineering and Geo-Sciences,*
  Technical University of Catalonia, Barcelona, Spain
[2] *Department of Physics, University of Las Palmas de Gran Canaria,*
  Las Palmas de Gran Canaria, Spain

ABSTRACT: The volcanic Canary Islands, in the Atlantic Ocean, have the characteristics of small islands in an environment varying from sub-humid to arid, and quite different from one island to another and even within the same island. These conditions produce specific water availability circumstances to be solved in each of the islands, through island Water Councils. The two most important islands, Gran Canaria and Tenerife contain about 90% of the 2 million inhabitants. Intensive groundwater exploitation for more than a century, and especially in the last half century, has produced a deep change in groundwater flow, the drying up of springs and the depletion of aquifer reserves. This has given rise to a special culture of water winning and use. To solve water problems the progressive and decisive introduction of seawater and brackish water desalination has been a key element, and more recently of reclaimed wastewater, although gradually. This forms a complex system that includes private water markets and public water offers. About 50% of water is for irrigation, at prices that occasionally may reach or exceed 1 €/m³. The up to now unsustainable situation has had the benefit of allowing the economic and social development from an agriculture-based economy toward tourism and services. Currently it is evolving toward a more balanced economy but with high water costs and some environmental, although bearable, damage.

Keywords: Canary Islands, intensive groundwater development, water cost, sustainability issues, water markets

1 INTRODUCTION

Water and its use have special characteristics in the Canary Islands when compared with the mainland. What is presented here is contained in diverse publications or is the result of the experience of the authors during several decades. References can be found in Cabrera *et al.* (2011) and in the respective websites of the seven Water Authorities [see *Consejo Insular de Aguas* of the respective island].

The Canary Islands (the Canaries) are a volcanic archipelago consisting of seven major islands and a few small isles and islets. They are located in the eastern Atlantic Ocean, between 27°37' and 29°25' North, and 13°20' and 18°10' West, in front of the Saharan coast of Africa. Along a length of 400 km from East to West, the islands are Lanzarote (LZ), Fuerteventura (FV), Gran Canaria (GC), Tenerife (TF),

La Gomera (GO), La Palma (LP) and El Hierro (HI) (see Figure 1). The main geographical characteristics are shown in Table 1.

Mean gross income in 2005 was close to 36 M€ or 17,000 €/inhabitant. The economic activity and employment are respectively 78% and 86% in services (dominated by tourism), 20% and 11.5% in industry (dominated by construction), and 2% and 2.5% in agriculture, feedstock and fisheries. There are large variations from island to island. What is presented below refers mainly to GC and TF, the most active islands, although the arid LZ and FV have important tourism activities.

The Canary Islands are an Autonomous Community of Spain, with their own Government and Parliament, which are responsible for water resources and the environment. Spanish legislation applies, but incorporating specific rules. Each island has its own local government (*Cabildo*), with important administrative, social and economic functions. Hydrologically each island is a separate Water District of Spain (*Consejo Insular de Aguas* or Insular Water Council), under the coordination of a Canarian Water Directorate.

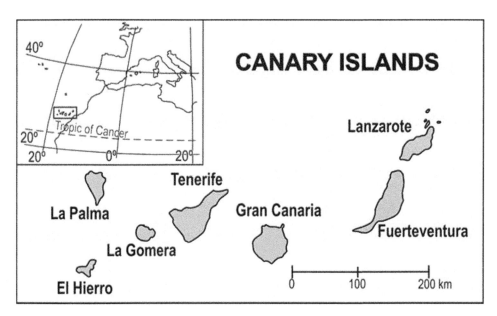

Figure 1 Situation of the Canary Islands.

Table 1 Basic data on the Canary Islands.

Island	LZ	FV	GC	TF	GO	LP	HI	Total
Surface area (km²)	850	1,650	1,570	2,050	710	370	270	7,470
Maximum altitude (m)	670	807	1,954	3,718	2,426	1,484	1,501	
Inhabitants (thousands)	142	103	846	907	23	87	11	2,119

2   WATER RESOURCES

The Canaries are in the arid Saharan belt. In spite of this the temperature is mild since the sea is relatively cold due to an oceanic upwelling area under the influence of the humid north-easterly, mid-latitude trade winds. When the trade winds intersect the islands they are forced to ascend, producing an important increase in rainfall in the north-easterly areas of the high islands, while the coastal and southern areas remain dry but good for out-of-season irrigated intensive agricultural production. Only LZ and FV remain arid due to their relative low altitude.

Due to the relatively high permeability of the outcropping volcanic materials, surface runoff is limited to some intense rainfall events. Storage reservoirs cannot be constructed, except when old, low permeability, volcanic formations crop out in the deeply eroded areas of GC and GO. Actually GC has numerous dams higher than 15 m in a small space, but of small total capacity due to the highly sloping land.

From the hydrogeological point of view, the islands consist of a low permeability core, covered and surrounded by younger, more permeable, heterogeneous volcanic materials and some derived sediments. Circumstances vary from island to island and from site to site. Groundwater flows towards the coast, and on the way down, under original natural conditions, spring areas appeared and fed some permanent flows in some tracts of the deep gullies. Table 2 summarizes the hydrological data.

Groundwater is the most important natural water resource in the high islands. Besides direct channelling of spring water, where it is still available, groundwater is intensively developed through horizontal and vertical works. Horizontal works consist of tunnels (water galleries) to intersect the aquifer formations; they are more developed in TF. Vertical works are deep, large diameter wells (Canarian wells), excavated using mining technology. Both of them may contain secondary works, mainly water galleries and small diameter, long horizontal boreholes to increase the chance of intersecting permeable layers. More recently, deeply penetrating, mechanically drilled wells have been introduced for new emplacements or to deepen existing Canarian wells.

Seawater desalination was introduced in LZ in the early 1960s, and in the early 1970s in GC. This water is now commonly used for urban supply, but also for irrigation of some cash crops. In addition, brackish groundwater is freshened by means of reverse osmosis, mostly for crop irrigation in small, private facilities.

Reclamation of treated urban waste water for agricultural uses started in the late 1970s, but development has been slow due to high salinity and variable quality.

Table 2 Basic hydrological data.

Island	LZ	FV	GC	TF	GO	LP	HI	Average
Average precipitation, P (mm/year)								
maximum	250	200	950	1,000	900	1,400	700	
island weighted average	156	111	300	740	425	368	373	323
minimum	90	60	100	200	100	300	200	
Surface runoff, ES (mm/year)	2.5	8	19	25	45	51	27	29
Recharge, R (mm/year)	4	9	57	185	191	188	101	105

This is currently improving and salinity is sometimes reduced by reverse osmosis treatment.

Water galleries produce water continuously, so in low demand seasons it may be wasted to the sea. To reduce the wastage of groundwater reserves, in TF this water is purchased by the water authority and stored in artificial reservoirs to be sold and distributed in high demand moments. Also some bulkheads have been constructed inside water galleries to regulate the discharge when favourable conditions are found.

Table 3 shows water resources data, Table 4 their evolution and Table 5 the current water demand for the different uses.

Table 3 Average water resources in the 2000–2010 decade, when groundwater extraction started to decrease and the use of reserves was dwindling. Values in hm³/year. [hm³ = cubic hectometre = million m³ = 10^6 m³].

Island	LZ	FV	GC	TF	GO	LP	HI	Total
Total Water								
Precipitation	134	184	466	865	140	518	101	2,408
Surface runoff	1.3	5	75	20	8	1.5	0.6	111
Recharge	3.3	14.2	87	239	63	265	27	642
Available Resources								
Surface water	0.1	0.0	24	0.0	1.4	0	0	25
Groundwater	0.5	2	100	180	4.5	58	2.4	347 (*)
Springs	0	0.0	0.1	5	6.7	10	0	22
Seawater desalination	19	12	60	19	0	0	0.5	110
Brackish water desalination	0.5	2	18	0	0	0	0	20
Reuse	0	0	12	8	0	0	0.4	20

(*): A significant fraction comes from groundwater reserves depletion.

Table 4 Evolution of average water resources used in the archipelago, in hm³/year.

Year	1973	1978	1986	1991	1993	2000	2009	Current trend
Groundwater	459	450	411	393	386	326	343	Decreasing
Surface water	25	19	20	21	21	25	27	Stable
Brackish water desalination	0	0	0	0	0	32	25	Stable
Seawater desalination	7	16	21	34	27	92	107	Growing
Reuse of urban waste water	0	0	0	0	0	21	19	Fluctuating
Total	491	485	452	448	444	496	516	Stable

Table 5 Water resources use in 2010.

Use	Agriculture	Urban	Tourism	Recreation	Industry	Other	Total
hm³/year	232	171	54	19	15	2	493
%	47	35	11	3.6	3	0.4	100

3   WATER DEVELOPMENT FOR AGRICULTURE

In historical times, the inhabitants, some thousands, did not irrigate their crops. After the progressive incorporation of the islands in the Crown of Castille, between the late 15th century and the early 16th century, population and trade grew fast, thus creating a high demand for goods and crops. The new settlers obtained irrigation rights from the Spanish Crown to capture and use spring water. Descendants shared the rights and formed societies (*Heredades*). As well as irrigating staple crops and fruit trees, water was used to cultivate sugarcane to produce sugar to be exported.

This situation lasted, with slow evolution, until the mid-19th century. Then non-irrigated cacti were cultivated for dye production, until they were replaced by artificial production. At the turn of the century intensive banana cultivation was introduced, mostly to be consumed in the mainland, peaking in the 1960s, and from then out-of-season cultivation of tomatoes and vegetables to be exported was intensively introduced mostly in the sunny, dry areas. Flowers and exotic plants are third in importance for consumption in the mainland and for exportation, even if they use less than 1% of the cultivated land. This dramatically increased water demand. Since available water resources were not enough, from the late 19th century private societies were formed to develop further resources. This started with the construction of dams where this was possible (in GC), and by drilling long water galleries (mostly in TF), and afterwards through Canarian wells when pumping machinery was made available (mostly in GC). The capital resources were obtained by means of shares to be paid through a proportional fraction of the water and the benefits from selling the excess water. Farmers have also received for exploitation some public water works built with public water funds, as some reservoirs in GC.

Groundwater development was already important by the 1920s and continued to grow until the late 1970s, after which the total production of groundwater stabilized, although activities to extend wells and galleries and to drill new boreholes continued. Since year 2000, groundwater production has decreased, mostly due to the high cost, decreasing irrigated surface area, introduction of less water-demanding crops, and other water sources being made available. The number of galleries and wells is shown in Table 6. Crops have evolved from bananas in the northern slopes, mostly for consumption in mainland Spain, to vegetables under cover in the sunny southern areas, where they can be grown early in the year, when European markets demand them at competitive prices. Feedstock is of secondary importance.

Table 6 Current number of groundwater sources in use, in rounded up figures.

Island	LZ	FV	GC	TF	GO	LP	HI
Large springs remaining	0	0	0 (*)	0	10	10	0
Water galleries in use	1	0	10	425	2	150	3
Canarian wells and boreholes in use	10	30	1,190	170	10	75	10

(*): There were at least 15 large springs and 7 permanent streams before 1930.

4 EFFECTS OF INTENSIVE GROUNDWATER USE

The intensive use of groundwater, mostly for agriculture, has produced a continuous water level drawdown, and consequently the progressive decrease of spring flow. In fact, many springs disappeared decades ago. This is accompanied by the periodical need to deepen wells and water galleries, dramatically increasing the cost of water. Currently, in high water demand seasons, water prices go up and may approach -occasionally exceeding- those of desalted seawater.

Due to the abrupt relief and poor accessibility of many areas, environmental changes were not noticed by the population, in times of poor environmental sensitivity. Actually most people are not aware that environmental degradation happened decades ago. Only groundwater developers noticed it, especially those tapping spring flow. This was the origin of numerous complaints and legal suits, documented as early as the 19th century but especially in the early 20th century. The value of the lost natural resources is not known, nor the burden of pumping from increasingly deep water tables, despite the social benefits obtained.

Total groundwater exploitation does not exceed total recharge in any of the islands. However there is a continuous water-table drawdown resulting from hydrodynamic conditions. Since a large part of groundwater exploitation is at mid- and high elevation, a significant fraction of recharge continues to be discharged diffusely into the sea along the coast. Thus, a fraction of abstracted groundwater resources come from depletion of reserves. Theoretically, to get a large fraction of recharge the water table has further deepened, approaching sea level. This is unfeasible due to the vertical decrease of permeability, inadequacy of existing wells and galleries, loss of yield and water quality, and the enormous operation costs. Fortunately, in some cases the local aquifer structure may help in approaching sustainable results with less drawdown, a moderate abstraction decrease and the adequate emplacement of wells. To carry out effective plans, hydrogeological knowledge has to be greatly improved and groundwater users' communities are needed, as well as an adequate, accepted and updated water plan, carefully enforced by a committed water authority.

5 GROUNDWATER QUALITY RELATED TO AGRICULTURE

Groundwater quality is highly variable, from low mineral content in the high, rainy areas, to brackish in the dry parts, due to climatic aridity. Seawater intrusion affects limited areas. The excess of Na over Cl due to the weathering of volcanic materials is not a serious problem for freshwater in permeable soils. However, there are rather large areas, especially in TF, where the deep-seated $Na–HCO_3$ water cannot be used for irrigation without being blended with other water resources or after costly treatment. High F contents are found in some areas, especially in TF. There are no serious problems of high B content. Soils and groundwater have rather high natural P content.

Agriculture is predominantly practiced below 800 m altitude, and the most important irrigated cash crops are below 300 m. So, a large part of the islands' area is free of intensive agriculture-derived contamination. In the areas affected by irrigation return flows, mostly valley bottoms, coastal plains and terraced slopes, high NO_3 content can be found, with many points exceeding the 50 mg/L NO_3 threshold, up to several hundred mg/L in some areas.

The large desalination plants dispose the brine into the sea through outfalls reaching the shelf break. However the small private desalination plants using brackish groundwater, many of them for irrigation, are relatively far away from the coast, and thus brine disposal is an issue of concern. In FV, a pipeline system to collect and dispose of the brine from dispersed agricultural wells was constructed by the *Cabildo*, which also promoted the construction of the wells and the associated desalination plants. The correct operation and maintenance of such pipeline systems, where they exist, is not always guaranteed.

Agriculture is currently an important part of the landscape and land conservation in northern areas, in the mid-altitude wide valleys (*vegas*) and in the spectacular terraced sides of the steep slopes. There, irrigation is socially and environmentally important. The recent abandonment of some traditional banana crop areas has had a noticeable impact on the landscape, but probably this will be soon forgotten.

6   WATER MARKETS AND RELATED SOCIAL ISSUES

The development of water resources since the 15th century was carried out mainly by the *Heredades* and other private societies, mostly for crop irrigation, as mentioned before. After supplying the needs of the associated farmers, the excess water was sold to other farmers and to towns. When demand grew, shareholding societies were created to develop and sell new groundwater resources, mostly in GC and TF. Thus, informal water markets developed, but remained a poorly transparent, private affair. Several such markets are still active, although dwindling. In theory water markets should allow optimal access to water at the minimum price. In practice things are otherwise since markets are dominated by a few *water-holders* (*aguatenientes*) who set the prices among themselves and impose them on others. The transportation of purchased water from the producer's well or reservoir to the buyer's point of use is carried at a price by *brokers* (*intermediarios*) who own, share or rent the canals and pipes, often without alternative for the buyer. The *intermediario* often do not feel responsible for water losses or thefts during transportation, and the quality supplied may not be the same as that of the source. This is the origin of complaints, but there is no way to set them. Besides, there are markets for well and gallery shares, although these are in recession. The number of persons with water shares represents an important fraction of population, but really most of the shares are in a few hands. But this widespread popular participation in water affairs makes people aware of the value of water.

This system is imperfect and with some abuses, but it works and play an important role for mid-altitude agriculture and town supply, where other water sources are not available. Currently, conditions in some coastal areas of GC and TF may be different since the public water administration, mainly in TF, is offering water at low altitude under more controlled water quantity and quality conditions. This water comes from storage of low-priced water obtained out-of-season and from reclaimed water, sometimes with salinity reduction by blending or reverse osmosis treatment. This helps to moderate the excesses of private water markets and tame water prices.

Actually the cost of water in constant money units has been approximately stable during recent decades.

Prices paid for irrigation water in high-demand seasons may approach or exceed that of seawater desalination. Thus, only producers of highly competitive crops may afford them, or will use desalinated sea water if the plots are at low altitude. Currently a privately promoted desalination plant exists in Gáldar, northern GC, which constitutes a new kind of private water enterprise, which compete (jointly with public water offer) with groundwater. This explains the decrease in groundwater abstraction and the start of aquifer recovery in some areas.

7   ADMINISTRATIVE AND SOCIAL ISSUES

Due to the early intensive development of groundwater, which was not foreseen in the Water Act of 1876, and the large number of legal complaints related to water, the Spanish Government decided in 1924 to enforce a Special Water Regime for the Canaries, later modified in 1948. Water Services were created to administer the restrictions on private water development through the requirement for permits to carry out new works. The Water Act of 1985, which declared all waters to be a public domain, took into account the Canarian situation through a Canarian Water Act that follows the principles of Spanish law. To deal with the continuous use of groundwater reserves, the Canarian Government at the time enforced the regulation of not allowing any change in the characteristics of wells and galleries of those owners who decided to remain in the private domain without getting a public concession. This produced an intense and widespread negative reaction and precipitated the fall of that Canarian Government. The newly elected one modified this rule by considering that works needed to sustain the flow cannot be considered a change of the characteristics. This tamed the situation but not the already unsustainable groundwater reserves depletion rate. Current taming is coming through the increased cost of exploitation and the progressive decline of agriculture, at the expense of further deterioration and social damage, but allowing a smooth transition toward a new situation. The incorporation of the European Framework Directive through the Spanish Water Act has created new stress, but currently with little public reaction since the social situation is now very different.

8   CONCLUSIONS

The Canary Islands present very different hydrologic characteristics but share some common circumstances. Water is scarce and costly but sustains the local economy, which has evolved from agriculture to services, with important tourism activities. Most water development, especially groundwater, has been made by private owners, but currently public and private desalinization of seawater and brackish water is important. Groundwater development has been carried out at the expense of using part of the groundwater storage, with serious water-table drawdown and the drying out of springs during the 20th century, and the loss of natural resources, but with little public reaction. Water is expensive, but at a sustained price in constant money units.

The formerly unsustainable development of groundwater is now being tamed as new water resources become available, although they are energy-intensive.

ACKNOWLEDGEMENTS

Dr. M.R. Llamas has carefully reviewed the original text, improving it largely and providing interesting comments.

REFERENCE

Cabrera, M.C.; Jiménez, J. & Custodio, E. (eds) (2011). *El Conocimiento de los recursos hídricos en Canarias: cuatro décadas después del Proyecto SPA–15* [The knowledge of water resources in the Canaries: four decades after Project SPA–15]. Asociación Internacional de Hidrogeólogos–Grupo Español. Las Palmas de Gran Canaria, Spain.

Glossary

Most of the definitions presented in the glossary have been taken from the *Water Footprint Manual* (Hoesktra *et al.*, 2011), the European *Water Framework Directive* (EC, 2000) and the *European Environment Agency online multilingual environmental glossary* (EEA, 2012). These are in line with the significant work of the UN Statistics Division on *System of Environmental-Economic Accounting for Water (SEEA-Water)* (UN Statistics Division, 2012).

Available groundwater resource It means the long-term annual average rate of overall recharge of the body of groundwater less the long-term annual rate of flow required to achieve the ecological quality objectives for associated surface waters, to avoid any significant diminution in the ecological status of such waters and to avoid any significant damage to associated terrestrial ecosystems.

Blue water Fresh surface water and groundwater, i.e. the water in freshwater lakes, rivers and aquifers.

Blue water availability Runoff (through groundwater and rivers) minus environmental requirements, such as river environmental flows or wetland needs. Blue water availability typically varies within the year and from year to year as well.

Blue water footprint Volume of surface and groundwater appropriated by humans to produce or consume a good or service. Consumption refers to the volume of freshwater used and then evaporated or incorporated into a product. It also includes water abstracted from surface or groundwater in a catchment and transferred to another catchment or disposed into the sea or saline water bodies. It is the amount of water abstracted from groundwater or surface water that does not return to the catchment from which it was withdrawn.

Blue water scarcity The ratio of blue water footprint to blue water availability. Blue water scarcity varies within the year and from year to year.

Body of groundwater It means a distinct volume of groundwater within an aquifer or aquifers.

Body of surface water It means a discrete and significant element of surface water such as a lake, a reservoir, a stream, river or canal, part of a stream, a transitional water or a stretch of coastal water.

Crop yield Weight of harvested crop per unit of harvested area. It can be measured in terms of dry matter.

Direct water footprint   The direct water footprint of a consumer or producer (or a group of consumers or producers) refers to the freshwater consumption and pollution that is associated to the water use by the consumer or producer. It is distinct from the indirect water footprint, which refers to the water consumption and pollution that can be associated with the production of the goods and services consumed by the consumer or the inputs used by the producer.

Economic water productivity (or apparent water productivity)   Economic value of the products produced per unit of water consumption or pollution. See also *water productivity*. It is a ratio of production value over water consumption, and differs from the marginal value (the productivity of the last unit of water), generally used to determine allocation efficiency.

Ecosystem green water requirements (or terrestrial ecosystem water requirements)   The amount of green water consumed by forest and other terrestrial ecosystems, which contributes to the supply of ecosystem services at a wide range of spatial and temporal scales.

Ecosystem water requirements   The amount of green water and blue water consumed by aquatic and terrestrial ecosystems, which contributes to the supply of ecosystem services at a wide range of spatial and temporal scales.

Environmental flow requirements (or aquatic ecosystem water requirements or ecosystem blue water requirements)   The quantity, quality and timing of water flows required to sustain freshwater and estuarine ecosystems and the human livelihoods and well-being that depend on these ecosystems.

Environmental security   When social systems interact with ecological systems in a sustainable ways, all individuals have fair and accessible access to ecosystem services, and mechanisms exists to prevent environmental degradation and crisis.

Evapotranspiration   This term represents the combination of evaporative losses from the soil surface and transpiration from the plant surface. These two phenomena occur simultaneously and there is no easy way of distinguishing between them.

Extended water footprint   The extended water footprint (EWF) refers to a set of indicators for economic and quantitative analysis of water resources. The EWF combines the contribution of the standard water footprint accounting in terms of water consumed/polluted with an economic perspective primarily based on the determination of the economic value of water.

Food security   Food security exists when all people, at all times, have physical and economic access to sufficient, safe and nutritious food to meet their dietary needs and food preferences for an active and healthy life.

Good groundwater chemical status   It is the chemical status of a groundwater body whose chemical composition: is not affected by salt intrusion; meets the established quality requirements; does not prevent that the associated surface waters achieve the established environmental objectives; and does not cause significant damages to the associated terrestrial ecosystems.

Good groundwater quantitative status   It is the status of a groundwater body where: the average annual abstraction rate on the long term is lower than the available water resources, and is not affected by anthropogenic alterations that can prevent that the associated surface waters achieve the established environmental objectives or can cause water salinization or other intrusion.

Good groundwater status It means the status achieved by a groundwater body when both its quantitative status and its chemical status are at least *good*.

Good surface water status It means the status achieved by a surface water body when both its ecological status and its chemical status are at least *good*.

Green water The precipitation on land that does not run off or percolate deeply recharging the groundwater, and is stored in the soil or temporarily stays on top of the soil or vegetation. It corresponds to the pedologic water, or water hold in the rootzone.

Green water footprint Volume of rainwater consumed during the vegetal production process. This is particularly relevant for agricultural and forestry products (products based on crops or wood), where it refers to the total rainwater evapotranspiration (from fields and plantations) plus the water incorporated into the harvested crop or wood.

Grey water footprint The grey water footprint of a product is an indicator of freshwater pollution that can be associated with the production of a product over its full supply chain. It is defined as the volume of freshwater that is required to assimilate the load of pollutants based on existing ambient water quality standards. It is calculated as the volume of water that is required to dilute pollutants to such an extent that the quality of the water remains above agreed water quality standards.

Gross value added (GVA) It is the value of goods and services produced in an economy at different stages of the productive process. The gross value added at market prices is equal to the gross output (value of production) minus the intermediate consumption.

Groundwater It means all water which is below the surface of the ground in the saturation zone and in direct contact with the ground or subsoil.

Groundwater governance It is the exercise of political, economic and administrative authority in the management of groundwater resources at all levels comprising the mechanisms, processes and institutions through which the citizens of the nation articulate their interests, mediate their differences and fulfil their legal rights and obligations to ensure sustainable and efficient utilization of groundwater resources for the benefit of humankind and dependent ecosystems.

Groundwater recharge The process by which external water is added to the zone of saturation of an aquifer, either directly into a formation or indirectly by way of another formation.

Indirect water footprint The indirect water footprint of a consumer or producer refers to the freshwater consumption and pollution *behind* products being consumed or produced. It is equal to the sum of the water footprints of all products consumed by the consumer or of all (non-water) inputs used by the producer.

Infiltration It refers to the downward movement of water into soils and may be defined for rain or ponded conditions.

Net Income It is equal to the income that a firm or a nation has after subtracting costs and expenses from the total revenue. Net income is an accounting term. It refers to the GVA plus subsidies and taxes, minus the consumption of fixed capital and salary payments, rentals and interests.

Pollution Pollution means the direct or indirect introduction, as a result of human activity, of substances or heat into the air, water or land which may be harmful to human health or the quality of aquatic ecosystems or terrestrial ecosystems directly depending on aquatic ecosystems, which result in damage to material property, or which impair or interfere with amenities and other legitimate uses of the environment.

Renewable resources   Natural resources that, after exploitation, can return to their previous stock levels by natural processes of growth or replenishment.

River basin district   It means the area of land and sea, made up of one or more neighbouring river basins together with their associated groundwater and coastal waters, which is identified under Article 3(1) of the Water Framework Directive (WFD) as the main unit for management of river basins.

Surface water   Surface water means inland waters, except groundwater; transitional waters and coastal waters, except in respect of chemical status for which it shall also include territorial waters.

Value of production   It is defined as the total economic value received for the commodities sold in the market.

Virtual water export   The virtual water export from a geographically delineated area (e.g. a nation or catchment area) is the volume of virtual water associated with the export of goods or services from the area. It is the total volume of freshwater consumed or polluted to produce the products for export.

Virtual water flow   The virtual water flow between two geographically delineated areas (e.g. two nations) is the volume of virtual water that is being transferred from the one to another area as a result of product trade.

Virtual water import   The virtual water import into a geographically delineated area (e.g. a nation or catchment area) is the volume of virtual water associated with the import of goods or services into the area. It is the total volume of freshwater used (in the export areas) to produce the products. Viewed from the perspective of the importing area, this water can be seen as an additional source of water that comes on top of the available water resources within the area itself.

Water abstraction   See *water withdrawal*.

Water consumption   The volume of freshwater used and then evaporated or incorporated into a product. It also includes water abstracted from surface or groundwater in a catchment and returned to another catchment or the sea.

Water demand   It is defined as the volume of water requested by users to satisfy their needs. In a simplified way it is often considered equal to water abstraction, although conceptually the two terms do not have the same meaning. In economic terms, demand is the willingness of users/companies to pay for a specific service or product.

Water footprint   The water footprint is an indicator of freshwater use that looks at both direct and indirect water use of a consumer or producer. The water footprint of an individual, community or business is defined as the total volume of freshwater that is used to produce the goods and services consumed by the individual or community or produced by the business. Water use is measured in terms of water volumes consumed (evaporated) and/or polluted per unit of time. A water footprint can be calculated for a particular product, for any well-defined group of consumers (e.g. an individual, family, village, city, province, catchment, given geographical area, state or nation) or producers (e.g. a public organization, private enterprise or economic sector). The water footprint is a geographically explicit indicator, not only showing volumes of water use and pollution, but also the locations.

Water footprint of a product   The water footprint of a product (a commodity, good or service) is the total volume of freshwater used to produce the product, summed over the various steps of the production chain. The water footprint of a product refers not only to the total volume of water used; it also refers to where and when the water is used.

Water footprint of the consumption of a geographically delineated area   It is defined as the total amount of freshwater that is used to produce the goods and services consumed by the inhabitants of a geographically delineated area. Part of this water footprint lies outside the boundaries of the area. The term should not be confused with the *water footprint within a geographically delineated area*, which refers to the total freshwater volume consumed or polluted within the boundaries of the area.

Water footprint within a geographically delineated area   It is defined as the total freshwater consumption and pollution within the boundaries of the area. The area can be, for example, a hydrological unit like a catchment area or a river basin or an administrative unit like a municipality, province, state or nation.

Water productivity   Product units produced per unit of water consumption or pollution. Water productivity (product units/m^3) is the inverse of the water footprint (m^3/product unit). Blue water productivity refers to the product units obtained per m^3 of blue water consumed. Green water productivity refers to the product units obtained per m^3 of green water consumed. Grey water productivity refers to the product units obtained per m^3 of grey water produced. The term *water productivity* is a similar term as the terms labour productivity or land productivity, but now production is divided over the water input. When water productivity is measured in monetary output instead of physical output per unit of water, one can speak about *economic water productivity*.

Water security   Water security is defined as the availability of an acceptable quantity and quality of water for health, livelihoods, ecosystems and production, coupled with an acceptable level of water-related risks to people, environments and economies.

Water use   Three types of water use are distinguished: a) withdrawal, where water is taken from a river, or surface or underground reservoir, and after use returned to a natural water body, e.g. water used for cooling in industrial processes. Such return flows are particularly important for downstream users in the case of water taken from rivers; b) consumptive, which starts with withdrawal but in this case without any return, e.g. irrigation, steam escaping into the atmosphere, water contained in final products, i.e. it is no longer available directly for subsequent uses; c) non-withdrawal, i.e. the in situ use of a water body for navigation (including the floating of logs by the lumber industry), fishing, recreation, effluent disposal and hydroelectric power generation.

Water withdrawal   It is the volume of freshwater abstraction from a surface or groundwater source. Part of the freshwater withdrawal will evaporate, another part will return to the catchment where it was withdrawn and yet another part may return to another catchment or the sea.

REFERENCES

EEA (2012). EEA multilingual environmental glossary. European Environmental Agency. Copenhagen, Denmark. Available online from: [http://glossary.eea.europa.eu/EEAGlossary/].

Hoekstra, A.Y.; Chapagain, A.K.; Aldaya, M.M. & Mekonnen, M.M. (2011). The water footprint assessment manual: Setting the global standard. Earthscan, London, UK. Available online from: [http://www.waterfootprint.org/?page=files/WaterFootprintAssessmentManual].

UN Statistics Division (2012). System of Environmental-Economic Accounting for Water (SEEA-Water). United Nations Statistics Division. Available online from: [http://unstats.un.org/unsd/envaccounting/seeaw/].

EC (2000). The EU Water Framework Directive – Integrated river basin management for Europe. European Commission. Available online from: [http://ec.europa.eu/environment/water/water-framework/index_en.html].

Author index

Subject index

Geographic index